Arena Birds

Arena Birds

Sexual Selection and Behavior

Paul A. Johnsgard

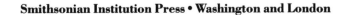

Smithsonian Institution Press • Washington and London

Designer: Janice Wheeler
Editor: Matthew Abbate
Production Editor: Rebecca Browning

Library of Congress Cataloging-in-Publication Data
Johnsgard, Paul A.
Arena birds : sexual selection and behavior / Paul A. Johnsgard.
   p.   cm. Includes bibliographical references (p.   ) and index.
ISBN 1-56098-315-9   1. Birds—Behavior.   2. Lek behavior.   3. Sexual selection in
animals.   I. Title.   QL698.3.J64   1994   598.256′2—dc20   93-8131

British Library Cataloguing-in-Publication Data available

Manufactured in the United States of America

∞ The paper used in this publication meets the minimum requirements of the American
National Standard for Permanence of Paper for Printed Library Materials Z39.48-1984.

Contents

Preface

For nearly as long as I can remember I have been especially fascinated by those species and groups of birds whose males attempt to achieve mating success within competitive arenas. In college, when I became interested in evolutionary aspects of behavior, I began to appreciate more fully the significance of the remarkable male plumages and behaviors that are typical of most arena-breeding birds. I was attracted successively to such avian groups as ducks, grouse, pheasants, hummingbirds, and bustards, and eventually came to recognize that the underlying reason for their interest to me was their unique mating systems and their associated remarkable mate attraction and mate choice signaling behaviors.

A few years ago I decided that I might as well put these interests into common perspective, and focus on the process, as well as the products, of sexual selection in birds. I asked the Smithsonian Institution Press if they might be interested in publishing a book dealing with sexual selection in arena-breeding birds and some of its varied results. These include the major female attraction mechanisms used by males, and especially the breeding assemblages of males and their associated brilliant plumages and elaborate visual displays. I also wished to discuss the use of site-specific "courts" by males of various bird species and the construction of bowers by bowerbirds for comparable reproductive signaling purposes, even though such sites and structures do not always qualify as typical arenas. Finally, I wanted to deal with the significance of these sexual signaling devices and related behaviors in influencing individual mate choice behavior by females, which is one of the basic tenets of Darwinian sexual selection, and one that has often been challenged historically.

Since this book deals primarily with avian arena behavior in promiscuous and polygynous birds, I have not felt it necessary to deal with other related albeit peripheral topics. I have thus not discussed reversed sexual selection in polyandrous birds, none of which are yet known to exhibit arena behavior. Nor have I discussed in detail the accumulating evidence that sexual selection operates effectively in monogamous birds, except for a few specific subgroups of waterfowl that exhibit a rudimentary form of arena behavior. I have relied upon Sibley and Monroe (1990) for my usage of both scientific and English nomenclature, and also for the choice of linear sequences of taxa listed in tables. However, I have made a few minor deviations from their technical and vernacular nomenclature, based on past taxonomic practices and personal subjective decisions.

The first, relatively technical chapter in this book deals necessarily with the basic theory of sexual selection as it is now visualized to apply to birds. The second chapter concerns the ways in which sexual selection is reflected in actual male spacing and dominance interactions, and mate choice behavior by females of typical arena-breeding birds. Persons not already familiar with the more theoretical aspects of these topics might find it helpful in reading these sections to refer frequently to the glossary, or perhaps they may want to go directly to the following, more descriptive chapters, leaving these two relatively theoretical chapters for later consideration.

As usual, I have had the assistance of numerous people in the course of assembling information for this book. These include librarians at the University of Nebraska, as well as those at the University of Kansas, University of Michigan, Yale University, and Oxford University (Edward Grey Institute), all of whom were extremely helpful to me. Various zoo personnel at the Omaha, San Diego (Balboa Park and Wild Animal Park), Bronx, London, Berlin, and Walsrode zoos were also of great help to me. I must especially mention David Rimlinger of the San Diego Wild Animal Park, Wolfgang Grummt of the Tierpark Berlin, and many long-time friends at the Wildfowl and Wetlands Trust (U.K.) in this regard. As usual, my old friend Ken Fink encouraged me to undertake the project from its very beginning, and his willingness to allow me to select from his superb collection of transparencies when assembling the color plate section was an important element in my decision to write the book. Ken also very kindly read an early draft of the manuscript. David Rimlinger, Curator of Birds of the San Diego Zoo, also sent me a large number of slides to choose from. Various manuscript chapters or sections were read by Drs. Bruce Beehler, Josef Kren, Frank McKinney, Robert Payne, William Scharf, F. Gary Stiles, and Larry L. Wolf. Several additional anonymous readers critically read various chapters.

I have used my own color transparencies for illustrating the color section whenever possible. However, in addition to those photos that were provided by Ken Fink and David Rimlinger I would like to acknowledge my appreciation for the use of photographs by Steffan Andersson for drawing references.

I have completed a large number of new pen-and-ink drawings specifically for this book, but on occasion have variously modified and used some of my older line drawings that have appeared in my earlier books as well. In many cases these drawings are based on my own photos or sketches, but in some they represent new drawings by me of previously published illustrations, as indicated in the captions. With the kind permission of his widow, Mary Lou Pritchard, I have also used two drawings from an earlier book that were done for me by the late C. G. Pritchard.

Arena Birds

1

Sexual Selection in Birds

Observations in biology have probably produced more insights than all experiments combined.   E. Mayr (1982)

In 1871 Charles Darwin finally dropped the other shoe; in *The Descent of Man* he expanded his theory of natural selection to specifically include humans, a subject that he had judiciously skirted in his monumental *Origin of Species* more than a decade previously. Almost lost in the whirlwind of controversy that accompanied the suggestion that even humans have evolved "from some lower form" was the major theoretical contribution of his new book, a laying out of the principles of sexual selection. Darwin defined sexual selection as being dependent "on the advantage which certain individuals have over others of the same sex and species solely in respect to reproduction." He believed that in some species males have acquired their present reproductively related or secondary sexual structures "not by being better fitted to survive in the struggle for existence, but from having gained [sexual] advantage over other males, and from having transmitted this advantage to their male offspring alone." Darwin thus distinguished this from his more confined view of natural selection, which he limited to the genetic effects on natural populations resulting from differences in *survival* (viability) rather than reproduction.

Darwin believed that sexual selection is driven by two quite different forces or conditions. The first of these is that some males are effectively able to dominate or "conquer other males" and thus gain unrestricted sexual access to females. In some cases, the development of male traits related to such male-male fighting may be carried to such a seemingly questionable extreme that, in Darwin's words, the reproductive advantages they generate must be "in the long run greater than those derived from rather more perfect adaptation to their conditions of life."

The case of males being able to dominate or defeat other males is now referred to as "intrasexual selection." Analogously but less frequently, intrasexual selection may include the effects of competitive or aggressive interactions among females (Selander, 1972). Such a situation is especially likely to occur in polyandrous mating systems, as in some phalaropes when mated females prevent other females from approaching their mates. However, formalized female–female dominance contests in an arena–like situation are still unreported for birds.

The second means by which males can reproductively benefit from sexual selection is by being among those that, in Darwin's words, "prove most attractive to the females." These more attractive males have a relatively greater reproductive access to females as a result of females choosing to mate preferentially with them. Cases involving heterosexual attraction are now generally referred to as "intersexual" (or epigamic) selection. Darwin concluded that, in either of these two situations, some males will be able to "leave a greater number of offspring than their beaten and less attractive rivals." Indeed, Darwin suggested that, of the two possible sources of sexual selection, "the power to charm the female has sometimes been more important than the power to conquer other males in battle."

Darwin predicted that sexual selection is most likely to be effective in those situations where there is an unbalanced sex ratio in the adult population, so that some males are unable to obtain mates. However, he suggested that even in monogamous species of birds having nearly equal sex ratios, the more vigorous and early-breeding females may possibly choose to mate with the most vigorous or most attractive males, and both sexes would thereby achieve an advantage in rearing their offspring. Nevertheless, among polygamous species the potential reproductive advantages accruing to successful males are obviously even richer than for monogamous species, and the opportunities for the evolution of secondary sexual traits that may influence individual reproductive success are even greater.

Concerning birds specifically, Darwin suggested that secondary sexual characters in this group are more diversified and conspicuous than in any other group of animals, and he remarked that, on the whole, birds would "appear to be the most aesthetic of all animals, excepting of course man." Concerning birds, he argued that if one can admit "that the females prefer, or are unconsciously excited by the more beautiful males, then the males should slowly but surely be rendered more and more attractive through sexual selection." At the time, Darwin had little evidence supporting the idea that female birds can actually make what would seem to be such highly aesthetic judgments in choosing mates. Later writers, even beginning with Darwin's contemporary and codiscoverer of natural selection A. R. Wallace, have tended to question this point of his argument in particular (Huxley, 1938). Yet it is now generally accepted among ornithologists not only that such capabilities are possible, but also that female choice tendencies in mating have been at least as powerful a driving force in the sexual selection of birds as have mating influences based on male dominance attributes.

Modern Views of Sexual Selection in Birds

The tremendous resurgence of interest in sexual selection in recent years has resulted in a burgeoning literature on the subject, and one that cannot possibly be reviewed adequately in a book of this type. Few people now would dare question the importance of sexual selection as a significant factor in avian evolution. Even fewer would question the role of male-male competition as a significant and fairly self-evident component of the overall mechanism of sexual selection, just as Darwin suggested (Table 1).

However, the role of intersexual selection by female choice is seemingly far more complicated and its effects more controversial. One of the most fundamental questions is not whether female choice traits exist, as was often questioned in the past, but

Table 1 • Models of Sexual Selection among Birds[1]

I. Intrasexual selection: male-male competition effects.
  A. Male interactions produce selection for maximum mass and aggressiveness, often resulting in positive sexual mass dimorphism ratios (larger males with weapons).
  B. Male interactions produce selection for maximum agility, especially aerial agility, often resulting in negative sexual mass dimorphism ratios (smaller, more agile males).
  C. Male interactions produce selection for delayed sexual maturation in that sex, especially when reproductive success depends on relative dominance experience (sexual bimaturism and bide-my-time sexual strategies).
  D. Male interactions result in selection associated with competition among sperm of different individuals to fertilize a limited number of eggs (sperm competition).

II. Intersexual selection: female choice effects.
  A. Avoidance of interspecific matings (reproductive isolation).
  B. Choice among available conspecific mates.
    1. Female gets male's resources and genes—most monogamous species. Courtship feeding sometimes occurs with polygyny; in harem polygyny females may also benefit from territorial resources controlled by male. In polyandry males provide incubation and parental care efforts.
    2. Female gets male's genes only—most polygynous and promiscuous species.
      a. Male heterosexual signaling attributes.
        aa. Male's appearance indicates overall gene quality (honest advertising of good genes, including advertisements of "handicaps").
        bb. Male's appearance not indicative of gene quality ("sexy son" effects and nonadaptive runaway process influences).

(continued)

Table 1 • (*Continued*)

---

    b. Active female choice attributes.
      aa. Female chooses a male trait that itself directly contributes to overall offspring viability (direct selection for viability-based traits).
      bb. Female chooses a male trait that may itself reduce viability but is correlated with improved overall viability ("handicap" hypothesis).
      cc. Female chooses a male trait that may reduce her own fecundity, but improves her male descendants' sexual attractiveness ("sexy son" hypothesis).
      dd. Female chooses a male trait that is neutral but correlated through genetic linkage with improved offspring viability ("hitchhiking" alleles).
      ee. Female chooses a male because of her behavior preferences for traits independent of offspring viability, and which may eventually reduce viability in male descendants (nonadaptive runaway process).

---

1. Classification partly after Maynard Smith (1987, 1991). See glossary for any unfamiliar terms in this and following tables.

rather how they evolve. Do such traits develop adaptively, because they favor the transmission of genes that promote survival (the "good genes" hypothesis)? Or do they develop in a nonadaptive or even maladaptive way (the "nonadaptive" hypothesis), such that female mating preferences may lead to the evolution of male traits that are not ecologically adapted to the environment (Kirkpatrick, 1982, 1987)?

It is equally interesting to consider female choice from the standpoint of the resulting evolved male traits. Are the male's traits that are being flaunted to the female actually indicative of his high genetic quality in attributes other than those directly associated with his mating success? That is, do his attractive traits represent "honest advertising" of good genes? In one model, males that are able to survive while producing and maintaining various seemingly costly traits—such as very bright colors or extremely long tails—may be advertising honestly by exhibiting them. Such phenotypically variable male traits are those whose expression correlates with the survivorship qualities and relative vigor of the individuals that carry them, thus truthfully reflecting their overall genetic quality (Kodric-Brown and Brown, 1984). By carrying and exhibiting such extravagant or physiologically costly traits, these males are thus perhaps variously "handicapped." However, by mating with such handicapped males the females can obtain other directly associated genes correlated with high viability that will be passed on to the following generation (Zahavi, 1975; Iwasa, Pomiankowski, and Nee, 1991). Thus, in this model females are indirectly choosing males carrying genes favoring overall viability, rather than merely choosing "sexy" males whose specific attractiveness traits have no survival value as such. Andersson (1986) has suggested that, even within the sexual selection constraints of a monogamous mating system, seemingly costly male ornaments, and associated female preferences for such costly traits, can "hitchhike" to high genetic frequencies by association with alleles that confer slight survival advantages, provided that such alleles arise often enough. He thus stressed gene-based overall viability differences, rather than mating success differences, as a selective factor driving the evolution of such costly male traits.

Alternatively, a male's attractive traits may not be directly indicative of his overall fitness or gene quality, but instead may simply be a nonadaptive end result of a positive feedback or so-called "runaway" evolutionary process, as originally proposed by Fisher (1930). This process results from the females showing increased mating preference for an advantageous genetically based male trait. As the genes for female preference increase, the male trait itself increases in a positive feedback process. For example, the longer the males' tails might become, the more the females would tend to prefer those males having the longest tails. The trait will therefore eventually become exaggerated to a point where its very presence becomes increasingly costly—in terms of classic natural selection—to the males that possess it. Pomiankowski, Iwasa, and Nee (1991) have recently suggested that such a female preference for a costly male trait can indeed be achieved if the mutation pressure on the male character is biased toward that character.

Kirkpatrick and Ryan (1991) have proposed that the evolutionary mechanisms for female choice among polygynous animals fall into two broad categories: those driven by direct natural selection (directly increasing the fitness of the females' descendants), and those driven by indirect selection. As examples of direct selection they include such possibilities as that (1) the chosen males might directly provide some significant resources for the females or their offspring. Or, (2) there may simply be some time- and energy-saving benefits resulting to those females who search for conspicuous males rather than spending more time looking for drab ones. Another possibility (3) is that females may be effectively choosing those males that have lower probabilities of transmitting diseases or parasites. There is also the possibility (4) that males may be chosen for a particular trait that is genetically associated with some phenotypically quite unrelated traits (a linkage known as pleiotropy) that result in improved offspring viability.

One variant of the good genes model of female choice is the total viability hypothesis, which states that females should mate preferentially with those males carrying the fewest deleterious genes in their genomes (Manning, 1984). A similar variant is the age indicator hypothesis. In this model, females may choose to mate with males carrying ornaments and weapons that are positively correlated with age, inasmuch as older males are likely to have a higher average fitness (having already survived longer) than younger ones, and presumably carry fewer undesirable genes. Age-related traits may thus become a valuable clue to females for assessing individual male fitness (Manning, 1985). Male traits such as body mass (in some relatively large bird species) and weapons such as tarsal spurs are examples of such age-related traits in birds. Among other things, this model predicts that males should sexually mature and acquire their secondary sexual characteristics later in life than females (sexual bimaturism); thus females can more readily assess the ages of their prospective mates. There is some evidence for sexual bimaturism in groups such as the lekking grouse (Wiley, 1974), although some of the evidence is contradictory with regard to the expected distribution of mating systems (Wittenberger, 1978).

Not all direct selection possibilities are readily distinguishable, especially in the field, from certain indirect selection effects. Such effects include the unstable "runaway" process proposed by R. A. Fisher: effects resulting from increasing female preferences for progressively more elaborate male traits that may be of no general adaptive value, and may ultimately even be a survival disadvantage to the males carrying them. In ethological terms, this process could result from the females innately responding preferentially to certain "superoptimal" male secondary sexual characteristics that may be exhibited by a relatively few males in the population. Such female preferences may increase more rapidly than the traits can evolve, at least until adequate counterselection pressures come into play. This model predicts that some male secondary sexual traits, while critical to attracting females and mating, may place serious costs on a male's overall viability. A hypothetical example might be the remarkably long secondary feathers of male great argus pheasants (*Argusianus argus*). These feathers are actually longer than the birds' primaries, and additionally support all of the approximately 200 eyelike iridescent ocelli that are directly exposed to and oriented toward the female during an elaborate wing-spreading courtship display. These feathers must place a severe burden on males in terms of their flight efficiency, in addition to the energy cost of replacing the feathers during molting. Direct evidence on these points is still lacking. Kirkpatrick and Ryan suggest that the runaway model of sexual selection is of greater historical interest than it is presently empirically supportable.

Among the alternative good genes but indirect selection models of female preference, Kirkpatrick and Ryan suggest that females' mating choices may result in them pairing with the most vigorous and healthy available males, as an indirect reflection of the males' overall fitness. Examples include the idea that genetically disease-resistant or parasite-resistant males may reflect these desirable traits through their exhibition of more perfect or elaborate plumage, or by performing their sexual signaling more strenuously than diseased or parasite-prone birds. Such behavior might effect a genetic correlation between the female's mating preference and the male's genetic basis for pathogen resistance, so that the development of increased mating preferences for these male phenotypes would also indirectly select for the evolution of improved re-

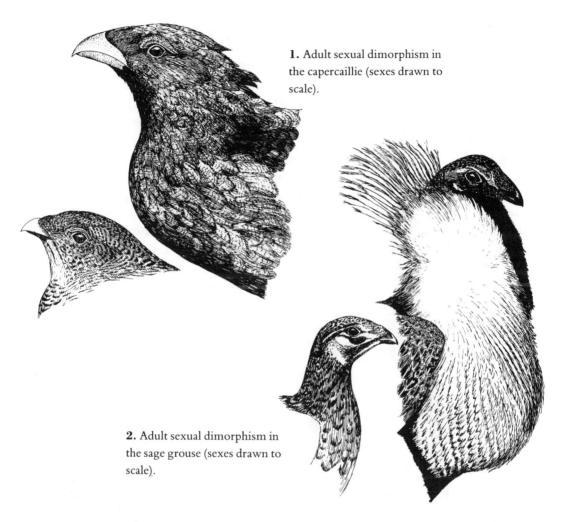

1. Adult sexual dimorphism in the capercaillie (sexes drawn to scale).

2. Adult sexual dimorphism in the sage grouse (sexes drawn to scale).

sistance to such diseases or parasites (Hamilton and Zuk, 1982). This hypothesis predicts that preferred mates within single species should have the fewest parasites, but among different species sexual selection should be most evident in those species that are especially subject to parasitic infections. There is an increasing body of evidence supporting the possibility that a direct relationship sometimes exists between mate choice by females and relative male parasite loads (Clayton, 1991; Loye and Zuk, 1991).

Secondary Sexual Traits in Monogamous and Polygynous Birds

The general anticipated result of sexual selection in polygynous bird species is that of increased sexual differences in three general categories. These include (1) adult mass (sexual size dimorphism), (2) adult plumage color and/or plumage length (sexual dichromatism), and (3) adult sexual behavior (sexual diethism). This last category includes heterosexual posturing, vocalizations, and similar mechanisms of intersexual advertisement.

For example, high levels of adult sexual dimorphism, dichromatism, and diethism occur in such polygynous grouse species as the Old World capercaillie (*Tetrao urogallus*) and its North American counterpart, the sage grouse (*Centrocercus urophasianus*) (Figures 1 and 2). By comparison with the highly polygynous and lek-forming sage grouse, males of the essentially monogamous white-tailed ptarmigan (*Lagopus leucurus*) are far smaller in overall size, and there is relatively little sexual size dimorphism or dichromatism apparent among adults (Figure 3).

More limited size dimorphism but a higher degree of sexual dichromatism is present in such forest-inhabiting pheasants as the Malay peacock-pheasant (*Polyplectron*

3. Comparative adult male traits of the polygynous, lekking sage grouse and the monogamous white-tailed ptarmigan (drawn to scale).

4. Adult sexual dichromatism in the Malay peacock-pheasant (sexes drawn to scale).

malacense) (Figure 4) and among the several montane forest species of Asiatic trag-opans (*Tragopan* spp.) (Figure 5). In the case of the tragopans it must be assumed that the elaborate and highly colorful lappets of adult males are primarily a reflection of sexual selection effects, rather than serving mainly as species-specific reproductive isolating mechanisms. This is because the various tragopans occupy generally al-lopatric geographic ranges along the Himalayan mountain chain from Pakistan to China. Furthermore, they show no tendency to reinforce these interspecific differ-ences in their limited areas of interspecific sympatry, where selection against hybrid-ization might perhaps be expected to occur. Although tragopans have generally been believed to form monogamous if short-term pair bonds (Johnsgard, 1986), the males' complex display behavior and the relatively high sexual size dimorphism present in the genus suggest that promiscuous mating systems may exist (Sigurjønsdøttir, 1981).

Such examples of sexual differences are characteristic of the typical patterns of sex-ual dimorphism (using this term broadly to include dichromatism and diethism) that will be described repeatedly in the chapters on specific bird groups that follow. How-ever, exceptions to these predictions and patterns do exist, and it is important to consider these real or apparent exceptions in the light of sexual and classic natural selection. That is, do they make evolutionary "sense," or do they weaken the general argument that has been advanced for using sexual selection as a general explanation for sexual dimorphism in birds?

One of the problems with the use of sexual selection as a general explanation for the presence of brightly colored male birds is that bright or ornate male plumages can also be found among species of some strongly monogamous bird groups. Such groups include, for example, most typical ducks (Anatinae), as well as most species of New World warblers (Parulini) and tanagers (Thraupini) and the Old World sunbirds (Nectariniidae). A variety of hypotheses have been advanced to try to account for these exceptional and generally unexpected cases, as summarized in Table 2.

As Darwin suggested long ago, sexual selection can help explain some of these seemingly anomalous situations, and more recently Kirkpatrick, Price, and Arnold (1990) have suggested several possible mechanisms by which sexual selection can op-erate in monogamous or basically monogamous birds. For example, it can occur (1) where early-breeding females that are in excellent breeding condition are able to ob-tain the best males, and thus enjoy the highest rates of reproductive success. Or, (2) unusually attractive males may even stimulate females to increase their reproductive effort. Additionally, males of species that are normally monogamous may increase their reproductive output by (3) occasionally obtaining second mates following the completion of their first breeding cycle, or (4) obtaining extrapair copulations, there-by effectively becoming polygynous. Recently Wagner (1992) has reported that in the strictly monogamous razorbill (*Alca torda*) both sexes gather in arena-like groups dur-

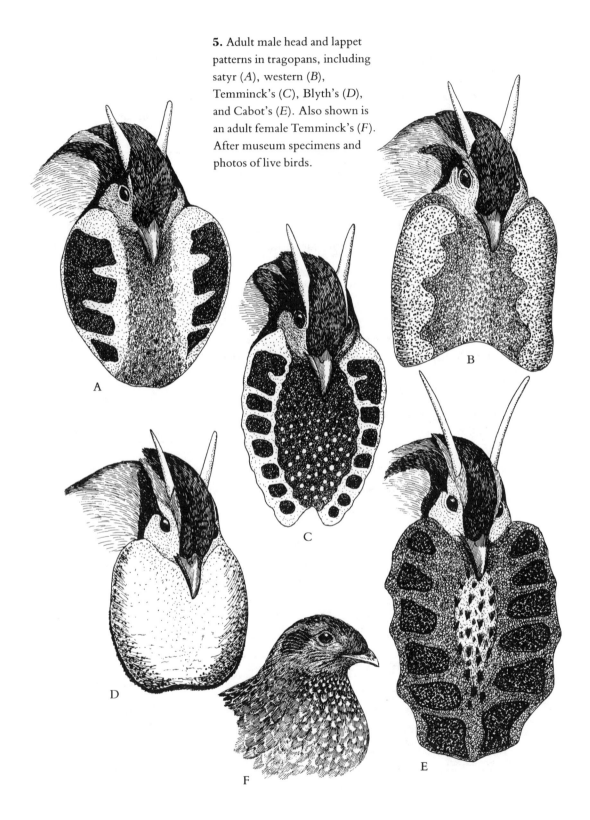

5. Adult male head and lappet patterns in tragopans, including satyr (*A*), western (*B*), Temminck's (*C*), Blyth's (*D*), and Cabot's (*E*). Also shown is an adult female Temminck's (*F*). After museum specimens and photos of live birds.

**Table 2 • Hypotheses That Relate to Male Showiness
among Normally Monogamous Species of Birds**[1]

I. Hypotheses that are directly related to sexual selection and differential male mating success.

    A. Differential access hypotheses: relative male showiness affects individual access to potential mates.

        1. Individual male fitness/vigor hypothesis: showiness may reflect a male's fitness, based on (1) its relative age, experience, and maturity, (2) its relative ability to provide good parental care, or (3) its general health, such as its relative parasite or pathogen load, allowing such males to mate earlier and produce more offspring.

        2. Fecund female hypothesis: healthiest and most fecund females will mate earlier, select males having the most showy traits, and produce more offspring (intergrades functionally with the preceding hypothesis).

        3. Sex ratio effects: operational sex ratio may have an excess of males, with only the more showy ones obtaining females.

        4. Facultative polygyny effects: females may not all breed simultaneously, thus allowing the more showy males to mate sequentially with additional females during a single breeding season.

        5. Facultative promiscuity effects: increased showiness may increase a male's chances for stealing extrapair copulations.

    B. Differential allocation hypothesis: mating with more attractive males may induce females to increase their reproductive output (Burley, 1986).

II. Hypotheses wholly independent of sexual selection effects.

    A. Predation deflection ("unprofitable prey") hypothesis: bright males might be signaling to predators that they represent unprofitable prey choices.

    B. Ecological constraints hypothesis: habitat-related showy colors may actually be cryptic in typical environments (e.g., the "bright" greens of trogons, turacos, and parrots) or may serve physiological (e.g., heat-regulatory) functions.

    C. Phylogenetic inertia hypothesis: postadaptive phenotypic traits such as male showiness may persist in a once-nonmonogamous population long after monogamy has developed and a need for male showiness has ceased.

1. Classification partly after Zuk (1991).

ing the prelaying period. There males compete strongly to obtain extrapair copulations, and the most aggressive and persistent males are those most likely to achieve these. Thus, under such circumstances sexual selection might operate even in such unlikely groups as the strongly monogamous alcids.

Sexual selection may also come into play among monogamous species where (5) adult sex ratios are skewed so as to produce a marked excess of males. This is the usual situation among anatine ducks, whose pair bonds are formed under intense male competition for mates, and usually last only until the female begins incubation. Sexual selection may not be highly effective for maintaining bright male plumages in most of these ducks, inasmuch as dull male nuptial plumages are typical of many island-dwelling or otherwise geographically isolated waterfowl species, where selection for reproductive isolation is now lacking but where sexual selection might still persist (Sibley, 1957). Yet in many tropical areas of Africa and South America, where breeding seasons may be long or irregular, males of many duck species remain brightly colored year-round. Thus they are continuously in breeding plumage, rather than undergoing periodic dull or "eclipse" plumages between annual breeding cycles. Presumably this results, at least in part, because of sexual selection pressures to maintain the males in a constantly "attractive" state.

Sexual selection may thus occur in monogamous species where long breeding seasons may allow early-breeding females to choose the "best" available males, as Darwin originally suggested, where the long breeding season allows a male to mate sequentially but monogamously with a succession of females during a single breeding season, or where opportunities for extrapair copulations exist. A few arguments completely independent of sexual selection have also been advanced to help explain some cases of male-only showiness that seem to defy simple explanation. These arguments often seem to stretch credibility, such as the phylogenetic inertia hypothesis and the predator deflection hypothesis (Table 2). Sigurjønsdøttir (1981) found no statistical evidence supporting a predation-based (unprofitable prey) hypothesis advanced by Baker and Parker (1979), at least with regard to galliform game birds. Baker and Parker had suggested that the bright colors of males of some species that do not remain with their young have evolved to reduce predation pressures, insofar as predators might learn that chasing such brightly colored prey would be relatively unprofitable, and that dull-colored prey might provide more attractive targets. The first field evidence supporting this explanation for bright coloration in male birds is apparently that of Götmark (1992).

An equally difficult theoretical situation arises in those cases where sexual selection might be expected to have produced male showiness in a group of birds but hasn't. Equally surprising are those situations where adults of both sexes exhibit exaggerated or relatively showy plumage development. This condition is particularly unexpected in those habitats where such conspicuous or elaborated plumage features would seemingly place showy females at a selective disadvantage. Again, a variety of ideas have been advanced to try to account for these seeming anomalies, as summarized in Table 3. Most of these are self-explanatory; a few will be discussed later in this book where appropriate.

Sexual Selection and Acoustic and Visual Male Signals

With regard to the role of sexual selection in promoting sexual diethism in acoustic traits, Searcy and Andersson (1986) have reviewed the possible relationship between sexual selection and song evolution in birds, frogs, and insects. For birds, they concluded that differences in song repertoire size among individual males may perhaps influence female choice and hence the individual males' relative mating success. However, few studies of this possibility have fully controlled for the possibility of other

Table 3 • Hypotheses That Relate to Sexual Monomorphism among Nonmonogamous Birds[1]

I. Hypotheses that relate to a lack of showiness or exaggerated plumage development in both sexes.
- A. Phylogenetic inertia hypothesis: assumes that the ancestral stock was monomorphic, and that sexual selection has not yet had enough time to produce sexual dimorphism.
- B. Behavior transference hypothesis: suggests that behavioral differences (sexual diethism) produced by sexual selection may substitute for morphological differences (sexual dimorphism) or coloration differences (sexual dichromatism).

II. Hypotheses that relate to the presence of showiness or exaggerated plumage development in both sexes.
- A. Social selection hypotheses: assume that intraspecific social selection pressures may favor monomorphism.
   1. Balanced selection: mutual sexual selection may operate in both sexes during breeding, favoring mutual displays and affecting the complexity of signaling devices in both sexes (Huxley, 1914).
   2. Selection for male mimicry by females.
      - a. Male mimicry may increase female fitness during competition with males while foraging (Bleiweiss, 1985).
      - b. Male mimicry may help to intimidate other females during competition for copulations on display grounds (Trail, 1990).
- B. Hypotheses independent of social selection pressures.
   1. Conspicuous plumage in both sexes may be the result of ecological constraints (see Table 2).
   2. Conspicuous plumages in both sexes may be predator-related, advertising unprofitable prey (see Table 2).

---

1. Adapted with modifications from Trail (1990).

likely influences on female choice (such as variations in territory quality). In terms of the possible role of male songs in influencing mating success through the effects of intermale song "contests," most evidence so far available comes from the obviously negative effects of muting experiments. In such cases, all-or-none (song vs. no song) differences rather than vocal quality variations are the operative experimental factors being measured. Relatively little information is yet available on how graduated variations in male song performance might influence social status or relative dominance among males.

The selective advantages inherent in both female choice and male dominance effects of song are more difficult to discover. Searcy and Andersson suggested that, in terms of female choice, they are most likely to involve such readily understandable advantages as enhancing reproductive isolation between species through the females' selective attraction to the songs of conspecifics. Quantitative male song variations might also reduce the energy costs of movement for females that are searching for mates, through their passive attraction to the most "intense" advertising signal available to them.

Adaptive advantages in female choice of specific male song traits might possibly also relate to enhanced opportunities for material benefits (assuming that females actually receive any direct benefits from males beyond simple fertilization, and that song quality is somehow directly related to these benefits), or to enhanced male fertility (again assuming a relationship exists between a male's song attributes and his overall fertility). Related to questions of a male's fertility are those of his overall genetic quality, and the possibility that genetically based male song variations might allow mate choices that influence heterozygosity and related phenomena. Finally, there is the possibility that female choice of elaborate male songs may simply be a result of runaway (Fisherian) selection, rather than being directly adaptive. Searcy and Andersson suggested that most or all of these hypotheses may apply to birds generally, but that few have been specifically established as significant operative factors in real-world situations.

Similarly, evolved male song interactions with other males may well have varied functions that are only indirectly related to sexual selection as such. A male may engage in male–male song contests because such behavior may (1) help to establish his species identity (functioning in reproductive isolation), (2) assist his resource-holding abilities (probably especially important for monogamous rather than polygynous species), (3) help announce and perhaps thus control the local density of conspecific males (perhaps thereby signaling relative individual fitness), or (4) proclaim his individual identity. Of these possibilities, the species identity explanation was judged the simplest and most attractive by Searcy and Andersson, and this role is independent of direct sexual selection functions. Good evidence supporting the possibility that males can dominate, or be dominated by, other males through their vocal abilities alone is still lacking.

In a related discussion of sexual selection and avian acoustic signaling behavior, Loffredo and Borgia (1986) tested the prediction that males of polygynous species of birds having leklike mating systems should exhibit convergently similar acoustic displays. They found polygynous species in a wide range of selected taxa to be more noisy, to use a broader range of sound frequencies, to be less melodic, and to use a greater number of nonvocal sounds than do their monogamous counterparts. By comparison, monogamous species are more prone to generate highly tonal sounds that facilitate individual recognition, and, in contrast to polygynous species, monogamously paired birds often participate in vocal duetting and antiphonal singing. These results confirmed Loffredo and Borgia's general prediction, that the kinds of sounds birds generate tend to maximize male fitness among species in each of the mating systems studied. The results also support the view that the acoustic attributes of such advertising signals are not arbitrary, but rather show broad similarities across taxonomic lines that may result from their comparable use by females in assessing the relative dominance of prospective mates.

As to the role of sexual selection in the evolution of male visual traits as secondary sexual signals, and associated female preferences for such specific traits, recent field evidence for that basic Darwinian tenet is finally beginning to accrue. One such example comes from Swedish field work (von Schantz et al., 1989; Göransson et al., 1990), which established that in a feral population of ring-necked pheasants (*Phasianus colchicus*) the most important predictor of harem size was tarsal spur length. Additionally, both the general phenotypic condition of the males and their overall estimated viability (based on year-to-year survival records of males that were at least two years old) were also significantly correlated with male spur length. This example, the first to establish a significant correlation between a sexually selected male trait and individual male viability, strongly supports the view that viability-based processes (rather than nonadaptive runaway processes) can contribute to the evolution of secondary sexual traits in males and the development of female choice tendencies for such traits. Additionally, female reproductive success was also correlated with male spur length, a surprising finding that Pomiankowski (1989) suggested might be accounted for if male spur length is actually a condition-dependent trait. Unlike many polygynous species, male pheasants establish territories within which females forage and perhaps nest; a male's condition may thus reflect the relative quality of his territory.

In a related field study, Wittzell (1991) likewise reported spur length to be the most important phenotypic factor to be correlated with male pheasant reproductive success, based on both observed harem sizes and estimated number of sired chicks. He additionally found spur length to be positively correlated with his estimates of male survival, the criteria of which differed somewhat from those used by Göransson et al. (1990). In a British study that used different and perhaps more reliable mate choice criteria (copulatory solicitation behavior), Hillgarth (1990b) was not able to establish a positive correlation between individual male spur lengths and female mating choices. However, mate choice correlations were established (Hillgarth, 1990a) with display rates of males and with their estimated parasite loads (males that called frequently and displayed for longer durations, and those that carried low levels of coccidial infections, were preferred).

In an important field study, Andersson (1989, 1992) examined female mating choice characteristics in the lek-breeding Jackson's widowbird (*Euplectes jacksoni*), a sparrowlike bird of eastern Africa. In this species, individual male success in obtaining copulations evidently depends on several factors, including display rate and lek attendance, which influence the frequency of female visits to a specific male, and relative male tail length, which affects the probability of a male copulating with a visiting female. Males with experimentally shortened tails received fewer female visits than did control males. Additionally, tail length in unaltered males was found to be correlated with a body-mass-to-tarsal-length index of overall body condition, which suggests that, like pheasant spur length, tail length in these birds may serve as a convenient and reliable visual indicator of male viability for females.

Several other recent studies have provided data supporting the view that females of various birds may likewise cue on male ornamental feather length traits as a basis for making mating choices (Cherry, 1990). In the Lawes' parotia (*Parotia lawesii*), a small bird-of-paradise, males' mating success was found to be correlated with the lengths of their ornamental head wires (Pruett-Jones and Pruett-Jones, 1990). Additionally, experimentally altered male tail lengths in two African sparrowlike birds, the long-tailed widowbird (*Euplectes progne*) and the queen whydah (*Vidua regia*), have been found to influence mate choice behavior by females (Andersson, 1982; Barnard, 1990). Surprisingly, a similar tendency to choose males bearing experimentally exaggerated male tail feather traits has been found in a monogamous swallow species (*Hirundo rustica*), and this tendency has likewise been attributed to sexual selection (Möller, 1988, 1991).

Sometimes feather color or male plumage pattern, rather than feather length, may be important in mate choice by females. For example, in the great snipe (*Gallinago media*) the amount of white on the outer tail feathers of the nocturnally displaying males may be an important aspect of their individual breeding success (Höglund, Eriksson, and Lindell, 1990). Similarly, the degree of ornamentation (number of eye-like iridescent ocelli) and the relative symmetry of ornamental markings in the train of adult male blue peacocks (*Pavo cristatus*) may provide a simple visual basis for mate choice in this species. Variations in visual characteristics of the peacock's train are age-correlated; therefore female selection for particular ornamental train attributes such as number of ocelli or train symmetry may result in selection favoring reproduction by older males (Manning, 1987, 1989; Manning and Hartley, 1991). Sexual selection for colorful patterns might also occur in monogamous species, since females have been found to prefer more colorful males in the monogamous house finch (*Carpodacus mexicanus*) (Hill, 1990). As will be discussed later, similar evidence regarding female mate choice and male plumage traits is now accruing for mallard ducks (*Anas platyrhynchos*) (Klint, 1980; Weidmann, 1990).

In some polygynous species, environmental "signal posts" established by resident males, rather than male secondary sexual traits as such, may serve as effective male dominance markers, and may also be selected for through female choice tendencies. Thus in bowerbirds (Ptilonorhynchidae), the elaborate structures built, maintained, variably decorated, and strongly defended by individual males may serve to inform females of their dominance status, and thus may also serve as indirect indicators of relative male quality (Borgia, Pruett-Jones, and Pruett-Jones, 1985; Borgia, 1986). Indeed, in the satin bowerbird (*Ptilonorhynchus violaceus*) the degree of bower ornamentation has been experimentally established to influence mate choice by females, which is strong evidence for the significance of individual variations in bowers and their decorations in terms of sexual selection value in attracting females (Borgia, 1985b; Borgia and Gore, 1986).

Female Mate Choice and Male Aggregation Strategies in Polygynous Birds

The relative costs and benefits to participating males of participating in competitive mating situations, especially aggregated ones, and to females of choosing their mates from the available males thus participating, are obviously important theoretical aspects of sexual selection. Clearly, males and females may employ quite different "strategies" in terms of the reproductive benefits that might potentially accrue from participating, and the risks or costs of engaging in this behavior. More specifically, what are the most important factors likely to influence mate choice in an adaptive manner when females are choosing mating partners in situations where several potential mates are simultaneously available, as is typically the case in arena species? Alternatively, what are the relative costs and benefits to males in participating in such competitive mating aggregations, as opposed to those of advertising solitarily?

These two important questions have been addressed by many writers, and some of their suggested answers are summarized in Tables 4 and 5. Posing the question in terms of female costs and benefits (Table 4) produces a balance sheet in which various risks are compensated for by adopting appropriate cost–cutting measures, and benefits are seemingly maximized by using some fairly simple rules that may be followed when females are forced to choose among several potential mates. Among the most widespread of these apparent rules are that (1) visiting larger leks is preferable to visiting smaller ones (maximizing choice opportunities and perhaps reducing individual predation risks), (2) long-distance cues (visual or acoustic) can be used for readily lo-

Table 4 • Costs and Benefits Influencing Female Mate Choice Strategies in Lekking and Other "Noneconomic" Avian Mating Systems[1]

Cost/Benefit Criterion	Potential Related Behavioral Strategies	Suggested Avian Examples
Potential Female Costs		
1. Difficulties of identifying the most fit males	a. Choose obvious age- or fitness-related male traits	Ring-necked pheasant, blue peafowl, Jackson's widowbird
2. Increased predation risk	a. Favor larger or denser leks (to reduce individual risk)	Lekking grouse, white-bearded and golden-headed manakins, Guianan cock-of-the-rock
	b. Favor safer (centrally located) males	Lekking grouse, ruff, great snipe
	c. Favor males providing protective shelters	Bowerbirds?
3. Travel time and energy expenditure	a. Use long-distance cues	Probably widespread
	b. Attract males to sites that females already use ("hot spots")	Village indigobird, green hermit, golden-headed manakin
4. Presence of competing females	a. Arrive early and mate quickly	Probably widespread, esp. lekking grouse
	b. Dominate other females using the same lek	Sharp-tailed and sage grouse
5. Danger of injury from other birds	a. Avoid male harassment by showing submissive signals	Probably widespread
	b. Employ intersexual mimicry	Calfbird
	c. Favor less aggressive males for copulation	No certain examples
6. Danger of other males disrupting copulations	a. Favor males that deter disruption by other subordinate males	White-bearded and long-tailed manakins; Guianan cock-of-the-rock
	b. Favor leks with stable male hierarchy	Prairie-chickens

(continued)

Table 4 • (*Continued*)

Cost/Benefit Criterion	Potential Related Behavioral Strategies	Suggested Avian Examples
Potential Female Benefits		
1. Access to many males from which to make best mate choice	a. Choose most dominant ("hotshot") males	Probably widespread; see Table 6
	b. Choose most conspicuous and/or elaborate males	Red junglefowl, blue pea-fowl, sage grouse, great snipe, long-tailed and Jackson's widowbirds
	c. Choose most actively displaying males	Sage grouse, Lawes' parotia
	d. Choose most healthy and parasite-free males	Ring-necked pheasant, red junglefowl, satin bowerbird, Lawes' parotia
	e. Copy other females' mating choices	Black and sage grouse, ruff, white-bearded and long-tailed manakins
	f. Go to larger and/or denser leks	Widespread? (see above)
2. Increased fertilization probability	a. Select mature males	See Table 7
	b. Select males having sufficient sperm (i.e., large testes)	Sharp-tailed grouse, long-tailed manakin

1. Adapted in part from Reynolds and Gross (1990) but excluding nonavian examples, and additionally including rock ptarmigan (Brodsky, 1988), sharp-tailed grouse (Nitchuk and Evans, 1978; Tsuji, Kozlovic, and Sokolowski, 1992), black grouse (Höglund, Alatalo, and Lundberg, 1989), sage grouse (Hartzler and Jenni, 1988), red junglefowl (Zuk et al., 1990a), blue peafowl (Manning, 1987; Petrie, Halliday, and Sanders, 1991), ring-necked pheasant (Göransson et al., 1990), great snipe (Höglund and Robertson, 1990), calfbird (Trail, 1990), satin bowerbird (Borgia, 1985b; Borgia and Collis, 1990), Lawes' parotia (Pruett-Jones and Pruett-Jones, 1990), Jackson's widowbird (Andersson, 1989), and long-tailed widowbird (Craig, 1988).

cating such leks, (3) leks should be visited early (and if necessary repeatedly) to avoid undue competition from other females, and (4) mating partners should be chosen that are both socially dominant and sexually experienced, to avoid disruption from other males and to achieve effective fertilization. In the absence of other apparent clues as to best mating choices, following the lead of other females may be an appropriate and commonly adopted mating choice strategy (Pruett-Jones, 1992).

Similarly, males probably incur significant costs or risks by participating in competitive mating aggregations (Table 5). The most important may be the reduced probabilities of mating for all but a few of the most successful individuals. Furthermore, nearby males in the cluster may directly interfere with or interrupt the limited copulation opportunities. Additionally, frequent fighting among the males over territory or dominance status may drain energy from and perhaps invite risk of serious

Table 5 • Costs and Benefits Influencing Male Clustering Strategies in "Noneconomic" Polygynous Avian Mating Systems[1]

Costs/Benefits	Specific Influences	Suggested Examples
Potential Male Costs		
1. Mate competition	Grouped males will reduce individual mating chances	Self-evident
2. Predation risk	Grouped males may attract predators	Sage and black grouse, Guianan cock-of-the-rock
3. Male sexual interference	Copulation efforts may be disrupted by other nearby competing males	Grouse spp., ruff, buff-breasted sandpiper, white-bearded and golden-headed manakins
4. Male competitive aggression	Male fighting may lead to serious injury or death	Ruff, capercaillie
5. Sperm limitations	Successful males may exhaust their sperm supplies during intensive mating activities	No direct evidence
6. Genetic effects on population	Restriction of mating to a few (alpha) males may reduce genetic diversity	No direct evidence
Potential Group Benefits		
1. Better signal generation	Grouped males may generate strongest and/or most stimulating sexual signals	Lekking grouse, black-and-white manakin

(continued)

Table 5 • (*Continued*)

Costs/Benefits	Specific Influences	Suggested Examples
2. Better predation avoidance	Grouped males may reduce predation risk by numbers or by increased vigilance	Lekking grouse, Guianan cock-of-the-rock
3. Information pooling potential	Males may get information on limited food resources by aggregating	Black grouse
Potential Individual Male Benefits		
1. Individual fitness filter	Alpha males may control most mating opportunities, transmitting "best" genes	Many examples
2. Individual fitness "badges"	Alpha males may acquire "reputations," thus avoiding constant challenges and disruptive fighting	Self-evident
3. Site inheritance possibility	Younger beta male(s) may outlive and inherit the territory of alpha male; thus are rewarded for cooperating with alpha male	Long-tailed manakin
Other Possible Influences		
1. Habitat constraints	Specific habitat attributes may be needed for optimum male display sites, such as high-visibility locations	Birds-of-paradise, forest and prairie grouse, ruff, bustards
2. Prior female distributional tendencies	Males may aggregate at sites that females often use for varied reasons ("hot spots")	Green hermit, village indigobird

1. In small part derived from Bradbury and Gibson (1983).

injury to participants. The attraction of predators to the assembled and conspicuous males may also pose a serious threat. This is especially true for birds on the site's periphery, which may be forced into locations placing them in greater risk than are more centrally located individuals. Perhaps increased male crowding and contacts may also facilitate the transmission of parasites or pathogens among the participants.

On the other hand, the payoffs for a few of the participating males may be great indeed. Not only might the grouped males be able to attract far more females than a single male could by displaying alone, but one or more of the males is likely to be able to dominate mating opportunities, and thereby pass on a greatly disproportionate number of his own genes to the next generation. Well-structured male aggregations should thus ideally function as effective individual fitness filters, reducing or preventing matings by less fit males, and concentrating these matings among the males whose genes are most likely to benefit the species as a whole.

It is this highly skewed male mating distribution that certainly lies at the heart of such male congregations. It is now well established that such skewed mating distributions occur among males of species representing virtually all of the arena-breeding bird groups (Table 6). In all of these species, and surely in many more, a few most successful or alpha males (column *a* in Table 6) have been found to obtain anywhere from a significant minority to a majority or sometimes even 100 percent of the mating opportunities for the entire group. Frequently, secondarily successful or beta males (column *b* in the table) share significantly in these activities too. However, only in a relatively few species does regular access to mating opportunities extend beyond about four or five birds in the entire male group, regardless of its size.

The characteristics that help to define alpha males are probably varied, and certainly include not only age but also less easily measured traits, such as social dominance, individual appearance, body mass, and perhaps also display rates. However, there is no doubt that calendar age is positively correlated with reproductive success in many arena-forming species, as is summarized in Table 7, and females of polygynous species that choose to mate with older males may automatically be obtaining mates that are of higher than average genetic quality (Weatherhead, 1984). Clearly there is not necessarily an indefinite linear age-dominance relationship, and quite possibly some species (such as lekking species of prairie grouse) might reach the peak of their vigor and sexual interest at three or four years of age only to decline again rather rapidly thereafter. On the other hand, it would seem that males of some passerine avian groups such as lyrebirds, manakins, bowerbirds, and birds-of-paradise may not become significant contestants for dominance and reproductive success until they are about five to seven years of age, and indeed they may not even acquire their fully adult male nuptial plumages until then.

Manning (1985, 1987) has argued that male age may sometimes serve as a direct fitness indicator in birds and mammals. This is the case because older males, having thus demonstrated their survival capacity, must possess a generally higher average

Table 6 • Relative Individual Mating Success Observed among the Most Successful Males in Arena-Breeding Birds

Species	Total Males	Total Matings	Proportion of Mating by Dominant Males (%)					Authority
			a	b	c	d	e	
Sage grouse	30	87	47	29	6	4	4	Wiley, 1973
	ca. 50	51	29	25	14	8	6	Lumsden, 1968
	20–50	444[1]	61	21	8	5	ca.1	Hartzler & Jenni, 1988
Black grouse	8–10	46[1]	28	24	24	15	6	Kruijt & Hogan, 1967
Capercaillie	5	23	92	8	—	—	—	Müller, 1979
Greater prairie-chicken	9	30	70	23	7	—	—	Robel, 1966
Sharp-tailed grouse	15+	17	76	?	?	?	?	Lumsden, 1965
Great snipe	13	29[2]	28	17	15	13	10	Höglund & Lundberg, 1987
Buff-breasted sandpiper	22	22	32	18	14	9	9	Payne, 1984
Ruff	21+	85	52	16	10	6	4	van Rhijn, 1991
	15	82	23	17	17	16	13	Bancke & Meesenburg, 1958
Calfbird	8	41	100	—	—	—	—	Trail, 1990
Guianan cock-of-the-rock	61	138[2]	30	8	8	7	7	Trail, 1985c
Golden-headed manakin	13	87	25	16	16	16	14	Lill, 1976
Long-tailed manakin	85	117[1]	67	?	?	?	?	McDonald, 1989b
White-bearded manakin	20	76	32	17	14	12	10	Lill, 1974a

(continued)

Table 6 • (*Continued*)

Species	Total Males	Total Matings	Proportion of Mating by Dominant Males (%)					Authority
			a	b	c	d	e	
Satin bower-bird	37[3]	212[2]	15	12	12	10	9	Borgia, 1985b
Lawes' parotia	23	103[2]	27	20	14	11	8	Pruett-Jones & Pruett-Jones, 1990
Lesser bird-of-paradise	8	25	96	4	—	—	—	Beehler, 1983d
Raggiana bird-of-paradise	5	35	43	20	11	9	—	Beehler, 1988
Green indigo-bird	11	31	61	22	6	3	3	Payne & Payne, 1977
Jackson's widowbird	36[4]	29	14	11	8	8	5	Andersson, 1989

1. Two or more years of data combined.
2. Percentages only approximate; estimated from graphic summary.
3. Comparison is of separate bower sites, not male groups.
4. Comparison involves results from four separate leks.

level of fitness than do younger ones, at least until aging effects begin to reduce their fertility. The size and complexity of many male ornaments and dominance-related weapons such as tarsal spurs are often positively correlated with age, and sometimes body mass itself may be age-related, as in males of some unusually large species of bustards and those of a few other groups. Thus, females choosing to mate with males showing such traits are also automatically selecting older than average males. The size or complexity of these age-related traits will thus gradually increase, until the reproductive advantages to males bearing such structures are counterbalanced by other more general considerations of survival. Certainly more information is needed on the characteristics of alpha males that help females to identify them and help them to maintain their positions of rank and unrestricted access to females without engaging in constant fighting.

Table 7 • Examples of Male Age-Related Dominance/Fitness Ratios and/or Age-Dependent Sexual Success Rates in Lekking Birds

Sage grouse: The male with the largest number of observed copulations (63) was at least three years old, and during three years a small proportion of older males performed most copulations (Hartzler and Jenni, 1988).

Black grouse: Of 449 observed copulations, juveniles obtained only 1.6 percent. Copulation success increased as males matured, at least to their fifth year, when over half of such males held central territories. Central territories were initially attained at 2–5 years of age, with 95 percent attained when 2–4 years old. Central territory holders who were 3–4 years old averaged 9.6 matings each; males of the same age not holding central territories averaged 2.3 matings (de Vos, 1983).

Capercaillie: The oldest and most aggressive males have the highest social rank, hold the largest territories, and are selected by females for mating in 90 percent of the cases (Müller, 1974). One alpha male was 13.5 years old at his death (Müller, 1979).

Greater prairie-chicken: Of 97 copulations, none were by juveniles, and most were by males at least three years old (Robel, 1966).

Sharp-tailed grouse: 13 percent of territorial males performed 93 percent of observed copulations. With one exception, these males held central territories, gained by overt aggression or by establishing peripheral territories as yearlings and gradually moving toward the center as the older birds died (Kermott, 1982). One alpha male was nearly three years old (Lumsden, 1965).

Swallow-tailed manakin: Relative male dominance and reproductive success are evidently age-related, with adult plumage acquisition requiring 2–3 years, and a linear dominance hierarchy existing (Foster, 1981).

Long-tailed manakin: Males under three years old had subdefinitive plumages and low social rank. Beta males were at least 8 years old; alpha males were at least 7–10 years old (McDonald, 1989a).

Golden-headed manakin: During first three years males progressed through "visitor" and "intruder" stages to "resident" status. Residency may persist for 6–8 years (Lill, 1976).

White-bearded manakin: One relatively old male was highly successful in mating over a three-year period (Snow, 1962).

Summary

With the passage of more than a century, it is evident that much of Charles Darwin's basic theory of sexual selection remains largely intact, and indeed his ideas on the subject are held in higher regard than at perhaps any time since they first appeared. Consider that the most critical elements of the theory—namely the presence of male dominance gradients within local populations and the capacity of females to make active mate choices from an array of nearly identical males comprising such populations—were largely based on the scant and often anecdotal information that was available to Darwin. He freely admitted that, because of a lack of firm data on these points, his arguments supporting the theory were necessarily "highly speculative."

This intellectual and personal gamble on Darwin's part becomes even more impressive when one remembers that he applied his sexual selection theory to vertebrates (and some nonvertebrates) generally, and that he specifically tried to account for the presence of many distinctively human features through his evolutionary (if not revolutionary) reasoning. The human attributes he considered included not only rather obvious sex-related morphological traits such as adult secondary sexual anatomical structures, but also such seemingly unique and presumably God-given mental faculties as a sense of aesthetic beauty.

Darwin thus used his sexual selection theory to support and supplement his earlier general views on organic evolution by natural selection in the animal kingdom generally. Furthermore, he assembled strong evidence supporting his belief that the human species itself, as well as many of the "higher" mental traits especially characteristic of humans, are direct evolutionary products of natural and sexual selection. It is hard to imagine the personal pains and social pressures that Darwin must have had to endure as a result of such intellectual courage. By thus exposing and defending his admittedly "highly irreligious" views, he effectively demolished any remaining doubts that contemporary scientists might have held about the common ancestry of humans and the animal kingdom in general. Just as Galileo and Copernicus had placed the earth in its own proper, if relatively obscure, position within the universe, Darwin had finally provided humanity with an honest view of our true relationship to the natural world around us, as well as an objective mechanism for understanding that world. With Darwin's help, biology had acquired the basis for becoming a modern science, and science as a whole was gradually freed of religious interference.

Having now reviewed basic sexual selection theory, the following chapter will elaborate on two aspects of avian social behavior that are most strongly impacted by sexual selection. One of these is the tendency for sexually active males to cluster to varying degrees, and to establish relatively stable male-to-male dominance relationships within such interacting groups. The other important attribute is the tendency for prenesting females to be attracted to such male assemblages, and to make definite mating choices from among the participating males, with their choices usually directly related to the already established male dominance relationships.

2

Arenas, Courts, and Leks: Dominant Males, Discerning Females

Arenas and Courts

In what was perhaps the first extended discussion of arena behavior in birds, Armstrong (1947) described "a definite and remarkable type of display which we shall call 'arena display.' In its characteristic form it takes place on an area frequented by a number of birds. It involves the precise localisation of display." Armstrong restricted the use of the term "arena" to the birds' general displaying area, whereas an individual's specific display site, "when tended, defended or kept bare of herbage and litter," was defined as a "court." Species of birds that participate in this kind of social display were called lek birds, using a Swedish term already in English use for describing the male gatherings of black grouse (*Tetrao tetrix*) and ruffs (*Philomachus pugnax*) on traditionally used display grounds. In using the term *arena* in this context, Armstrong was essentially accepting the terminology proposed earlier by Mayr (1935), who defined arena behavior as a type of territoriality in which the males establish a mating station having no connection with feeding or nesting sites. Armstrong thus emphasized the congregational and locational aspects of this type of mating behavior, rather than the absence of any specific breeding resources within a male's territory.

Gilliard (1962) later defined an avian arena in much the same manner as had Armstrong, namely as the collective display site of a local population. He also similarly defined a male's court as the specific display site of an individual male. Later (1969) Gilliard stated that arena behavior involves a group of males gathered around a traditional mating site, the arena. He suggested that in bowerbirds, arena behavior represents male courtship reshaped by males that have been emancipated from nesting responsibilities to include certain "non-discardable" nesting tendencies. In this way, the bowers of bowerbirds were regarded as representing behavioral relics of the males' inherent nest-building instincts.

In his comparative review of grouse mating behavior, Hjorth (1970) identified the general area where most male grouse display as their "display ground," and the portion of the display ground where several males show territorial activity as an "arena." With regard to black grouse (*Tetrao tetrix*), de Vos (1983) defined a territory as that part of the home range of an individual male grouse where it is dominant over all other individuals of the species. A "display ground" was defined as the area where the display sites of one or several males are located; it was specifically called an "arena" if the display sites of several males are closely situated.

Snow (1982, 1985) seemingly avoided the use of "arenas" and "arena behaviors" in discussing these phenomena. Rather he defined an avian "display ground" as a communal male display site. He called the display site of an individual male a "court" (especially when it is located on the ground; otherwise he referred to it as a "display perch"). In his discussion of manakin mating behavior, Snow (1963c) referred to avian group or "collective" displays as those where two or more males are associated spatially, including those cases in which males are sexual rivals, and true "communal" displays as those where participating males take an equal part in joint display performance. Later Schwartz and Snow (1978) defined a manakin's "display area" as an area claimed by an individual male as a territory, in which he displays, and into which he accepts females and occasionally also males (for communal displays).

Leks and Lekking Behavior

Although the term "lek" has long been used for the arena displays of black grouse (Selous, 1909–10), Armstrong (1947) referred more generally to "lek" species as those that participate in localized arena display.

In a general discussion of avian lekking behavior, Snow (1985) defined a lek as a communal display ground (or arena) where several males congregate for the sole purpose of attracting and courting females, and to which females come for mating. Snow noted that display grounds where participating males are well separated but within earshot of one another may be described as "dispersed" or "exploded" leks, and that the degree of male dispersion in some species may perhaps vary with population densities. Hjorth (1970) restricted his use of the term "lek" to refer to the collective male grouse displays performed within an arena. He also clearly restricted the term "lek" to use as a verb rather than a noun.

Oring (1982) stated that use of the term "lek" should be limited to those situations in which grouped males establish small, closely positioned mating courts. Females visit such groups only for purposes of choosing mates and mating but subsequently nest away from the arena, and they alone provide parental care. This definition, which clearly includes the important element of active mate choice by females, is similar to one that had been used previously by Bradbury (1977, 1981). Oring suggested that although many leks also exhibit "traditional" (year-to-year) use, this is not a necessary or even a desirable part of the definition. Most recently, Höglund (1989) clearly defined lekking species as including those in which males provide females with no resources except their gametes, in which displaying males are clumped spatially, and in which females can exert active mate choice based on criteria other than territorial quality and parental care potential. This tripartite definition not only obviously included typical or classic leks, but also specifically included exploded leks having intermediate degrees of spacing.

I am using the term "lek" to designate a clumped and socially structured (by relative individual dominance) avian arena, in the same general context and with the same criteria as were used by Höglund (1989). As noted earlier, Hjorth (1970) defined "lek" as the aggregated advertisement behavior of male grouse, rather than the environmental location of such behavior. Such usage may well reflect the traditional Swedish meaning of the verb *leka* (to flirt in a playful manner), but in this book "lekking" is used to refer to the clustering behavior itself. "Lek" is correspondingly used only as a noun, to refer to a lekking site (a clustered male arena). Contrary to Hjorth, avian leks are also clearly social congregations rather than simple aggregations. That is, the participating males exhibit clear internal social structuring, and the male clustering and female attraction attributes result from social interactions of the participants, rather than from any collective ecological attraction to a particular environmental substrate. To avoid the implicit oxymoronic difficulties of a "dispersed lek" (Payne, 1984), I am using "exploded lek" to describe those clustered display sites whose collective members are generally out of view, but within auditory range, of one another. I am also including mobile (or "detached") leks within the concept of the lek. These do not have a precise site or location, and thus may not have any territorially dependent dominance attributes.

There is also no clear evidence yet for lekking behavior in any polyandrous birds. This should not be surprising considering that, although behavioral role reversal of the sexes in incubation and brood care is easily achieved, such reversal is not possible in terms of actual gamete characteristics and production rates. Thus, female birds will always produce the rarer, larger, and therefore the more valuable of the two types of gametes, which represent the limiting factor influencing a species' reproductive success rates. Among phalaropes, which practice female-access polyandry, both sexes congregate at temporally varied aquatic sites where they feed, display, and copulate. Emlen and Oring (1977) described this mating system as one resembling an explosive breeding assemblage, with competition sometimes occurring among the females for access to males during the critical period of copulation and nest initiation. However, these courting groups are evidently not socially structured, the sites are not localized nor are they used only for social display, and actual short-term pair bonds (lasting until male incubation of the clutch begins) are formed, so few if any of the important criteria for defining arena behavior are met by the phalaropes.

The term "lek" to designate arena-like sexual congregations has gradually come into more general use not only by ornithologists for birds, but also in reference to sexually oriented assemblages in several other animal groups, especially mammals (Wiley, 1991). These occur in at least six species of ungulates (Gosling, 1986; Fryxell, 1987; Gosling and Petrie, 1990; Clutton-Brock, 1991; Balmford, 1992), a few bats

(Bradbury, 1977; Attenborough, 1990; McWilliam, 1990), the walrus (Fay, Ray, and Kibalchick, 1984), and a marsupial (Cockburn and Lazenby-Cohen, 1992). In some of these species the locations in which the males display are mobile rather than fixed, but in most the basic criteria for lekking behavior described earlier are certainly met. Lekking among mammals is mostly confined to herbivores that have relatively precocial offspring and are fairly mobile, conditions that favor the possibility of male emancipation from paternal involvement. Among the ungulates, lekking is especially typical of grazing species that live in large, mixed herds and occur at least locally at high density levels. There are two general types of ungulate lekking, those in which the leks are visited by herds of females and those visited only by estrus females (Clutton-Brock, Green, and Hasegawa, 1988).

Lekking behavior, or at least competitive male mating aggregations, may also occur in some other vertebrate groups, including the mating assemblages of various snakes (Gregory, 1974), amphibians (Howard, 1978; Kruse, 1981; Sullivan, 1983; Willson, 1984), and fish (Endler, 1983; Loiselle and Barlow, 1978; Willson, 1984). The concept of lekking has even been extended to include a variety of insect mating clusters. However, these insect assemblages may not always fulfill all the criteria mentioned earlier (such as active female choice behavior) for typical lekking (Bateman, 1948; Cicero, 1983; Thornhill and Alcock, 1983; Bradbury, 1985; Knapton, 1985; Alcock, 1987; Shelly, 1989).

A Summary of Relevant Terminology

Given the varying use of such common terms as "arena," "court," and "lek" even among ornithologists, it might be well to define some related terms as they are used in this book, both to provide reading clarity and perhaps to offer some insights into the limits and interrelationships of phenomena. It may be initially useful to clearly distinguish process (behavior) from the physical environment (habitat or substrate) within which the behavior occurs, and occasionally to take into account the results of these behaviors on the physical environment. I have thus provided a tentative functional classification of male behaviors in nonmonogamous birds (Table 8), as well as a similarly tentative classification of their associated environmental display sites (Table 9). Both outlines provide some implicit definitions of terms that are at occasional variance with previously used definitions, or at least indicate a clear personal preference in those situations where conflicting usages have been applied in the past.

To elaborate on some of these, I have elected to use the terms "arena behavior" and "arenas" in a sense that is comparable to that of "lekking" and "leks," namely to encompass the male advertisement behaviors (arena behavior) and their environmental locations (arenas) of all those nonmonogamous (polygynous or promiscuous) avian species that show some apparent degree of male clustering, whether only slight (the males forming "exploded" areas) or substantial (forming classic "clumped" arenas).

Table 8 • A Classification of Male Advertisement Behavior and Behavioral Interactions in Polygynous or Promiscuous Birds

I. Males display solitarily; monogamous or polygynous species.
   A. Dispersed, noncontiguous variably resource-inclusive territories. Males often display throughout the territory, but may not actively defend all of it.
   B. Often contiguous, mostly smaller and nest-centered territories of variably colonial species. Males often display at the nest site (especially if favorable nest sites are limited) or at other restricted display sites.

II. Males display competitively in small or large arenas; polygynous or promiscuous species. Mostly display-site-only territories, to which females are attracted sexually, and from which they choose mating partners.
   A. Exploded leks; contiguous territories, the males within hearing range of but often visually isolated from each other: some grouse, bustards, hummingbirds, manakins (*Neopelma, Tyranneutes, Pipra pipra*), cotingids (bellbirds, umbrellabirds), some birds-of-paradise (parotias, king).
   B. Highly clumped male display sites; the males in close proximity to one another (the intermale distances much smaller than intergroup distances) and usually in visual contact.
      1. Male displays occur within exclusive-use territorial boundaries (typical leks).
         a. Terrestrial leks: lekking grouse, ruff, great snipe.
         b. Arboreal leks: some birds-of-paradise (*Paradisaea*), cotingids (calfbird, red-cotingas, etc.), and some manakins (*Manacus, Pipra*, etc.).
      2. Male displays not confined within exclusive-use territories.
         a. Stationary leks, coordinated ("communal") male displays.
            aa. Participating males overtly competitive: some manakins (*Pipra serena*).
            bb. Participating males seemingly "cooperative": *Chiroxiphia* manakins.
         b. Mobile display sites, males competitive but nonterritorial, or defending "moving territories" around females.
            aa. Terrestrial arenas: great bustard, wild turkey, *Corapipo* manakins.
            bb. Aquatic arenas: most typical ducks (Anatinae).
            cc. Aerial and submarine courtship chases: many ducks (most dabbling ducks, pochards, and sea ducks).

In my view, an individual male's display location may best be called simply a terrestrial display site (if it is seemingly not environmentally constrained), a display perch (if elevated), or a stage (if its location is seemingly otherwise facilitated by specific environmental features or is spatially limited by habitat constraints). If the display site has either been adaptively modified by the "owner" or has actually been constructed by him, the site becomes defined as a "court." Such a court might be a simple ground court (if the site is kept free of vegetational debris or visually obstructing overhead

**Table 9 • A Classification of Male Avian Display Sites,
Based on Site Characteristics**

I. Display sites in which the environment has not been adaptively altered by resident males.
  A. Arenas: aggregated male display sites that do not have evident unique environmental features making them especially suitable for display; ducks, lekking grouse, ruff, buff-breasted sandpiper, great snipe, bustards.
    1. Clumped leks: display sites are close together; males usually display while in direct visual contact, and their common territorial boundaries are often quite apparent.
    2. Exploded leks: display sites somewhat clumped, the males often isolated visually but not acoustically, and their territorial boundaries often indefinite.
  B. Stages: display sites clearly influenced by environmental constraints (especially those affecting visual/acoustic signaling effectiveness), thus modifying the spatial distributions of individual males, including logs, mounds, boulders, etc.; ruffed grouse, argus pheasants, some manakins, bowerbirds, and birds-of-paradise.
  C. Display perches: elevated display sites in trees, shrubs, vines, etc. that have not been obviously modified by the owner; many manakins and cotingids.
II. Display sites in which the environment has been adaptively modified by the participating males.
  A. Courts: display sites modified by clearing and/or local site alteration.
    1. Terrestrial or primarily terrestrial courts.
      a. Ground courts: level display sites variably cleared or trampled by resident males; peacock-pheasants, argus pheasants, Albert's lyrebird, Guianan cock-of-the-rock, golden-collared manakin, tooth-billed and Archbold's bowerbirds, parotias, Jackson's widowbird.
      b. Mound courts: elevated sites that have been piled up by resident males; brush turkey, superb lyrebird.
      c. Bowers: display structures that have been constructed by local site alteration plus brought-in materials; bowerbirds.
    2. Arboreal courts in shrubs, vines, and trees: twigs and leaves clipped so as to influence visibility at an arboreal display perch; some birds-of-paradise (*Paradisaea, Cicinnurus*), some manakins (*Pipra*).

vegetation), a mound court (if the dirt has been piled up by the bird itself), an arboreal court (if the associated vegetation surrounding an arboreal display perch has been clipped or removed), or a bower (if the display structure has actually been "assembled" or constructed in some way using materials brought in by the owner).

Of all these environmental modifications by males, the most difficult to account for ethologically as to its evolutionary origin is the avian bower. In some species such a "bower" might actually be a rudimentary nest (Marshall, 1954), or may at least reflect the male's "non-discardable" nesting instincts (Gilliard, 1969). It may now simply operate as a heterosexual, species-specific visual lure to attract the attention of females, perhaps supplementing the male's own visual or acoustic signals and serving as a comparative visual marker to denote an individual male's relative dominance status. The bower may also provide some degree of overhead protection from aerial predators for visiting females (Borgia, Pruett-Jones, and Pruett-Jones, 1985). It might conceivably even function as a kind of copulatory trap, which physically restricts the female from making a ready escape during copulation attempts by the male.

A Survey of Arena-Forming Bird Species of the World

In what was one of the first major reviews of avian mating systems, including both classic leks ("male-dominance polygyny: clumped") and exploded leks ("male-dominance polygyny: intermediate dispersion"), Oring (1982) suggested that exploded leks provide a continuously graded intermediate situation between dispersed male dominance polygyny and classic leks, and suggested male spacing at distances of at least 50 meters as a criterion of intermediate dispersion. By that criterion he listed some 43 species, representing 11 avian families, that he considered to exhibit male dominance polygyny, the males showing either clumped or intermediate dispersion characteristics.

In an important general and early survey of arena behavior in birds, Payne (1984) provided a comparative review of individual male and female breeding success data, sexual dimorphism, and related phenomena in a rather large number of arena-breeding species and groups, but he did not attempt to provide a complete survey of arena behavior as it appears throughout the entire class Aves. He did, however, conclude that in most lekking species males are larger than females, and that in a few groups (such as grouse, cotingids, and birds-of-paradise) overall adult body mass and/or sexual dimorphism may be more apparent in those species exhibiting lekking behavior or showing intense sexual selection among males than in monogamous species. However, reversed adult sexual dimorphism is apparent among the smaller species of several groups (manakins, hummingbirds, and bustards) in which males perform active aerial displays, suggesting that the male's associated agility in such species may be more important than his fighting prowess. Payne concluded that intrasexual competition and fighting among males may explain most instances of evolved sexual dimorphism in lekking birds. On the other hand, active female choice of male mates must certainly have played some role in groups such as hummingbirds and manakins, in which adult males are frequently smaller than females but nevertheless show strong sexual dichromatism.

Höglund (1989) has recently provided a virtually complete analysis of the possible relationships between the occurrence of sexual dimorphism and sexual dichromatism and that of lekking behavior in birds. His survey included a total of 117 taxa (species or genera) of birds that he regarded as lekking, representing 11 families. These birds were defined as including all those species in which male display distribution is clumped, and those displaying out of sight but "within earshot" of one another. Using this criterion, bowerbirds were excluded from his analysis. Otherwise it is a quite complete listing, and to a large degree provided a basis for my own tabular listing of arena-breeding birds (Tables 10 and 11). My list of more than 150 species or genera includes not only all lekking species known to me, but also some apparently dispersed-territory species or groups (such as bowerbirds) that may not qualify as arena species as such, but that establish courts, construct bowers, or are close relatives of known arena species and warrant inclusion for that reason alone. It is probable that some additional arena-breeding species will be found, especially among the relatively little-studied manakins. However, it seems unlikely that more than about 2 percent of the world's extant species of birds (about 9,700, according to Sibley and Monroe, 1990) will ever be found to qualify as arena breeders.

Rather surprisingly, Höglund's (1989) analysis caused him to conclude that the presence of sexual plumage dichromatism and sexual mass dimorphism is not directly and significantly correlated with the occurrence of lekking behavior. Höglund judged, furthermore, that lekking probably did not precede the evolution of these dimorphism types any more often than the reverse sequence of events, at least in those phylogenies for which he could hypothesize a probable temporal sequence for the two. He judged relatively larger males to be especially characteristic of ground-displaying species, whereas the evolution of small and agile males may have been favored in groups or species having aerial and/or arboreal displays. Finally, he concluded that although in some groups the evolution of sexual dichromatism may indeed have been favored by the selective effects of lekking behavior, in others sexual selection may instead have produced significant sexual diethism, such as the evolution of sexually distinctive acoustic displays.

Effects of Natural Selection vs. Sexual Selection among Lekking Birds

It is certainly apparent that the species in Tables 10 and 11 include many of the world's most stunningly beautiful birds, as well as those having some of the most remarkable and spectacular mating behaviors, as will be described individually in the following chapters. It is also of some interest that the list includes some of the avian groups that are most prone to exhibit interspecific (sometimes intergeneric) hybridization under wild conditions. This is but one of several paradoxical facets of arena behavior. Another of these is the basic "lek paradox" (Borgia, 1979), that of females evidently preferring those males having the most elaborate and conspicuous structural and behavioral traits, traits that seemingly would be selected against in terms of their survival under natural conditions (Darwin, 1871). This paradox may prove somewhat

Table 10 • Dispersion, Dimorphism, Dichromatism, and Displays in Males of Nonpasserine Arena-Breeding Birds[1]

Species or Genus	Dispersion	Dimorphism		Displays	Sources
		Mass	Color		
Grouse[2]					
Sage grouse	Lek	+	+	G	Wiley, 1973
Blue grouse	D/EL	+	+	G/T	Hjorth, 1970
Capercaillie	Lek	+	+	G	Hjorth, 1970
Black-billed capercaillie	Lek	+	+	G	Andreev, 1979
Black grouse	Lek	+	+	G/A	Hjorth, 1970
Caucasian black grouse	Lek	+	+	G/A	Hjorth, 1970
Ruffed grouse	D/EL	+	+	S	Hjorth, 1970
Greater prairie-chicken	Lek	+	+	G/A	Ballard & Robel, 1974
Lesser prairie-chicken	Lek	+	+	G/A	Hjorth, 1970
Sharp-tailed grouse	Lek	+	+	G/A	Lumsden, 1965
Pheasants[2]					
Tragopan spp.	D	+	+	S?	Johnsgard, 1986
Polyplectron spp.	D	+	+	S	Johnsgard, 1986
Crested argus	D	+	+	S	Johnsgard, 1986
Great argus	D	+	+	S	Davison, 1981
Blue peafowl	D	+	+	G	Ridley et al., 1984
Wild turkey	ML	+	+	G/J	Watts, 1968
Ocellated turkey	D	+	+	G	Steadman et al., 1979
Bustards[2]					
Houbara bustard	D/EL	+	+	G	Collins, 1984
Little bustard	EL	+	+	G/A	Schulz, 1985
Great bustard	Lek/ML	+	+	G	Gewalt, 1959
Australian bustard	EL?	+	+	G	Fitzherbert, 1978
Great Indian bustard	EL?	+	−	G	Ali & Rahmani, 1984

(*continued*)

Table 10 • (*Continued*)

Species or Genus	Dispersion	Dimorphism		Displays	Sources
		Mass	Color		
Kori bustard	EL?	+	−	G	Hellmich, 1988
Denham's bustard	EL?	+	−	G/A	Tarboton, 1989
Black-bellied bustard	EL	+	+	G/A	Schulz, 1986
Rufous-crested bustard	EL?	+	+	G/A	Johnsgard, 1991
Bengal florican	EL?	−	+	G/A	Rahmani, 1990
Lesser florican	EL?	−	+	G/A	Rahmani, 1990
Sandpipers					
Buff-breasted sandpiper	Lek/D	+	−	G	Pruett-Jones, 1988
Ruff	Lek	+	−	G	Hogan-Warburg, 1966
Great snipe	Lek	−	−	G	Lemnell, 1978
Parrots					
Kakapo	Lek	−	−	S	Merton et al., 1984
Hummingbirds[2]					
Band-tailed barbthroat	EL/Lek	+	−	A/T	Stiles & Wolf, 1979
Green (Guy's) hermit	EL/Lek	−	+	A/T	Snow, 1974
Long-tailed hermit	EL/Lek	−	−	A/T	Stiles & Wolf, 1979
Reddish hermit	EL/Lek	−?	+	A/T	Snow, 1973a
Little hermit	EL/Lek	−	−	A/T	Snow, 1974
White-tipped sicklebill	EL/Lek	+	−	A/T	Stiles & Wolf, 1979
Rufous saberwing	EL?	−	+	A/T	Skutch, 1972
Brown violet-ear	EL	−	−	A/T	Ffrench, 1980
Green violet-ear	Lek	+	+	A/T	Fjeldså & Krabbe, 1990

(*continued*)

Table 10 • (*Continued*)

| Species or Genus | Dispersion | Dimorphism | | Displays | Sources |
		Mass	Color		
White-eared hummingbird	EL/Lek	+	+	A/T	Bent, 1940
Blue-throated goldentail	EL/Lek	+	+	A/T	Skutch, 1972
White-bellied emerald	Lek	−	−	A/T	Atwood et al., 1991
Violet saberwing	EL/Lek	+?	+	A/T	Wolf, 1970
Scaly-breasted hummingbird	EL/Lek	+?	−	A/T	Wolf, 1970
White-necked Jacobin	Lek?	+	+	A/T	Belt, 1874
Violet-headed hummingbird	EL/Lek	+?	+	A/T	Payne, 1984
Blue-chested hummingbird	EL/Lek	+	+	A/T	Stiles & Wolf, 1979
Charming hummingbird	EL?	−	+	A/T	Stiles & Skutch, 1989
Cinnamon hummingbird	EL?	−	−	A/T	Stiles & Skutch, 1989
Rufous-tailed hummingbird	EL?	+	+	A/T	Skutch, 1972
White-tailed emerald	EL?	+	+	A/T	Stiles & Skutch, 1989
Coppery-headed emerald	EL?	+	+	A/T	Stiles & Skutch, 1989
Snowcap	EL?	−	+	A/T	Stiles & Skutch, 1989
Amethyst-throated hummingbird	EL?	?	+	A/T	Skutch, 1967
Crimson topaz	EL/Lek	+	+	A/T	Davis, 1958
Marvelous spatuletail	Lek	+	+	A/T	Fjeldså & Krabbe, 1990
Wine-throated hummingbird	EL?	?	+	A/T	Skutch, 1972

(*continued*)

Table 10 • (*Continued*)

| Species or Genus | Dispersion | Dimorphism | | | Sources |
		Mass	Color	Displays	
Calliope hummingbird	D/EL	+	+	A/T	Tamm et al., 1989
Broad-tailed hummingbird	D/Lek	+?	+	A/T	Barash, 1972
Trogons					
Apaloderma spp.	EL?	?	+	T	Brosset, 1983

1. Adapted, with many additions and modifications, from Höglund (1989). Dispersion symbols: D = dispersed males, EL = exploded lek, ML = mobile lek. When two dispersion types are shown, the first is believed more typical. Dimorphism symbols: + = dimorphic, − = nondimorphic, r = reverse dimorphism. Display symbols: A = aerial display, G = ground or log display, J = joint male display, S = stage or arena display, T = tree or shrub display. Aerial displays include display jumps, but exclude static wing flapping.
2. Species sequence and vernacular names as per Johnsgard (1983a, 1983b, and 1991), except that "prairie-chicken" is used rather than pinnated grouse. Following Höglund (1989) I have attributed aerial displays to all hummingbirds, although true courtship flights in lekking hummingbirds have been questioned (Atwood, Fitz, and Bamesberger, 1991). Available descriptions of hummingbirds also often make it impossible to judge the degree of male dispersion, which in any case seems to vary considerably.

illusory, when the potential benefits of appropriate female choice of the most dominant, sexually experienced, or vigorous male (such as increased safety from copulation disruption, improved chances of mating with disease-free mates, and greater assurance of fertilization) are weighed against the relatively low costs incurred by females in their nearly simultaneous examination of a large sample of readily available and competitively interacting males (Reynolds and Gross, 1990).

The hybridization enigma is more complicated. Can one explain the fact that, although males of lekking species have some of the most elaborate nuptial plumages and complex sexual displays of all birds, nearly all these species are notoriously prone to hybridization? At least in theory, the elaborate male plumages and behavioral signals should serve not only as intraspecific signals helping to identify the most vigorous and fittest males, but also as effective premating reproductive isolating mechanisms that reduce or avoid possibilities of hybridization. The general argument is that, in species whose pair bonds are not forged gradually, and in which incipient "mistakes" in mating choices that could result in interspecific mating can be corrected

**Table 11 • Dispersion, Dimorphism, Dichromatism, and Displays in Males
of Passerine Arena-Breeding Birds[1]**

| Species or Genus | Dispersion | Dimorphism | | Displays | Sources |
		Mass	Color		
Lyrebirds					
Superb lyrebird	D	+	+	S	Kenyon, 1972
Albert's lyrebird	D	+	+	S	Curtis, 1972
Tyrant flycatchers					
Ochre-bellied flycatcher	EL	−	−	A/T	Snow & Snow, 1979
McConnell's flycatcher	EL	−	−	A/T	Willis et al., 1978
Gray-hooded flycatcher	EL	−	−	A/T	Willis et al., 1978
Speckled mourner	EL	−	−	T	Stiles & Skutch, 1989
Thrush-like schiffornis	EL	−	−	T	Trail, 1990
Cotingas[2]					
Black-necked red-cotinga	Lek	r	+	A/T	Trail & Donahue, 1991
Guianan red-cotinga	Lek	r	+	A/T	Trail & Donahue, 1991
Black-and-gold cotinga	Lek	−	+	T	Snow, 1982
Dusky piha	Lek	−	−	T	Snow, 1982
Screaming piha	Lek	−	−	T	B. Snow, 1961
Rufous piha	D/EL	−	−	T	Skutch, 1989
Red-ruffed fruitcrow	Lek	−	−	T	Wünschmann, 1966
Bare-necked umbrellabird	EL?	+	+	T	Snow, 1982
Long-wattled umbrellabird	EL?	+	+	T	Snow, 1982
Amazonian umbrellabird	EL?	+	+	T	Sick, 1954
Calfbird (capuchinbird)	Lek	−	−	T	Snow, 1972; Trail, 1990
Three-wattled bellbird	EL?	+	+	A/T	B. Snow, 1977b
White bellbird	EL?	+	+	A/T	Snow, 1973b
Bearded bellbird	EL?	+	+	A/T	Snow, 1970
Bare-throated bellbird	EL?	+	+	A/T	Snow, 1978
Guianan cock-of-the-rock	Lek	+	+	S/T	Trail, 1985c
Peruvian cock-of-the-rock	Lek	+	+	S/T	Benalcazar & Benalcazar, 1984
Sharpbill	EL	−	−	A/T	Stiles & Whitney, 1983
Manakins[2]					
Crimson-hooded manakin	EL	−?	+	J/A/T	Snow, 1963b

(continued)

Table 11 • (*Continued*)

Species or Genus	Dispersion	Dimorphism		Displays	Sources
		Mass	Color		
Band-tailed manakin	EL	−?	+	J/A/T	Robbins, 1983
Wire-tailed manakin	EL	−?	+	J/A/T	Schwartz & Snow, 1978
Red-capped manakin	Lek	r	+	A/T	Skutch, 1969
Golden-headed manakin	Lek	−?	+	A/T	Lill, 1976
Round-tailed manakin	Lek	−?	+	A/T	Robbins, 1985
White-crowned manakin	Lek	r	+	A/T	Stiles & Skutch, 1989
Blue-crowned manakin	Lek	−?	+	A/T	Skutch, 1969
White-fronted manakin	EL	−?	+	J/A/T	Prum, 1985
Long-tailed manakin	Lek	−	+	J/A/T	Foster, 1977
Blue-backed manakin	Lek	−	+	J/A/S	Snow, 1963a
Swallow-tailed manakin	Lek	−	+	J/A/T	Foster, 1981
Golden-winged manakin	EL	−	+	J/A/T	Prum & Johnson, 1987
Pin-tailed manakin	EL	−	+	A/S/T	Snow & Snow, 1985
White-throated manakin	Lek/ML	−	+	A/S/G	Prum, 1986
White-ruffed manakin	Lek/ML	−	+	A/G	Skutch, 1967
White-collared manakin	Lek	−	+	A/T	Stiles & Skutch, 1989
Orange-collared manakin	Lek	−	+	J/S/A/T	Stiles & Skutch, 1989
Golden-collared manakin	Lek	−	+	A/S/T	Chapman, 1935
White-bearded manakin	Lek	−	+	A/S/T	Lill, 1974a,b
Fiery-capped manakin	Lek	−	+	J/A/T	Robbins, 1985
Striped manakin	Lek	−	+	J/A/T	Sick, 1967
Club-winged manakin	EL	−	+	A/T	Willis, 1966
Grey-headed piprites	EL	−	−	T	Stiles & Skutch, 1989
Bulbuls					
Yellow-whiskered greenbul	Lek	−	−	A/T	Brosset, 1982
Bowerbirds (genera)[3]					
Scenopoeetes (1 sp.)	D	−	−	S/G	Gilliard, 1969
Archboldia (1 sp.)	D	+	+	S/G	Gilliard, 1969
Amblyornis (4 spp.)	D	+	−/+	S/G	Gilliard, 1969
Prionodura (1 sp.)	D	+	+	S/G	Gilliard, 1969
Sericulus (3 spp.)	D	+	+	S/G	Gilliard, 1969
Ptilonorhynchus (1 sp.)	D	−	+	S/G	Gilliard, 1969
Chlamydera (4 spp.)	D	−/+	−/+	S/G	Gilliard, 1969

(*continued*)

Table 11 • (*Continued*)

Species or Genus	Dispersion	Dimorphism Mass	Color	Displays	Sources
Birds-of-paradise[3]					
Standardwing	Lek	+	+	T	Frith, 1992
Black-billed sicklebill	D	−	+	T/A?	Beehler, 1987a
Pale-billed sicklebill	D	+	+	T	Beehler & Beehler, 1986
Superb bird-of-paradise	D	+	+	S	Beehler & Pruett-Jones, 1983
Western parotia	D/EL	+	+	S	Bergman, 1958
Carola's parotia	D/EL	+	+	S	Frith, 1968
Lawes' parotia	EL	+	+	S	Beehler & Pruett-Jones, 1983
Wahnes' parotia	EL	+	+	S	Beehler & Pruett-Jones, 1983
Magnificent riflebird	D	+	+	T	Cooper & Forshaw, 1979
Victoria's riflebird	D	+	+	T	Cooper & Forshaw, 1979
Paradise riflebird	?	+	+	T	LeCroy, 1981
Magnificent bird-of-paradise	D	+	+	S	Beehler & Pruett-Jones, 1983
Wilson's bird-of-paradise	D	+	+	S	Frith, 1974
King bird-of-paradise	EL	−	+	T	
Arfak astrapia	?	+	+	G/T	Frith & Frith, 1981
Splendid astrapia	D?	−	+	T	LeCroy, 1981
Ribbon-tailed astrapia	D?	+	+	T	LeCroy, 1981
Stephanie's astrapia	Lek	+	+	T	Healey, 1978
Huon astrapia	?	+	+	T	LeCroy, 1981
King-of-Saxony bird-of-paradise	D	+	+	T	Healey, 1975
Twelve-wired bird-of-paradise	D	+	+	T	Coates, 1990
Red bird-of-paradise	Lek	+	+	T	Frith, 1976
Lesser bird-of-paradise	Lek	+	+	T	LeCroy, 1981
Greater bird-of-paradise	Lek	+	+	T	LeCroy, 1981
Raggiana bird-of-paradise	Lek	+	+	T	Frith, 1982
Goldie's bird-of-paradise	Lek	+	+	T	LeCroy, 1981

(*continued*)

Table 11 • (*Continued*)

| Species or Genus | Dispersion | Dimorphism | | Displays | Sources |
		Mass	Color		
Emperor bird-of-paradise	Lek	+	+	T	Draffan, 1978
Blue bird-of-paradise	D	+	+	T	Coates, 1990
Whydahs and Widowbirds					
Village (green) indigobird	EL	+	+	A/T	Payne, 1973
Pin-tailed whydah	EL	+	+	A/T	Shaw, 1984
Eastern paradise-whydah	D?	+	+	A/T	Nicolai, 1969
Broad-tailed paradise-whydah	D?	+	+	T	Nicolai, 1969
Jackson's widowbird	Lek	+	+	S	Andersson, 1989

1. In part after Höglund (1989), but with many additions and modifications. Symbols as in previous table.
2. Species sequence and vernacular names of cotingas and manakins after Sibley and Monroe (1990), who also merged the cotingas, sharpbill, and manakins into a single enlarged family Tyrannidae. Male dispersion patterns of cotingas are based on Snow (1982) and other sources. I have followed Höglund (1989) in assuming that aerial displays occur in all manakins. Probably many of the still-unstudied species of manakins also represent lekking species, since males of all typical manakins so far studied in detail are known to display in leks or exploded leks.
3. Species sequence and vernacular names of birds-of-paradise as in Sibley and Monroe (1990). Generic sequence and species numbers of bowerbirds as in Cooper and Forshaw (1979). Male dispersions mostly after Diamond (1986). Probably some additional birds-of-paradise will be found to qualify as arboreal stage, or at least arena, species. All of the listed birds-of-paradise and bowerbirds are currently believed to be polygynous (Cooper and Forshaw, 1979).

prior to fertilization, it is especially important that species-specific social signals are in place to facilitate premating sexual and species recognition. Such signals presumably help to confirm proper species identification immediately prior to copulation, thus avoiding or reducing the chances of disadvantageous hybridization (Sibley, 1957).

Not only must the males of arena-breeding species shoulder the burden of carrying these species-specific identity badges, they must additionally advertise them clearly and effectively over a generally long breeding period. This advertisement process may expose the often colorful and/or loud males to significantly increased probabilities of predation, and correspondingly reduce their time available for foraging, resting, or other important activities. Effective male signal advertisement may obviously be influenced by the species' normal daily activity patterns (diurnal vs. nocturnal or crepuscular signaling requirements), its visual capabilities (color vs. monochromatic vision limitations, as well as interspecific variables in visual sensitivity and acuity), similar hearing and sound production constraints (vocal vs. mechanical sound production abilities, optimum frequency ranges, etc.), and general ecological factors (such as the most appropriate signals for visual or acoustic transmission through particular habitats, and the possibility of predators being attracted by their signaling behavior).

Sibley (1957) has argued that the most pronounced male signal characters should be found among those avian groups where the effects of sexual selection for maximizing male mating success combine with the effects of natural selection for reducing disadvantageous hybridization. Such additive effects should occur among those polygynous bird groups having numerous species that occur sympatrically and that also occasionally hybridize, resulting in natural selection against hybridization in addition to classic sexual selection. The groups that Sibley used to bolster this argument were the birds-of-paradise, hummingbirds, grouse, pheasants, manakins, and the typical surface-feeding (*Anas*) ducks. All of these are groups that generally show high levels of sexual dichromatism and elaborate and species-specific male sexual signals, and include numerous instances of sympatry among fairly closely related species. All of these groups also readily qualify as arena breeders, and hybridization is fairly to highly prevalent in all these groups except the manakins (Gray, 1958). The paucity of manakin hybrids known thus far may reflect the fact that most of these birds are small and elusive forest species of Central and South America whose biologies still remain largely unstudied, or perhaps the incidence of hybridization among closely related manakin species has been reduced as a result of geographic separation and ecologic isolation (Parkes, 1961).

Hybridization has been especially well documented among the grouse assemblage, which is an economically important bird group that has been exhaustively studied by both wildlife management biologists and ornithologists. Among grouse, hybridization is clearly more common among the lekking species than the nonlekking forms, and is also more common among polygynous than monogamous species (Table 12). Sibley suggested that the strongly specific male traits found in the lekking species of grouse evolved primarily under the influence of sexual selection. He further stated that hybridization may often result from situations of historically recent (secondary) contact between pairs of closely related species, because the degree of genetic relationship between the two species is still actually very close. Natural selection has not yet resulted in effective reproductive isolation between them, even though there may already be substantial interspecific differences in male appearances and male sexual signals. Indeed, many of the well-documented North American grouse hybrids that have been reliably reported are regarded as intergeneric rather than simply interspecific, even by current taxonomic standards (Figure 6). Sibley regarded this surprising situation as an artifact of human error in traditional taxonomic evaluation, which treated male plumage and related sexually limited traits as generic-level attributes rather than as species-level social signals. Of the approximately 30 reported interspecific hybrid combinations within the grouse assemblage as summarized by

Table 12 • Reported Natural Interspecific Hybrids among Lekking Grouse[1]

Hybrid Combination	Mating Types	Hybrid Frequency
Intrageneric hybrids		
Sharp-tailed grouse × greater prairie-chicken	Lek × lek	Usually under 3% of total population
Capercaillie × black-billed capercaillie	Lek × lek	Locally to 12% of total population
Black grouse × capercaillie	Lek × lek	Over 200 specimens known
Intergeneric hybrids		
Willow ptarmigan × black grouse	Monogamous & dispersed × lek	Over 30 specimens known
Rock ptarmigan × black grouse	Monogamous & dispersed × lek	Several specimens known
Willow ptarmigan × capercaillie	Monogamous & dispersed × lek	Several specimens known
Hazel grouse × black grouse	Monogamous & dispersed × lek	Several specimens known
Blue grouse × sage grouse	Promiscuous & dispersed × lek	Few specimens known
Sage grouse × sharp-tailed grouse	Lek × lek	Two specimens known
Blue grouse × sharp-tailed grouse	Promiscuous & dispersed × lek	One specimen known

1. Adapted from Johnsgard (1983a), where a summary of hybrids among the nonlekking grouse species can also be found.

Gray (1958), at least eight involve pairs of species that are still generally regarded as belonging to separate grouse genera.

Sibley (1957) also noted many hybrids among the birds-of-paradise, and observed that all of these reputedly hybridizing species are both polygynous and sexually dimorphic. There are indeed many cases in which newly discovered and described "species" of *Paradisaea* were later found simply to represent hybrids between two al-

6. Hybridization records in wild North American grouse (lines connect reported parental types; intergeneric combinations ["I"] are also indicated), including greater prairie-chicken (*A*), sharp-tailed grouse (*B*), sage grouse (*C*), blue grouse (*D*), ruffed grouse (*E*), and spruce grouse (*F*). Drawing by C. G. Pritchard.

Table 13 • Natural Intrageneric Hybrids Reported among Species of Paradisaea[1]

Hybridizing Species	Hybridization Rate	Hybrid Form
Raggiana × lesser	fairly common	*"mixta"*
Raggiana × greater	occasional	*"luptoni"*
Raggiana × emperor	occasional	*"maria"*
Emperor × lesser	occasional	*"duivenbodei"*
Raggiana × blue	one record	*"bloodi"*

1. Based on Gilliard (1969), and using the original but now invalid names by which these hybrids were first described. Indicated hybridization rate as judged by Coates (1990). All but the blue bird-of-paradise are typical lekking species. The lesser bird-of-paradise has reportedly also hybridized intergenerically with *Ptiloris, Cicinnurus,* and *Seleucides.*

ready known species (Table 13). Additionally, as in the grouse, several probable hybrids have been reported among seemingly well-defined genera of birds-of-paradise, especially when these genera have been based on the often remarkable differences in male nuptial plumages (Figure 7). Gray (1958) listed 10 genera and 13 species of birds-of-paradise that have been implicated in possible hybridization, within a family that contains a total of about 17 generally accepted genera and nearly 40 extant species. Of the 18 total reputed hybrid combinations that she summarized (including some combinations that would now be considered as intraspecific), the majority are actually intergeneric rather than simply interspecific, on the basis of the taxonomy used by her. The parental identities of some of these often strange-looking hybrids are certainly admittedly questionable (Fuller, 1979), but the general point that was made by Sibley, that of a high proclivity for hybridization among seemingly distantly related birds-of-paradise species, appears to be well established.

It thus seems quite evident that, in spite of their exposure to highly species-specific male plumages and sexual signaling behavior, females of many lekking species such as grouse and birds-of-paradise sometimes make serious "mistakes" during mate selection. There is little doubt that in such polygynous or promiscuous groups the "responsibility" for proper species affiliation resides exclusively with females. It is the females who are selectively attracted to the aggregated males by the latter's collective signaling, except in the special case of mobile arenas, where it is the males who are attracted to unmated females. Males of lekking species are notorious for their tendency to display sexually toward birds sometimes only remotely similar to females of their own species, or even toward crude models of females. Indeed, greater prairie-chicken (*Tympanuchus cupido*) males will often display sexually toward, or may even attempt to mate with, such unlikely female substitutes as straw hats that have been placed on their territories!

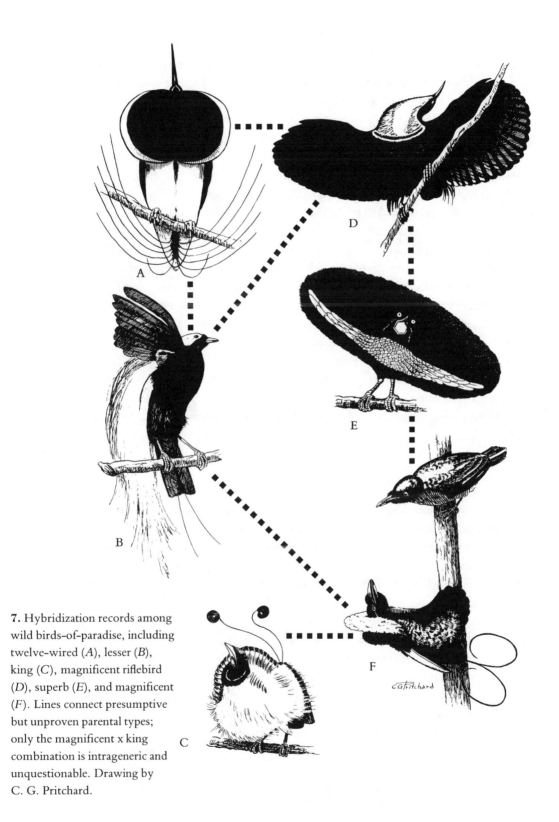

7. Hybridization records among wild birds-of-paradise, including twelve-wired (*A*), lesser (*B*), king (*C*), magnificent riflebird (*D*), superb (*E*), and magnificent (*F*). Lines connect presumptive but unproven parental types; only the magnificent x king combination is intrageneric and unquestionable. Drawing by C. G. Pritchard.

The surprisingly high frequency of hybrids among many lekking bird groups makes it clear that mistakes in species recognition during mating must be made regularly by both sexes. Reasons for such serious mistakes are perhaps diverse. One possibility is that there might be an absence or a relative rarity of nearby males of the right species, and thus a female might be attracted to a nearby lek of a closely related species, as may occasionally occur in some grouse (Sibley, 1957). Mating activity in leks is typically done quite hurriedly, and often in direct competition with other females. Possibly females take a "follow-the-leader" approach in choosing a mate, once they have been accidentally attracted to a lek of the wrong species. Or perhaps a female's sensory system may respond to the heterosexual signals of some other larger or more brilliantly arrayed species as a kind of supernormal stimulus, resulting in an unfortunate error of innate judgment.

It is improbable that females of any lekking species rely on imprinting or other early learning opportunities to achieve their male recognition abilities, inasmuch as young females among most lekking birds are unlikely to encounter their fathers during any possible learning-sensitive periods of immaturity. Presumably females must rely on innate species recognition capabilities, or at least on innate predispositions, when forced to choose between potential mates of differing species. In any case, it is clear that although females are seemingly usually able unerringly to locate and mate with the most fit among numerous competing males on a conspecific lek, they sometimes make glaring mistakes at the species level, or even occasionally at the generic level.

Some Final Observations on Lekking Birds

Life is seemingly dominated by choices, even if these are made as a result of innate tendencies rather than through rational consideration. Among lekking birds, males must choose either to participate or not to participate in leks. If participating, they then may have to choose their participation level or strategy, and thereby strongly influence their probable reproductive success. Males choosing to participate in leks may, for example, choose to spend their time fighting with other males, and thereby perhaps gain first rights of mating based on their sheer strength and male-male dominance. Or they may perhaps choose to avoid direct fights with other males and possibly even cooperate with clearly more dominant males. In so doing, they may avoid wasted energy in fights, possibly including direct injury or even death. Perhaps they will be able to steal occasional matings at opportune moments, as when dominant males are otherwise occupied, or perhaps they can survive long enough to outlive the dominant male and eventually inherit his mating rights.

Females must in turn initially choose which leks to visit; this may mean choosing between interspecific and intraspecific mating opportunities, should two lekking species occur in the same area. Thereafter females must make a passive or active mating choice among those males that are present within their chosen lek. Females exercising an active mate choice and selecting one of the better available males may benefit in two possible ways. The genes they acquire from this male may result in their offspring being generally better fitted for their own struggle for survival, or in their sons in turn being more attractive to females (Avery, 1984).

Because of the potential powerful influence of active female choice, several general models of lekking behavior have been proposed and seem to be variably supportable. One of these models, the female preference model, holds that females are attracted to larger rather than smaller assemblages of males, inasmuch as this behavior facilitates rapid and effective mate choice (Bradbury, 1981). Wiley (1991) termed this explanation the "male buffet" model. Choosing larger groups of males might also have some benefit in terms of reducing the probabilities of predators approaching unseen, but the very presence of numerous competing males may seriously disrupt the female's mating efforts. These disruptive influences could eventually result both in the establishment of male dominance hierarchies and in increased spatial separation of participating males (Foster, 1983). Furthermore, individual lekking males, except for the relatively few most dominant or most sexually attractive individuals, gain no inherent advantages from participating in such clustering, and at least the peripheral ones may indeed expose themselves to significantly increased dangers of predation. The cost of male clustering might only be overcome if substantial reproductive benefits accrue, namely if a substantial number of females are thereby attracted to the group. Male clustering trends would presumably continue until leks have become sufficiently large, and leks have become far enough apart, that females can no longer exert a preference for the largest lek without enlarging their own home ranges. Theoretically, females should visit only a single lek, the largest available one, prior to mating, and the distance between leks should approximate the diameter of a female's home range.

Another model for lek evolution, the hot spot model (Bradbury and Gibson, 1983), suggests that male clustering results from sequential attraction of males to locations or pathways regularly used by females for various purposes unrelated to reproduction. In this model, early-arriving males successively settle into areas separated by distances at least as great as a female's home range, in order to avoid sharing females. However, later males must join already established males at mating hot spots, or else must display solitarily. Unlike the female preference model, the hot spot model holds that leks should be separated at distances of no more than the diameter of a female's home range, and that females are likely to visit more than one lek before making a mating choice. In neither model are visiting females constrained to make identical mating choices (Beehler and Foster, 1988). Hot spots are of course most likely to develop in situations where several females have overlapping home ranges. A somewhat similar model was proposed earlier by Parker (1978), who suggested that leks could have evolved as males positioned themselves so as to maximize their mating opportunities, as in the hot spot model, and females visited such assemblages in order to minimize their searching time for males. Wiley (1991) similarly suggested that females may tend to return to a prior year's mating location to mate; thus such site fidelity tendencies might make it profitable for males to establish their territories at such preferred mating locations.

Beehler and Foster (1988) have pointed out that both these models, that of general female attraction to larger leks for their own reproductive benefits and that of general male-benefiting clustering behavior in hot spot areas regularly used by females, fail to take into account the possible important influence of individual males in female mating choices. These authors have proposed a third possible model for lek development, namely the "hotshot" model. This model emphasizes individual male-male interactions in facilitating lek development and maintenance, and secondarily considers female mate choice responses based on available male phenotypes. Thus, certain "hotshot" males (dominant males that are unusually successful at both attracting and fertilizing females) serve as a nucleus for the formation of leks. Leks would form as less successful and perhaps less experienced males gather around these more successful males, inasmuch as such associations might eventually lead to successful mating for some of the newcomers as well. Although females might initially choose their mates randomly, successful females would probably tend to mate with previous partners, and inexperienced females might copy mating choices of their mothers or of successful females.

Such male clustering around hotshots would be facilitated by (1) the development of a stable hierarchical system of male social dominance within this lek, as well as (2) marked associated differences in individual male mating success, and (3) a long-term deferred mating strategy by novice males, which must be sufficiently long-lived on average to make a deferred mating strategy worthwhile. There must also be (4) a large pool of available nearby females (and/or relatively large, overlapping female home ranges) for facilitating male mating inequality, and (5) relatively simple mating rules or "default strategies" available to females when choosing their mates. Such mate choice rules would be especially important for females visiting large and tightly clustered leks.

Visiting females must easily and quickly be able to identify (by a male's territorial position or his signaling behavior) the most dominant individual of any lek and mate specifically with him, rather than trying to identify the best available mate through possibly subtle phenotypic traits alone (Beehler and Foster, 1988). Obviously a simple female choice strategy based on relative male dominance would automatically favor mating with the largest and/or most highly aggressive males, whereas choosing a mate based on his perceived phenotypic "beauty," or perhaps on his relative aerial agility, might lead to selection for quite different male traits.

Wrangham (1980) suggested that, beyond the evident reproductive advantages accruing to some males by lekking, three possible types of lekking benefits might also accrue to participating females. These include direct genotypic benefits to females (the good genes hypothesis), possible phenotypic benefits (such as reduced predation rates near nest sites), and Wrangham's own preferred least-costly male hypothesis. This hypothesis states that lekking is favored whenever excessive male copulation attempts tend to interfere with female reproductive success, so that females may benefit reproductively from the absence of males on female home ranges. Female preferences for thus mating with the "least costly" males favor the males adopting a strategy of patiently waiting, rather than actively searching, for females. Real proof that the presence of nearby sexually active males actually hampers female reproductive success among any true lekking birds is still apparently lacking. However, the frequency of seemingly violent forced copulation (rape) behavior among many duck species might lead one to believe that sexually aggressive males may indeed sometimes interfere significantly with female reproductive success, and rape avoidance may represent an important if still unmeasured benefit for pair bonding among these ducks.

It has also been suggested (Queller, 1987) that lekking behavior might evolve simply through innate female preferences for visiting larger rather than smaller male assemblages without invoking any specific selective advantage to such clustering behavior, through a positive feedback process similar to the nonadaptive runaway process described for male secondary sexual characters in the previous chapter. Such a trend might perhaps be facilitated through the increased signaling effectiveness of grouped males through stimulus-pooling effects, which could result in a superoptimal stimulus. Wiley (1991) described this advantage of clustered males as representing one of disproportionate attraction, especially in the case of synchronized or coordinated male displays. This explanation would seemingly not address the important question as to why such strong social dominance and reproductive success gradients invariably occur among males participating in avian leks, or the associated aspects of active female choice within such assemblages.

It seems apparent that each of the available models has certain limitations when considered alone. It is also very clear that, as Darwin so aptly anticipated, both intrasexual interactions (the outcome of male-male dominance interactions) and heterosexual factors (females being individually attracted to specific males on the basis either of relative strength and status or phenotypic "beauty") evidently play important roles in the development and organization of arena–type mating systems in birds.

3

Ducks: Inciting Females and Mobile Arenas

The ducks (Anatinae) of the world have never been regarded as one of the bird groups performing arena displays, and because at least most of them form short-term pair bonds they certainly cannot be considered typical arena birds. However, they do possess at least several attributes typical of arena species. First, the available males usually display competitively in localized groups, over courtship periods sometimes lasting up to five or six months, during which time females apparently actively choose their mates from among the actively competing males. Secondly, the adult sex ratio in ducks consists of a variable excess of males, thus increasing male competition for the reduced supply of females. Third, bright colors, ornamental male plumages, and fairly elaborate male displays comparable to those of typical arena birds are typical of nearly all the duck species that display competitively and form only temporary pair bonds (or perhaps no pair bonds at all). Finally, male ducks are largely emancipated from any required participation in brooding and feeding their offspring, although many tropical species do participate in brood protection to varying degrees.

Males of most ducks do not establish resource-based territories, as least as a part of or prerequisite to male advertisement. Instead, like typical arena species the males simply advertise themselves. After pair bonding they may establish a small "moving territory" around their mates to defend them from rape, or perhaps (in a few species) may establish rather long-term resource-based territories. Resource-based territories exist in some rather specialized riverine-adapted ducks (*Merganetta, Hymenolaimus, Anas sparsa, A. waigiuensis*) that aggressively defend food-rich stretches along generally fast-flowing rivers or mountain streams. The shoreline-based resource-inclusive territories in the notably aggressive steamer ducks (*Tachyeres*) of coastal South America provide comparable examples. Nearly all these species, however, are rather sedentary birds with long-term pair bonds and biparental brood care and with mating strategies that more closely resemble those of geese and swans than typical ducks.

I would thus suggest that the social competitive display of most ducks can be regarded as quasi-arena behavior. Their usual display sites consist of areas of water where the still-unmated (often wintering or migrant) birds typically assemble, providing the appropriate environmental "arena" for such competitive display. However, this kind of generalized stage or arena is usually not limited in distribution, so social displays may occur on many different aquatic sites during the pair-forming period, especially in migrant species. Aquatic display may actually be interspersed with or supplemented with aerial display (courtship flights) between successive aquatic sites. Courtship flights represent a highly mobile and extremely competitive type of social display, the possible importance of which in influencing mate choice is still completely unknown, and set the ducks distinctly apart from all typical arena birds. No other arena or lek species are known to have simultaneous competitive aerial displays involving the entire assemblage of participants; an early suggestion that this might be the case for the pintail snipe (*Gallinago stenura*) has now been discredited (Byrkjedal, 1990). A few comments on aerial courtship in ducks are thus perhaps in order.

Courtship Flights and Other Aerial Chases of Ducks

Although many writers have described apparent courtship flights in ducks, they have often neglected to distinguish true courtship flights from attempted rape flights, territorial expulsion chases, and similar phenomena. Titman and Seymour (1981) reviewed the information on heterosexual pursuit flights in six species of typical surface-feeding ducks (*Anas*), and judged that they could be classified as either "three-bird flights" or "attempted rape flights." Three-bird flights (involving a mated pair and male "chaser") are most prevalent in relatively territorial species. In these, the resident male chases a pair out of its territory (either to remove both birds from the locality or perhaps to try to catch and rape the female). Attempted rape flights, involving a poorly protected or unprotected female and varying numbers of males, may thus at times begin as a three-bird flight, but additional males may join in, and the "motivation" of the participating males may shift from aggressive to sexual during its course.

In addition to these kinds of variably aggressive pursuits, true courtship flights involving several males competing for the attention of unmated females certainly occur in many species of ducks. By virtue of the number of males involved (up to a dozen or more) they can readily be separated from typical three-bird flights, but are not so easily distinguished from attempted rape flights. Often, however, females being chased by males during attempted rape flights utter distinctive "repulsion" notes (Lorenz, 1951–53) and show every sign of evasion and apparent fear. However, during courtship flights females typically utter "inciting" calls, which seem to function in showing their specific mate preferences. Males often utter courtship calls and, rather than trying to physically catch and rape the female, will attempt to fly beside or even slightly in front of her. In some species courtship flights are far rarer than in others; Lorenz mentioned that he had heard a female mallard inciting while in flight only once in his life. Indeed, mallard courtship flights are most common early in the pair-forming period. However, in species such as the northern pintail, American wigeon, and gadwall, courtship flights are extremely common during both early and middle stages of social display. In all these three species the males frequently utter loud burp-whistles during courtship flights and, especially in the northern pintail, males often fly ahead of the inciting female and try to orient the back of the head toward her (Figure 8, below). This same turning of the back of the head by males toward inciting females is evidently an extremely important male component in the pair-bonding process during aquatic courtship in many species of ducks (Johnsgard, 1965). It is possible that the exposed speculum and upper wing covert patterns of flying ducks, such as the contrasting white wing coverts of the male American wigeon (Figure 8, above), may also play some visual sex-signaling role during courtship flights. These wing areas are otherwise normally hidden by the flank feathers during aquatic courtship, except for brief (but conspicuous) speculum exposure during directed "mock preening" displays. It is possible that such wing patterns are equally significant during courtship flights, and at least in the case of the American wigeon would serve as an age-related visual signal, distinguishing first-year males (which lack immaculate white wing coverts) from older ones, which otherwise are visually very similar to first-year males.

Among deep-water and more diving-adapted ducks, their diving adaptations (especially their more posterior foot placement) cause them to take flight with greater difficulty than surface-feeding ducks, and these generally heavy-bodied birds have considerably less agility in flight than do the surface feeders. However, in the North American diving duck species such as canvasbacks (*Aythya valisineria*) and redheads (*A. americana*) aerial courtship flights also commonly occur, particularly early in the pair-forming period, when they often alternate with subsurface dives. Such aerial and subsurface chases were described in some detail for canvasbacks by Hochbaum (1944). Hochbaum also described "nuptial flights" of already-mated birds, during which the male attempts to catch the upper tail coverts or tail feathers of the female with his bill. Such tail-pulling behavior also occurs in redheads (Figure 11B), and Weller (1967) stated that although such behavior may rarely be seen during spring migration, it is most common among isolated pairs during the prelaying and laying period. Thus it probably plays little if any role as a pair-forming display, and instead may simply be a copulation intention activity.

Aquatic Courtship and Related Pair-Forming Behavior in Anatine Ducks

Early German ethologists writing on the social display of waterfowl (such as Oskar Heinroth and Konrad Lorenz) referred often to such activities as "social courtship play" (*Gesellschaftspiel*), implying that the activity might serve some social functions

8. Courtship flights of the American wigeon (above) and northern pintail (below). After photos by the author and Thomas Mangelsen.

other than pair bonding alone. Lorenz (1951–53) actually believed that in mandarin ducks (*Aix galericulata*) the males do not court particular females, and the display is "an affair of the males," even sometimes occurring in the absence of females. However, he generally believed that the social play of males seems to have "a very active role in choosing the mate," during which the competing males may display in general concert or, through a progressive gradient to the opposite extreme, a single male may display in a very directed way toward an individual female. Lorenz did not suggest that individual differences in display behavior or male plumage features might influence female choice of mates, although it is clear that he recognized that females show individual preferences toward particular males, as exhibited primarily through their mate selection or "inciting" behavior. Female incitement of male sexual competition may be fairly widespread in animals, and its potential significance as a factor in sexual selection has been well documented (Cox and LeBoeuf, 1977).

The role of inciting by female ducks in the pair-forming process can scarcely be overstated, and its importance in waterfowl courtship was clearly perceived by Lorenz. He stated that inciting is found in a homologous form in all "Anatinae" (= Anatini in most modern classifications, such as Johnsgard, 1965), in shelducks (= Tadornini), and also in the perching ducks "Cairininae" (= Cairinini). Lorenz was convinced that inciting as it exists in these diverse groups represents a real "phylogenetic order," and that it is necessary to examine the sexual behavior of the shelducks and sheldgeese, and indeed even to some extent the true geese (Anserini), in order to understand the courtship ceremonies of the surface-feeding ducks. Lorenz might well have taxonomically extended his "phylogenetic order" argument to include the pochards (Aythyini) and even the sea ducks (Mergini), but apparently not the stiff-tailed ducks (Oxyurini).

Social Displays of Shelducks and Perching Ducks

According to Lorenz, and as documented more fully by Johnsgard (1965), inciting in shelducks and sheldgeese is functionally agonistic in female motivation and resulting male response. Thus, in the sheldgoose species called the Egyptian goose (*Alopochen aegyptiacus*) a female incites by placing her body in line with her mate or potential mate, and makes threatening movements with her bill and outstretched neck toward the symbolic "enemy," often another male. This pattern of inciting behavior is illustrated here for the New Zealand shelduck (*Tadorna variegata*) (Figure 9A). Inciting is accompanied by loud calls, and the bill is always directly pointed toward the enemy, regardless of the latter's actual position. Thus, inciting in the sheldgeese and shelducks has not been ritualized to the degree that the head movements are fixed, regardless of the position of the enemy. In most true ducks the female's head movements as well as her vocalizations are highly stereotyped, and in some species the inciting movements have been nearly lost, or at least modified, to such a degree that they might easily be overlooked or mistaken for other behaviors. Similarly, whereas in shelducks and sheldgeese the female's mate invariably threatens or even attacks the indicated "enemy," such male responses have been ritualized in other groups to become relatively innocuous symbolic activities, such as bill pointing or chin lifting.

9. Social display in the New Zealand shelduck (*A*), the female inciting her mate (right) against an intruder male; mandarin duck (*B*), showing male burping (right); and American wood duck (*C*), showing male mock preen (left) and burping (right). Male tracheae of the shelduck and wood duck are also shown. After photos by the author.

A

B

C

Establishment and defense of a territory through aggressive encounters is critical to breeding success in shelducks; Patterson (1983) reported that in the common shelduck (*Tadorna tadorna*) the territory is "owned" by the female. The female evidently selects the territorial site, and her presence is needed for territorial establishment. Initial mate choice is probably attained through inciting behavior, with most pairs maintaining pair bonds from year to year. Interestingly, in at least this species of shelduck the largest males are not necessarily the most dominant; here successful pairing may lead to social dominance, rather than the reverse. The critical role of the female's behavior in achieving successful breeding by shelducks is thus apparent, although both sexes help defend the territory.

To the degree that females use inciting as a means of showing preference toward a male, a simple signaling mechanism for female-choice matings is of course available. Various authors have differed on the importance of female choice as an aspect of social display in ducks. Hochbaum (1944) believed that active male-male competition rather than female choice determines pair-bonding success in the canvasback (*Aythya valisineria*). It is also true that, especially in shelducks and sheldgeese, fights among males often result from female inciting; and should the intruding male win out over a female's initial choice, she may very well abandon this bird in favor of the winner. In such a case "female choice" is really a direct reflection of male-to-male competition, selecting for male strength and aggressiveness. It is not surprising, therefore, that in shelducks and sheldgeese male-male fighting rather than elaborate male displays is the usual pattern of behavior during competitive encounters, whereas in most of the structurally more "advanced" duck groups female inciting is more likely to initiate male courtship displays directed toward the female.

Among the structurally rather diverse perching ducks, the female inciting behaviors and the usual male responses are quite varied. In such species as muscovy ducks (*Cairina moschata*) and comb ducks (*Sarkidiornis melanotos*), the male responses consist largely of male-male aggression; these species correspondingly exhibit considerable sexual mass dimorphism but little sexual dichromatism. At the opposite end of the spectrum are generally rather sexually dichromatic species in which the males are only slightly larger than the females and rarely fight as a result of inciting. Instead, female inciting serves as a trigger for heterosexual display in the Lorenzian sense of "social play." Lorenz used two representative species, the mandarin duck and American wood duck (*Aix sponsa*), as examples of this advanced type of social display among perching ducks, but he mentioned that the Brazilian teal (*Amazonetta brasiliensis*) exhibits some similar characteristics. A more comprehensive survey of perching duck social behavior was also presented by Johnsgard (1965). Lorenz distinguished between the general social display patterns of mandarin ducks and American wood ducks in that, whereas he believed that male wood ducks tend to direct their displays toward specific females during social courtship, in mandarin ducks the female must take a more active role in choosing a mate. Lorenz thought this difference helps to explain the mandarin's more highly differentiated and showy male plumage. He also believed that the male wood duck exhibits an unusually large number of not very highly differentiated displays, suggesting that this situation might represent a more primitive condition.

Both mandarin ducks and American wood ducks engage in prolonged social courtship, which is often marked by extensive periods of female inciting and male posturing. Among mandarin ducks many of the male displays include crest raising and hackle spreading, as well as erection of the ornamental inner secondary "sail" feathers along the back. In this posture the male may mock-preen, momentarily lifting the sail feather enough to expose its lower iridescent vane, may perform a burp-whistle while vertically extending its neck (Figure 9B), or may perform various more elaborate "display-shakes." In both mandarins and wood ducks the male's trachea is inflated to form a bony chamber, or bulla, at the junction of the trachea and bronchi, or syrinx. Although the mechanism of sound production in such bullae remains uncertain, larger bony bullae are clearly associated with whistled notes during neck stretching. In true geese such bony bullae are lacking, and the sex differences in vocalizations among adults are often slight. In shelducks and sheldgeese the bullae of males are quite small (Figure 9A), and the associated differences in the adult sexes' vocalizations are less extreme than in most true ducks.

Among American wood ducks the male's displays are quite similar to those of mandarin ducks, although male plumage differences result in quite different visual effects. For example, the secondaries forming the distinctive "sail feather" in mandarins are not so specialized, and during the mock-preening display (Figure 9C, left) the entire inner wing, rather than this single feather, is briefly lifted out of the flanks and its iridescent surface is oriented toward a particular female. Like mandarin ducks, male wood ducks also perform burp-whistles while raising the crest and stretching the neck vertically (Figure 9C, right). Males often crowd closely around females during courtship, and are especially prone to swim immediately ahead of an inciting female and turn the back of the head toward her. In this distinctly heterosexual (rather than intrasexual) aggressive response to inciting, a fundamental shift away from the shelduck pattern of male hostile behavior is already apparent. Sexually selected male behaviors and structures that primarily attract a female's sexual attention, rather than increase a male's opportunities for winning an aggressive encounter with another competing male, are now clearly apparent.

It has been suggested (Dilger and Johnsgard, 1959) that the remarkably elaborate plumages of American wood ducks and mandarin ducks evidently did not evolve as antihybridization devices (reproductive isolating mechanisms), since no congeneric sympatry now occurs in these species. Instead, they perhaps represent supernormal stimuli developed under the combined impact of sexual selection resulting from intense and prolonged male-male sexual competition, and the tendency of the birds to display under crepuscular conditions.

Social Displays of the Surface-Feeding Ducks

The classic observations by Lorenz (1951–53) on the behavior of 14 species of ducks of the genus *Anas* and their near relatives provide the basis for understanding this group. Nearly all of the remaining approximately 20 species of these surface-feeding ducks were observed and described later by Johnsgard (1965). It is impossible to summarize the behavior of all of these adequately here, and only some general traits and trends will be attempted.

In all surface-feeding ducks thus far studied, females utter inciting notes and perform associated head movements that seem to stimulate male competitive display but rarely lead to overt male-to-male fighting. In most and probably all of these species the combination of female inciting and male turning the back of the head seems to constitute a major pair-bonding mechanism. Females, especially unmated females, of most if not all species utter "decrescendo calls" that also help to maintain contact be-

10. Social display in the American wigeon (*A*), showing males calling and wing-lifting during female inciting; speckled teal (*B*), showing male grunt-whistle (left) and bridling (right); falcated duck (*C*), showing male chin lifting (left), burping

tween pair members. Likewise most if not all additionally utter "repulsion" calls when sexually threatened by a strange male (Lorenz, 1951–53).

Among surface-feeding ducks, male displays tend to be rather diverse and fairly complex, typically consisting of simultaneous calls and posture that at least in part serve to emphasize their species-specific traits. Males of virtually all species perform a turning the back of the head, symbolic or "mock" drinking (an apparent appeasement gesture), ritualized or mock preening (primarily directed toward specific females), and a preliminary shaking movement that seems to draw attention to a male about to display. There are also various "burp" calls associated with neck stretching that are performed either independently or as part of a more complex posturing. In all the species of *Anas* so far observed, males have asymmetric bony tracheal bullae. Furthermore, in the species having unusually well-developed burp-whistle notes these bony bullae are fairly large (Figure 10). However, in the monotypic marbled teal (*Marmaronetta angustirostris*), which behaviorally links the surface-feeding ducks with the pochards, the bony surface of the bulla is interrupted with the same kind of membranous "windows" as are typical of pochards, which seem to have a softening acoustic effect on the male's vocalizations.

(center), and crest raising (right);
and mallard (*D*), showing male
head-up-tail-up (left), down-up
(center), and grunt-whistle
(right). Male tracheae are also
shown. After photos by the
author.

Bisexual surface-feeding duck displays performed between mates or prospective mates include not only the female inciting and simultaneous male turning-the-back-of-the-head sequence but also mutual mock drinking, mutual chin lifting (in some species such as the wigeons), and mutual head pumping, the last-named serving as the group's primary precopulatory display. These mutual or at least mate-directed displays, which appear to function more as pair bond maintenance mechanisms than mate choice mechanisms, are much more uniform across broad taxonomic groups than are the more species-specific courtship postures that are apparently critical in mate choice (competitive social display) situations.

Male displays that occur in social display situations are quite similar within taxonomic subgroups, so that Lorenz was able to suggest phyletic relationships within the surface-feeding duck group. These have been supported by later research. Thus, among the three wigeons a lifting of the folded wings (exposing the speculum and part of the white wing coverts) is typical, and is accompanied by whistles and chin lifting (Figure 10A, left, right). Female inciting among wigeons also is more strongly marked by chin lifting than by lateral bill pointing (Figure 10A, middle). Few other more elaborate male displays occur in the wigeons, which are unusual among *Anas* species in this regard, but which also represent a taxonomic extreme within the genus (or may be placed in a separate genus *Mareca*).

In the "green-winged" teal species group, including the Chilean or speckled teal (*Anas flavirostris*), a considerable number of male displays (essentially the entire *Anas* repertoire) are present, which are performed in rapid and sometimes almost unbroken sequence. These include the grunt-whistle (Figure 10B, left), a modified body shake during which the bill is dipped in the water and quickly retracted again as the bird rears upward. Another is bridling (Figure 10B, right), in which the head is drawn backward along the back as the bill is pointed approximately toward the courted female, although the head is moved toward the side of the body farthest from the female. In both of these displays a whistle is uttered at the peak of posturing. Several other obviously ritualized male displays (burping, head-up-tail-up, down-up, nod-swimming) also occur in this situation, as well as some less obviously ritualized preening, drinking, wing-flapping, and head- or body-shaking movements.

Male falcated ducks (*Anas falcata*), a species that taxonomically links the wigeons and green-winged teals, frequently perform crest raising, burp-whistling, and chin lifting with associated calling (Figure 10C). They additionally perform grunt-whistle and head-up-tail-up displays, the former with an exaggerated tail-raising and head-shaking component. The turning of the back of the head is especially evident in this species because of its flowing crest.

The social displays of male common mallards (*Anas platyrhynchos*) are probably the best understood of any species of waterfowl, and continue to be a rich source of new information on waterfowl pair-bonding behavior. (Because these studies are especially important in the area of mate choice behavior, they will be mentioned in greater detail later in this chapter.) In addition to several seemingly minor male ones, mallards perform three distinctly ritualized social courtship displays, all of which have associated whistle calls and are to varying degrees oriented toward specific females. They are typically stimulated by specific female activities, including inciting but especially "nod-swimming." This activity, during which the female assumes a posture similar to the precopulatory position while swimming rapidly, typically sets off one or more male displays, including the head-up-tail-up, down-up, and grunt-whistle (Figure 10D). Lorenz believed these three displays to be functionally "of equal value," and their sequence of performance "a matter of chance." Not surprisingly, more recent research has proven this not to be the case, but it is certainly true that all three of these displays are performed at a fairly high frequency during intensive social display.

Social Displays of Pochards and Sea Ducks

Like Lorenz's for the displays of surface-feeding ducks, Hochbaum's work (1944) on the canvasback provides a fundamental set of terms and basic descriptions for the social courtship displays of the diving duck group that are inclusively called pochards (tribe Aythyini). Additional terminology and interpretation was provided by Weller (1967) for the redhead (*Aythya americana*). As noted earlier, Hochbaum questioned whether canvasback pairs are formed by active female choice, suggesting that male competitiveness is more likely to be significant. He believed that the most important step toward attaining a mate in this species is the male's ability to attain and maintain a position at the female's side, which he judged could only be done through severe competition with other males during courtship. However, more recent authors have reported that females clearly play a major role in this species' mate selection (Bluhm, 1985).

Weller (1967) believed that female redheads try to maintain their positions "near a preferred male" and avoid attentions of intruding males by their inciting behavior and direct aggressive responses. Interestingly, Weller determined that male head throws were performed approximately three times more often by extra drakes than by preferred or so-called "pair-drakes," whereas the kinked-neck call was performed about twice as often by pair-drakes as by extra drakes. These two displays are among the most frequent and most conspicuous of the male courtship displays of pochards, including such North American species as canvasbacks, redheads, and ring-necked ducks (*Aythya collaris*) (Figure 11A,C,D, left). A low-forehead, seemingly threatlike "sneak" posture is also present in most or all pochard species, but it is often quite inconspicuous and is of uncertain function. Additionally, neck stretching (Figure 11D, right) is often performed by males, and may perhaps serve as a low-intensity avoidance or threat display, as a pair-greeting display (Weller), or only as an outward

11. Social display in the canvasback (*A*), showing male head throw (left), sneaking (center), and kinked-neck call (right); redhead, showing tail pulling during aerial chasing (*B*), along with head throw, sneaking, and kinked-neck surface displays (*C,* left to right); and ring-necked duck (*D*), showing head throw (left), sneaking (left–center), and neck stretching (right). A male trachea of the redhead is also shown. After photos by the author except for *B,* which is after a photo in Weller (1967).

indication that the bird is in the "courtship phase" of the reproductive cycle (Hochbaum). Neck stretching is clearly an integral part of inciting among female pochards, suggesting that at least in females it has threat and/or appeasement components.

Among pochards, rather low-amplitude and low-frequency but generally far-carrying calls are produced by males during the kinked-neck, head-throw, and sometimes the neck-stretching displays. In apparently all pochards the male has an inflated tracheal bulla with large membranous "windows" incorporated with the syrinx, and the tracheal tube itself is often variously inflated along part or much of its length, as illustrated for the redhead (Figure 11C). As with surface-feeding ducks, the acoustic mechanisms of male vocalizations are still quite obscure, but it seems highly probable that differences in male tracheal and syringeal structures contribute significantly toward both heterosexual and interspecific differences in duck courtship vocalizations.

A more active following, or overt chasing, of the female during aquatic courtship is typical of pochards than of surface-feeding ducks. These surface chases may often be interspersed with dives and, less frequently, with aerial chases as well. It is possible that such underwater and aerial chases are simply different manifestations of the same basic mate-choosing activity (Weller, 1967), and these may influence pair bonding by testing a given male's physical capacity to remain in the most favorable position, namely immediately beside the female (Hochbaum, 1944). However, in canvasbacks the females clearly retain the primary role in mate selection, based on their smaller relative abundance in adult populations and also their greater reproductive investment, as indicated by the great amount of time and energy females devote to the mate selection process (Bluhm, 1985; Lovvorn, 1990).

Once pair bonds have formed, males typically swim immediately beside their mates, and thereafter usually take the initiative in forcing other males to keep their distance. Such male protection also probably serves to facilitate relatively undisturbed foraging by their mates, allowing the hens to acquire the necessary energy reserves prior to egg laying.

Among typical sea ducks (tribe Mergini) the general pattern of social pair-forming behavior is much the same as in the pochards. For example, in the scoters such as the surf scoter (*Melanitta perspicillata*) courtship is marked by energetic surface and underwater chases, during which several males try to remain close to the female, who may perform inciting as well as direct efforts to escape their attentions (Figure 12A, left). Males may chase one another away from the female, and may perform various displays including calling while neck-stretching, tail cocking, and various racing or running movements across the water generally oriented toward the female. Underwater chases are common, and aerial chases also occur occasionally. No detailed studies of pair-forming behavior in scoters yet exist, so possible mechanisms of mate choice must remain speculative.

Courtship in the four species of eiders is somewhat better understood, especially in the case of the common eider (*Somateria mollissima*). Probably because of their rather large size and relatively heavy bodies, eider social display seems rather slow and ponderous. However, it consists of the usual combination of female inciting and male performance of a variety of ritualized postures and associated calls, termed "cooing movements" (McKinney, 1961). The three major cooing movements (two are shown in Figure 12B) are performed in markedly different frequencies and are interspersed with other less highly ritualized postures, such as a slightly modified wing-flapping (Figure 12B, right) display. All of the male displays are performed while the competing males attempt to hold their positions as close as possible to the female, which results in a good deal of chasing and threatening of other males, as in scoter displays. The female may sometimes dive to avoid these close attentions, and is quickly followed by the males. The periods of actual pair bonding (about half the birds initially pairing during fall, followed by a second spring period of pairing, at least in Scotland) seem to correspond with periods of intense social courtship and also with elevated testosterone levels in males. This suggests that these may be important behavioral and physiological elements influencing individual male mating success (Gorman, 1974).

Among goldeneyes, the social behavior of the common goldeneye (*Bucephala clangula*) is especially well studied (Dane and van der Kloot, 1964). An astonishing array of as many as 14 mostly highly ritualized and stereotyped postures are performed during social display by males, who crowd close to females that appear to initiate display through inciting and other apparently stimulative postures. Accompanying these ritualized display postures are various aggressive interactions, with the most intense period of aggressiveness coinciding with the period of most intensive courtship display, and with most of the aggression occurring in courting groups and involving competing males (Nilsson, 1969). Most of the male displays involve varying amounts of neck stretching and associated vocalizations, and in the common goldeneye not only is the tracheal bulla greatly inflated but the tracheal tube is also unusually enlarged, forming a retracted cone that can be considerably stretched and expanded in volume during neck-stretching displays (Figure 12C). There are, for example, two major types of head throws involving neck stretching, one of which lacks an accompanying kick (Figure 12C, right). The other (which occurs in two temporally discrete fast and slow forms) has a louder call and an accompanying double kick that throws water well up into the air behind the displaying bird (Figure 12C, left). Relative frequencies of these numerous displays vary somewhat under varied social conditions, such as the number of males or females present and the distance to the nearest other male or female in the courting group, suggesting that some functional differences probably occur among the displays. However, Dane and van der Kloot (1964) judged that the displays of both sexes could only rarely be interpreted as specific stimulus-response signals, and mused that even the species' seemingly "unordered sequences" of display could perhaps be selected for as a sexual isolating mechanism.

12. Social display in the surf scoter (*A*), showing male chasing (left) and neck stretching (right); common eider (*B*), showing cooing displays (left and center) and ritualized wing flapping (right); common goldeneye (*C*), showing head-throw-kick (left), masthead (center), and head throw displays (right); and hooded merganser (*D*), showing crest-raising (center) and head throw (right–center) displays. Male tracheae of the eider, goldeneye, and merganser are also shown. After photos by the author.

Social display among the mergansers is rather similar to that of the goldeneyes, although it is generally not so well documented. At least in the hooded merganser (*Mergus cucullatus*) and indeed the two other North American mergansers, social display is clearly competitive. Several males usually follow a single female while performing a variety of calls and ritualized postures, while the female responds with occasional inciting displays (Johnsgard, 1961b). The male courtship displays include frequent crest raising (Figure 12D, center) and an occasional head throw (Figure 12D, right-center) that is somewhat like that of the common goldeneye, and has a similar accompanying throaty call. Interestingly, the male's tracheal bulla is also distinctly[*] similar to that of a goldeneye, and wild hybrids between these outwardly rather dissimilar species have been reported, suggesting that a closer phyletic relationship might exist between them than is indicated by their bill shapes and male breeding plumages.

Social Display in the Stiff-tailed Ducks

The stiff-tailed ducks are a group of special behavioral interest. Although they possess diving mechanisms somewhat similar to those of sea ducks, they are even more highly adapted for diving than are these, and are correspondingly less well adapted for both land locomotion and sustained flight. Stiff-tailed ducks have an additional remarkable attribute, which is that males of most or all species are able to variously inflate their neck, by inflatable tracheal or esophageal mechanisms. Neck inflation may evidently function either as a strictly visual male display or, more often, in conjunction with vocal or mechanical sound production. Mechanical sound production involving water movement, by foot splashing, bill dipping, head immersion, or other similar methods, seems to be very common among stiff-tailed ducks, and perhaps its frequency relates to the fact that stifftails tend to inhabit reedy, overgrown ponds, where acoustic signals may be ecologically more useful than visual ones. This same situation might help account for the fact that plumage iridescence is lacking in all stifftails, and contrasting plumage patterns are largely limited to the male's head region, their bodies being mostly unicolored brown or gray. In virtually all species the elongated tail feathers are cocked during male display, and in both sexes are held at or below the surface preparatory to escape-diving, which provides a visual clue to help assess dominant-submissive relationships between birds.

However, stifftails are in some behavioral respects a seemingly rather "primitive" group, possibly diverging from the other ducks early in evolution, and as a result seem to lack (or have secondarily lost) some of the behavioral mechanisms of pair bonding that are otherwise characteristic of all the true ducks considered so far. They additionally have few if any clearly ritualized and mutual precopulatory displays, which results in their copulations often having a rapelike appearance.

Correlated with the fact that pair bonds are typically weaker in stifftails than in typical anatine ducks, ritualized female inciting behavior is apparently wholly lacking in this group. In some stifftails, especially the musk duck (*Biziura lobata*), pair bonds seem to be absent altogether, producing an apparently promiscuous situation exactly comparable to that found in many arena- or lek-breeding bird species. Additionally, males of several stifftail species differ from most true ducks in that they are distinctly territorial. Males maintain strongly defended temporary or permanent territories in which they advertise sexually. These territories may provide females with foraging as well as nesting opportunities. Such resource-based territories may thus attract pre-nesting females not only for their potential nesting sites but perhaps also for the resident male's protection from sexual harassment by other males, which may tend to collect near the territories of especially dominant males. For such reasons the stifftails are of special interest as models illustrating possible patterns of arena and lek behavioral evolution in birds generally.

Although some have questioned whether the black-headed duck (*Heteronetta atricapilla*) should be included within the stiff-tailed duck group, most current evidence from a variety of sources suggests that it should. However, its unique adaptations for obligatory egg parasitism set it ecologically and behaviorally apart from all other ducks, and have almost certainly affected its general social behavior as well. Both sexes are thus emancipated from parental responsibilities, which should weaken any need for strong pair bonding. Judging from limited observations in the wild and captivity, males do not appear to defend definable territories, but do seem to establish short-term pair bonds with individual females, probably especially during their egg-laying period (Carbonell, 1983). However, pair-bonding tendencies in females seem to be quite weak, and females lack obvious inciting behavior, which suggests that the importance of female behavior as an active mate choice mechanism may be rather limited. Instead, like many female stifftails, they seem to be only variably tolerant of a particular male remaining in their close vicinity, and the "preferred" male in turn prevents other males from closely approaching "his" female associate. Certainly no great ecological or social advantage would obviously derive to males from maintaining resource-based territories, nor to females from maintaining monogamous pair bonds in this socially parasitic species.

Most of the typical stifftail species are part of the genus *Oxyura*. Of these, the masked duck (*O. dominica*) seems to be the least specialized in most anatomical respects, and indeed it has often been separated as a monotypic genus (*Nomonyx*). Unfortunately the social and sexual behavior of this highly secretive and tropically distributed species is still essentially unknown. It is known that the adult male has both an inflatable esophagus and also a trachea that has two associated but rather small air sacs. These inflatable structures clearly affect the appearance of the displaying male (Figure 13A, right), and probably also influence the acoustic nature of his calls,

which have recently been described as a repeated pigeonlike cooing (Rod Hall, pers. comm.). Otherwise, male displays in this species appear to be infrequent and inconspicuous, suggesting that neither territorial advertisement nor nonmonogamous matings are likely to be well developed.

Much more is known of the social structure and sexual behavior of the ruddy duck (*Oxyura jamaicensis*), especially in the case of the North American race. Gray (1980) reported that in her California study area males showed no territorial behavior, but instead exhibited mate protection behavior against other males. Males varied in their individual pair-bonding tendencies from promiscuity (in most males) to short-term monogamous pair bonds that persisted through the incubation period. The latter occurred in situations where mate protection behavior to reduce harassment by other males was a seemingly important factor. The fairly long breeding season in California may also allow and indeed favor the development of successive polygynous or promiscuous matings, as compared for example with Manitoba with its short breeding season, where loose monogamy is reportedly more typical in this species. At the Wildfowl and Wetlands Trust in southwestern England, where breeding seasons are quite long and male display occurs over a six-month period, no pair bonds are formed by ruddy ducks (Carbonell, 1983).

D

E

13. Social display in the masked duck (*A*), showing male calling posture; North American ruddy duck (*B*), showing middle (right) and end (left) of bubbling sequence; maccoa duck (*C*), showing independent vibrating trumpet call (left) and water-flick (right); white-headed duck (*D*), showing kick-flap (left) and sideways piping (right); and musk duck (*E*), showing male whistle-kicking (left), as a young male watches (right). After photos by the author except for *A,* which is based on photos by Rod Hull.

Male sexual signaling in ruddy ducks, regardless of their individual degree of pair bonding, consists primarily of the remarkable "bubbling" display, during which the inflated tracheal air sac is repeatedly tapped by the male's lower bill, producing a low-pitched drumming noise and also forcing bubbles out of the breast feathers. This display is directed toward rival males as well as to nearby pairs, and also toward individual females (especially mates or potential mates). In all these varied situations the displaying male's body is usually oriented toward the object of the display (Figure 13B, left and right). Other more clearly threatlike displays may be directed toward rival males. Although females lack ritualized inciting behavior, their hostile responses toward displaying males often result in these birds in turn threatening or attacking other nearby males, and thus possibly serve as a kind of indirect inciting (Gray, 1980).

In the somewhat larger African maccoa duck (*Oxyura maccoa*) the males are strongly territorial, defending rather fixed boundaries against intrusion by all other males throughout the year. Females are allowed to move at will through these territories to forage, and also to select the best nesting sites and/or the fittest males. Promiscuity prevails, and successful males may succeed in attracting several females to nest within their territories (Siegfried, 1976). Thus, a male's displays are directed toward advertising his territorial status. These displays include stereotyped postures and noise-producing movements, low-pitched vocalizations, and (in common with most other male stifftails) bright bill coloration but relatively little plumage differentiation. One such male posture and call is the independent (or territorial) "vibrating trumpet call" (Figure 13C, left), in which the male faces a male rival and calls with a belching note, while his neck is enlarged and his bill is directed toward the rival. Another very similar display is the vibrating trumpet call, oriented instead toward a courted female, in which the male's bill is moved progressively downward during calling until it touches the water. Water flicking (Figure 13C, right), head dipping, and a vigorous "sousing" sequence are all displays that involve rather strong splashing movements with the bill, head, and entire forepart of the body respectively, and these movements probably have both acoustic and visual signal values. Such splashing displays are apparently more frequent and perhaps more important in the case of the rather closely related Argentine (*O. vittata*) and Australian (*O. australis*) blue-billed ducks (Johnsgard, 1966; Johnsgard and Nordeen, 1981). Of these two species, the Australian blue-billed duck is apparently polygynous, with males developing short-term sexual associations. The Argentine blue-billed duck may be somewhat more monogamous, judging from limited observations of captive birds (Carbonell, 1983). In these species splashing displays seem to be more important than vocalizations for male self-advertisement or possible territorial advertisement.

In its social structure and male advertising behavior, the white-headed duck (*O. leucocephala*) of Eurasia is similar in many respects to the maccoa duck. At least under captive conditions the males apparently form no pair bonds; instead they display to any and all females that approach them, and threaten or attack other nearby males (Matthews and Evans, 1974; Carbonell, 1983). The male's advertisement postures are rather different from those of the maccoa duck, but include a vocalized "tickering-purr" note while orienting laterally toward the female in a "sideways-hunch" posture. This lateral posturing sometimes gives rise to an explosive "kick-flap" (Figure 13D, left), with a loud splashing of the feet accompanied by an extremely rapid wing lifting and a quick dipping of the bill into the water. The kick-flap display is associated with "sidewise-piping," during which the lateral hunched posture is held while a loud piping note is uttered and the opened bill is pointed at the female (Figure 13D, right). Although extended observations in captivity strongly indicate that the white-headed duck is promiscuous, recent observations on wild birds in Spain suggest that there the birds may establish pair bonds, with males reportedly holding territories and defending nesting females within these territories. In this way the species may be somewhat intermediate in its social structure between the relatively monogamous stifftails, such as the black-headed duck and the Australian blue-billed duck, and the promiscuous maccoa duck and musk duck (*Biziura lobata*). It is also quite possible, as is evidently the case with the North American ruddy duck, that the strength of the white-headed duck's pair bond may vary regionally or temporally under differing environmental conditions.

The Australian musk duck, the largest and most aberrant of the stifftails, is also the species of waterfowl whose social structure is most like that of typical arena birds. The males are evidently wholly promiscuous, or at least strongly polygynous. Adult males are extremely large and powerful birds, weighing on average about 2,400 grams, or more than 1.5 times the average female weight, and some probably rather old males are known to have exceeded 3,200 grams, or about seven pounds. Their highly aggressive temperaments also fit the mold of sexually selected male traits based on male-to-male competitive interactions. Adult males actively patrol and defend their territories, admitting only females and young birds, and territorial contests between adult males are notably savage and often prolonged. Adult males also seasonally produce a distinctive musky odor from secretions of the uropygial gland (of unknown social function), and have both an inflated sublingual sac and a pendant throat lobe that during the breeding season can be enlarged and made turgid through changes in blood pressure. Territories are held by adult males at least through the breeding season, and the nearly constant male display activity (occurring both day and night, and with sporadic display extending through every month of the year) provides a long-distance means of attracting females. Females as well as younger males are attracted to displaying males, which sometimes results in a leklike social grouping. The young males remain some distance from displaying adults (Figure 13D, right), but watch them closely if discreetly. Females seem drawn as if magnetically to displaying males, and in some instances males have been observed to mount and copulate with them in a rapelike manner as soon as they venture close enough. It is not known whether females remain within a particular male's territory to nest there, but there is no indication that the male later participates in nest or brood defense.

Advertising displays of male musk ducks involve various energetic kicking movements, with associated pouch enlargement, tail cocking, and other posturing, which in the most intense display stages (Figure 13D, left) are supplemented by whistled notes (Johnsgard, 1966; Fullagar and Carbonell, 1986). The splashing sounds produced and the accompanying whistled vocalizations carry well over water and make effective long-distance acoustic signals; no bright colors or contrasting plumages of any kind are present on males. In many respects the male musk duck's secondary sexual characters, social system, dispersion pattern, and advertising behavior more closely resemble those of the large, solitary bustards (*Ardeotis*) than those of typical ducks.

Mate Choice Mechanisms and Tendencies in Mallards

Not surprisingly, the species of duck for which the greatest amount of information on mate choice mechanisms is available is the mallard, including its domesticated derivative forms. Generally, biologists have believed that female mate choice in ducks is most likely to be influenced by one or both of two mechanisms. First, females may detect and respond to variations in species-specific male plumage and related external morphological traits, whether innately or on the basis of experience. Second, females may perceive variations in individual male social behavior, including not only qualitatively different performance of species-specific displays but also quantitative variables in male-to-male dominance behavior and differences in the incidence of male display activities that are directed specifically toward females.

Mate choice studies involving the mallard are both numerous and highly instructive; Williams (1983) has effectively summarized much of the relevant information. Indicative of the probable role of individual male behavioral variables on mate choice by females, Bossema and Kruijt (1982) reported that nearly all the female mallards they studied mated with the males that courted them most intensively, regardless of the males' actual plumage color (including even domestic white variants). Bossema and Raemers (1985) later reported that larger males were more attractive to females than were smaller ones, and those males with a high tendency to "address" their social display toward particular females were more attractive to such females than were males less active in this respect. When given a choice of mates among groups of males that had either been fed *ad libitum* or had been raised on restricted diets, female mallards exclusively courted males of the well-fed group. Females additionally preferred those males showing high display activity levels and those that exhibited a "high plumage" aspect (Holmberg, Edsman, and Klint, 1989). The females also showed a tendency to court males of intermediate age (those roughly 18 months old, which was also the age class having the highest male display rates), and marginally tended to prefer males of "smaller" body size, as measured by sternum length. The significance of this last, rather surprising finding needs some explanation, since it seemingly conflicts with the results of Bossema and Raemers. The sternum length measurement was evidently positively correlated with male age variations, and the "larger" males were those at least 30 months old. Thus, the females may have been simply preferring to court smaller, younger, and more active males, rather than larger, older, and less active males.

Evidence for possible learned or innate female preferences involved in making proper species-specific mate choices, as well as a female's relative capacity to recognize and select the most fit or most "high-plumage" individual males, is of considerable interest in terms of sexual selection. Klint (1980) reported that female mallards can discriminate between normal wild-type and domestic white-plumaged male mallards in competitive courtship situations. The females consistently chose males of the typical or wild type, regardless of their own varied early experiences with these color types, suggesting that an innate and specific species recognition mechanism may be present. However, females that had been reared in isolation from hatching until about four months old showed no such preference for normally colored males, or even preferred an abnormally colored male, suggesting that social imprinting may also be important in mate choice. More recently, Weidmann (1990) reported that female mallards, even including inbred individuals, preferred to mate with "good-looking" males that most closely approached the wild type in plumage and softpart coloration and pattern, and that had few or no plumage blemishes. Such female choice tendencies could be experimentally modified either by making normally "attractive" males more "ugly" through plumage alteration, or by making previously "unattractive" males more "beautiful" through staining and bleaching techniques.

Brodsky, Ankney, and Dennis (1988) tested mate choice characteristics of both female mallards and the very closely related (some would argue conspecific) American black ducks (*Anas rubripes*). They used birds that had been selectively reared from hatching with males either of their own species or of the other. As adults, these females preferred to associate with the type of male with which they had been reared, at least under confined, relatively noncompetitive conditions. This obviously acquired social preference is of special interest, inasmuch as adult male black ducks are much less colorful than are breeding-condition male mallards. However, under unrestrained and directly competitive conditions, the females associated with the most dominant individual males (which were always mallards), regardless of their earlier social experiences and preferences.

These results are of particular ethological, ecological, and taxonomic interest in terms of species recognition and mate choice characteristics of female ducks. Male and female social displays in mallards and American black ducks are essentially identical. Not only are male mallards in breeding plumage more brightly colored than male black ducks, but there are slight quantitative differences in the two species' male display performance characteristics. Nevertheless, hybridization between these two incipient species is fairly frequent both in the wild and under captive conditions (Johnsgard, 1961a). A significant tendency among females of both species to select male mallards over male black ducks in competitive situations might significantly impact the hybridization incidence in regions of geographic overlap, especially in common wintering areas, where most of the pair-bonding behavior occurs. This female behavior would obviously favor the introduction of mallard genes into the much smaller black duck gene pool in such situations, thereby weakening whatever other reproductive isolating mechanisms (such as ecological differences) might have been developing between them, and eventually threatening the genetic integrity of the American black duck.

Williams (1983) concluded that the then-available information on mate choice in mallards indicated that male plumage and display differences in mallards and related ducks operate as species-specific isolating mechanisms. Males learn female traits through sexual imprinting, but females have an innate capacity to recognize species-typical male plumage characteristics. Females evidently also respond to cues that signal an individual male's interest in them, as well as his ability to protect and defend them. Williams stated that little evidence existed at the time of her writing to suggest that females pay any attention to the actual performance of male displays, although the more recent above-mentioned studies by Holmberg, Edsman, and Klint (1989) and others indicate that relative intensity of male directed display as well as relative male social dominance may play highly important roles in mate choice by females.

4

Galliform Birds: Stages, Courts, and Classic Leks

The large order Galliformes provides a great diversity of mating systems and social structuring, from extended monogamous pair bonding to totally promiscuous relationships. All species have highly precocial chicks, setting the ecological stage for male emancipation from participation in chick rearing (although necessary male defense of the female during nesting may influence pair-bonding tendencies). Typically, the small and/or relatively vulnerable species of many galliform birds such as the quails and partridges (Odontophorinae and Perdicinae) are monogamous, with strong male participation in nest and chick defense, whereas the largest species of phasianids (pheasants, peafowl, turkeys) are much more prone to have nonmonogamous mating systems, and are also more highly sexually dimorphic and dichromatic (Johnsgard, 1986). Monogamy in the African guineafowl (Numididae) and New World cracids (Cracidae) is not obviously size-related, and so other ecological constraints must be present in these tropical groups.

Although arena-like displays are likewise best developed in some of the typical pheasants and grouse, there is some tendency toward them even in some species of the seemingly most anatomically generalized galliform group, the megapodes (Megapodidae). Thus, in contrast to most megapodes, males of the brush turkey (*Alectura lathami*) are apparently polygynous, with each adult tending and defending an incubation mound that also serves as a display site, and from which other males are aggressively excluded. Females are permitted near the mound only when they adopt a solicitous stance, and egg laying (of a previously fertilized egg) often immediately follows copulation. Only males with mounds are effective in attracting females, who are fertilized for subsequent eggs during egg-laying visits. Females leave the site soon after copulation and egg laying are completed. Thereafter, females play no further role in reproduction (Jones, 1988). Basically this species' male advertising behavior pattern seems to represent a simple "stage" level of localized male display. Many males are expelled from their mounds by more dominant ones, and some males are able to build (or usurp), maintain, and defend two mounds, thus probably increasing their mating opportunities. Individual differences in mound building and

14. Male sexual display sequence in Temminck's tragopan, including preliminary head shaking and lappet exposure (*A*), wing-beating and clicking phase (*B*), and rearing phase, as seen from behind (*C*) and by female (*D*). Also shown is rearing phase of Cabot's tragopan display (*E*). *A–D* after photos by the author and David Rimlinger; *E* after a photo in Guangmei, Ronglun, and Zhengwang (1989).

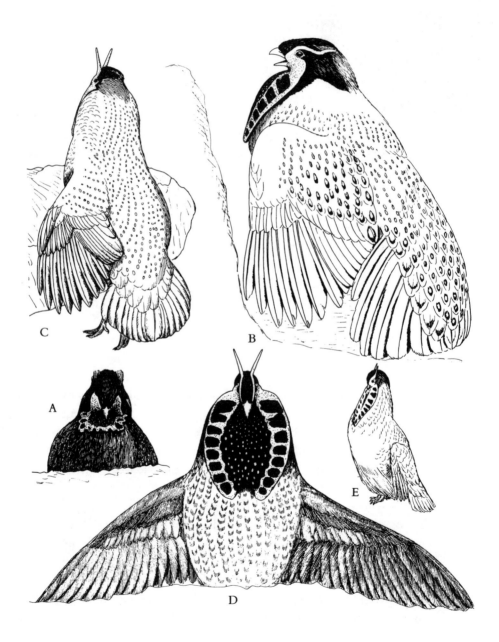

defense success may be reflections of individual male fitness (Troy and Elgar, 1991). Even in the mallee fowl (*Leipoa ocellata*), which is generally considered to be permanently monogamous, at least one case of polygyny has been recently reported (Weathers, Weathers, and Seymour, 1990).

Stage and Arena Displays in Pheasants

In the majority of the nearly 50 species of typical pheasants, nonmonogamous mating systems prevail, and in most of them rather pronounced sexual dimorphism and dichromatism are typical. Male pheasants are variably larger than females (generally with mass ratios of 1.1–1.5:1), and usually have one or more pairs of tarsal spurs that are primarily used in male-male fighting. Adult males also exhibit variably elaborated plumages, including colorful or iridescent body feathers, crests, elongated and elaborate tails, and variably large areas of colorful skin exposed on the head, especially around the eyes (Johnsgard, 1986). Large male size and elaborate plumages reach their extremes in the peafowl and the argus pheasants, and elaborate arena-like displays for male advertisement are also highly developed in these groups. However, even in some of the more typical, somewhat more partridgelike pheasants such as the tragopans (*Tragopan*) the rudiments of male arena displays can be discerned.

Although the tragopans have usually been considered (on the basis of observations in captive situations) to have short-term monogamous pair bonds probably lasting until incubation begins, the evidence for such pair bonding is highly suspect. Both the fairly substantial adult sexual dimorphism (1.0–1.7:1) and sexual dichromatism that exists in all five species (Figure 15) might lead one to suspect that polygynous matings may be fairly common in these birds. Regardless of the level of polygyny that normally exists in tragopans, the male advertisement displays offer a rather striking example of the possible role of the environment in shaping arena-like displays.

The male advertisement displays of only one species of tragopan, the Temminck's (*T. temmincki*), have so far been fully documented and described (Rimlinger, 1984; Johnsgard, 1986). However, what is known of the comparable displays of the other species suggests that in at least three of them there is a good deal of similarity, and the fourth is likely to be similar as well (Islam, 1991). While various male-to-female lateral displays may be performed frequently and are not environmentally constrained, the most elaborate and highly ritualized of the displays, the frontal display, is essentially dependent upon the presence of a suitable stage, consisting of a natural hiding place, such as a log or large rock, with a clear view in front of this site to a position where a female might occasionally stray. Indeed, when aviaries have only very low hiding places, captive male tragopans may attempt to perform their frontal displays by crouching or even lying flat on the ground, rather than standing in their normal posture. If no such hiding places are available in the area, frontal displays may not be attempted at all. However, by providing the aviary with an appropriately sized visual barrier, such as logs or large rocks, it is sometimes possible to initiate male advertisement display rather soon after (Kamal Islam, pers. comm.). In this respect, tragopans seem to qualify as stage-dependent species.

The frontal display of the Temminck's tragopan is a complex series of events usually lasting about 30–40 seconds. It begins with a male walking behind a rock or log and peering over the top at the female, which is usually 10–15 feet away. The male then begins to twitch his head, simultaneously expanding his gular lappet and erect-

15. Sexual signals of peacock-pheasants and argus pheasants, including head detail of male bronze-tailed peacock-pheasant (*A*), tidbitting by male Palawan peacock-pheasant (*B*), lateral display by male Rothschild's peacock-pheasant (*C*), lateral display by male crested argus (*D*), tail ocelli of Palawan peacock-pheasant (*E*), and tail ocelli of Malay peacock-pheasant (*F*). After specimens and photos by author except for *D,* which is after a photo in Huxley (1914).

ing his bluish "horns" (Figure 14A). His tail is then spread and his wings begin to beat rhythmically, and a prolonged sequence of clicklike calls is begun. At this stage the male is crouched behind the rock, so that for a brief time he is normally completely invisible to the female. This calling segment of the display (Figure 14B) usually lasts about 14 seconds, during which period the wingbeats are rather shallow but the tail is well spread, and a call is uttered at the top of each wingbeat cycle. The intervals between the calls gradually speed up, from about one per second to four per second as the calling phase progresses, which in turn leads directly to the climax phase. In this phase (Figure 14C, D) the male suddenly hisses and stretches vertically upward. As he does so, the lappet is maximally flattened and spread over his breast, the "horns" are maximally raised, the bird is standing on his tiptoes, and his wings are fully extended, with their tips touching the ground, perhaps to help in balance. From the female's view, the male's sudden appearance is almost apparition-like, with little of the male's body below the lappet normally visible to her because of the intervening rock. There is a brief pause at the climax of the display, after which the male sometimes sinks silently back down again behind the rock. However, at times the male suddenly runs around (or over) the rock toward the female in his full display posture, accompanied by loud hissing. The male may then chase the female for a distance, or he may abruptly stop at the point where she had just been standing. If she remains in place the display may terminate with mounting and an attempted copulation. Indeed, quite possibly the entire purpose of this bizarre display sequence is to mesmerize the female long enough to allow the male to approach and mount her.

It is of considerable evolutionary interest that this general male display sequence is apparently very similar in several of the tragopan species. However, because of some differences in lappet color and pattern, differing degrees of wingflap coordination and relative wing synchronization during the clicking phase, and acoustic differences in clicking, male species specificity is achieved (Islam, 1991). Only in a few geographic locations do two species of tragopans possibly now occur sympatrically (involving Cabot's and Temminck's, and Blyth's and satyr tragopans respectively). It thus seems unlikely that direct selection for display divergence as a reproductive isolating mechanism has influenced display complexity, or has favored the evolution of these interspecies differences in male tragopan traits. Instead, the varied patterns and behaviors that are now present may largely reflect the varied effects of sexual selection in five different tragopan populations. It also seems probable that the frontal displays of male tragopans have evolved in close conjunction within the normal environmental range of tragopan habitats, which are typically rich in logs, rocks, and boulders. As such, this stagelike display setting is probably readily adaptable to any number of usable sites within a male's home range.

No comparable stagelike displays occur in the red junglefowl (*Gallus gallus*), nor its domesticated variant, the domestic fowl. In the Burmese red junglefowl nonrandom mating occurs, at least in captive populations. There, a social hierarchy similar to that of the domestic fowl exists, with alpha-level or other high-ranking males usually showing the highest level of sexual activity, but with female peck-order rank apparently playing little if any direct role in mate discrimination by males. Thus, the opportunities for classic sexual selection on the basis of relative male dominance exist in junglefowl (Lill, 1966). Recently, Ligon et al. (1990) have provided more detailed documentation of sexual selection based on male-male competition, and the role of testosterone levels in male dominance status in red junglefowl. They reported that comb size was the only measured morphological trait that correlated with the winning of controlled fights between males. Since the comb size is strongly influenced by blood testosterone levels, it closely reflects the individual male's physical condition or vigor. As such, comb size might be a reliable signal of relative male dominance, perhaps providing a basis for active female choice of most-fit males. Additionally, there is evidence that sexual selection based on female choice exists in red junglefowl, since Zuk et al. (1990a) reported that in mate choice tests, females preferred unparasitized males over males that have been parasitized by an intestinal nematode, at a preference ratio of about 2:1. The parasitized males showed a number of diminished external ornamental male traits (duller combs and eyes, shorter combs and tail feathers, etc.) compared to unparasitized males, which may perhaps provide the visual basis for such female discrimination.

Among polygynous ring-necked pheasants (*Phasianus colchicus*) the sexes are somewhat dimorphic and highly dichromatic, the adult males having ornamental and partially iridescent plumages, elongated tails, colorful facial skin, and sharp tarsal spurs, the spurs being mostly used for male-male fighting. Unlike the situation in junglefowl, there is not yet any clear evidence from captive birds that females preferred unparasitized males over those that were experimentally parasitized with coccidian parasites. However, there was a correlation between display rate, coccidian parasite load, and female choice among male pheasants in the wild (Hillgarth, 1990a).

Ring-necked pheasants have a system of polygyny somewhat like that of junglefowl. However, in pheasants adult males accumulate "harems," or variable numbers of individually guarded females, within their rather large territories, rather than forming peck orders within rather stable flocks. Sexual bonding is based on mate guarding by males, which probably protects the females both from possible injury as well as harassment by other males. Advertisement display by males, consisting of calling and loud wing flapping, occurs in varied parts of their territories, so like junglefowl they clearly do not qualify as arena species. However, experienced males have larger harems than do first-time territory holders, and males with shorter (experimentally cut) tails have smaller than average harems. These results suggest that variations in overall individual male quality may be more a reflection of male behavioral or visual quality than of territory quality (Ridley and Hill, 1987). Thus, sexual selection seemingly operates at least in part through active female choice of specific male traits.

In a study of feral birds in Sweden, a variety of structural and behavioral variables among 81 males were studied and compared with harem size variations (Göransson et al., 1990). The best single predictor of male reproductive success was found to be tarsal spur length, which is apparently a reflection both of an individual's age and the environmental conditions affecting relative food availability. Harem size differences among males were evidently more a result of active female choice than of male-male competition, although the correlation between harem size and male tail length variations was quite low. The stronger spur length correlation was believed to reflect general male viability rather than specifically determining male dominance differences, since a separate study involving the experimental manipulation of male spur length did not alter their dominance relationships.

In a related study (Wittzell, 1991), the same pheasant population was studied from the standpoint of differences in individual male reproductive success relative to male variables in spur length, wing length, tarsal length, and tail length, as well as similar individual variables for females. Among males, there was a positive correlation between individual harem size and male wing length, tail length, and spur length. There were also correlations between the number of chicks individually sired by males and variations in their wing length, spur length, and total body mass. However, only the spur length measure formed a significant positive gradient when the other variables were held statistically constant. Similar correlations between female morphometric traits and their individual reproductive success were not conclusive. In short, sexual selection involving active female recognition and choice of the fitter males evidently occurs in ring-necked pheasants, and results in differential male reproductive success.

Comparatively little is yet known of the mating strategies of the Bornean wattled pheasant (*Lophura bulweri*), a jungle-dwelling pheasant that has only moderately well developed tarsal spurs but exhibits very substantial sexual dimorphism and plumage dichromatism. Judging from very limited observations of wild birds, males probably are polygynous or promiscuous. They are probably not stage-dependent like tragopans, nor harem-forming like ring-necked pheasants, but instead may be relatively mobile and nonterritorial. At least in captivity, the visually stunning but nearly silent displays of the male are performed in small, open areas that are partly surrounded by bushes. This tactic may force the female to observe from close range and may increase the element of visual "surprise" in the male's displays (Rimlinger, 1985). These consist of rapid about-face postures, with the spinelike tips of the disk-shaped tail audibly scraping the ground. The male's grotesquely enlarged blue facial wattles then contrast strongly with his velvety black body plumage and his immaculate white tail, while his ruby-red iris is visually exaggerated by the accompanying exposure of a surrounding vermilion eye ring (Johnsgard, 1986).

In the similarly forest-adapted peacock-pheasants (*Polyplectron*) an interesting gradient of sexual dichromatism and male display behavior exists, along with a probable shift from monogamous to nonmonogamous mating systems. Together with the previous information on sexual selection based on female choice in junglefowl and typical pheasants, this clinal variation provides a hypothetical mechanism to explain the evolutionary development of arena and lek behavior in pheasants and grouse. There are seven currently recognized species of peacock-pheasants (Johnsgard, 1986), which exhibit interesting interspecific differences in appearance as well as considerable differences in their degree of sexual dimorphism and dichromatism. The least dimorphic species is the apparently monogamous bronze-tailed pheasant (*P. chalchurum*), in which the plumage differences between the sexes are also quite minor. Males lack both definite crests and iridescent ocelli (Figure 15A), but do have a glossy bronze-colored band crossing the middle of the tail. In the probably polygynous Rothschild's peacock-pheasant (*P. inopinatum*) the male is slightly crested, and variably distinct iridescent ocelli are located on the tail, tail coverts, wing coverts, and secondaries (Figure 15C).

These ocelli and general male plumage specializations are more highly developed in the gray peacock-pheasant (*P. bicalcaratum*) (Figure 16A, B) and in the Malay peacock-pheasant (*P. malacense*) (Figure 15F). Sexual dimorphism is slight, but sexual dichromatism is well developed in both of these apparently polygynous species, especially the Malay peacock-pheasant, which has large numbers of well-developed ocelli (Figure 4). Male ocelli are more restricted in distribution but are highly conspicuous in the Palawan peacock-pheasant (Figures 15B, 16D), which has a highly developed male crest and head plumage pattern (Figure 16C) and extensive body feather iridescence. Males of this species are the most colorful of all the peacock-pheasants, and their lateral displays of their ocelli-laden tails are among the group's most highly ritualized and spectacular. Whereas the gray peacock-pheasant has a frontal display (Figure 16B) that is probably derived from functional food presentation or "tidbitting" behavior, in the Palawan the bill is hidden from the female's view and a highly compressed lateral posture is assumed. This hides much of the male's head below his eyes, and the ocelli on the entire upper surface of the tail and tail coverts are maximally exposed to her view (Figure 16D). This exhibition of the eyelike ocelli of male peacock-pheasants and their near relatives the peafowl and argus pheasants may mesmerize the female into a state of tonic immobility, as suggested as a possible function for male tragopan displays, perhaps thus making copulation more likely, or at least enhancing a particular male's conspicuousness and perhaps also his sexual attractiveness (Davison, 1983b).

Perhaps more interesting from the standpoint of the evolution of arena behavior than trends in postural display and feather patterning are some apparent "stage display" variations among peacock-pheasants. Limited observations of captive bronze-tailed pheasants suggest that the rather nondescript and dull-colored male may perform his rather simple wing-flapping, hissing, and "freezing" display while standing in essentially the same poorly illuminated location during each display bout. However, males of the more colorful Rothschild's peacock-pheasant would repeatedly and rather dramatically step out from behind a leafy bush and into the female's direct view just before displaying in a posture that effectively exposed their iridescent ocelli in the sunlight (Johnsgard, 1986).

Nothing has yet been reported on possible display localization sites of the male gray peacock-pheasant, nor of some other very closely related forms such as the Germain's

16. Sexual displays of peacock-pheasants, showing crest raising (*A*) and frontal display (*B*) in the gray peacock-pheasant, and head detail (*C*) and lateral display (*D*) in the Palawan peacock-pheasant. After photos by the author.

17. Frontal displays of male blue peafowl (*A*) and great argus pheasant (*C*). Also shown are tail covert ocellus of the male peafowl (*B*) and secondary feather ocelli of the male argus (*D*). After specimens and photos by the author.

peacock-pheasant (*P. germaini*). However, specific "display areas" or "arenas" have been reported for both the Malay peacock-pheasant and the Palawan peacock-pheasant. In the little-studied Palawan peacock-pheasant the male display arenas have been described as consisting of neatly swept areas about a meter in diameter, usually situated in a remote part of the forest and on rolling or flat land. Individual males may apparently maintain from one to as many as 16 of these areas, which presumably serve as primary display sites. In the Malay peacock-pheasant the display sites are from only about 0.4 meters up to nearly 2 meters in maximum dimension. They usually are located on human-made or game trails. Such locations suggest that open spaces are needed for effective display, by either reducing physical obstructions or providing a suitable visual backdrop. The display site itself might thus function as an epigamic signal. Display sites are tended and kept clean by their male owners and may be used for up to two months during a single year, as well as in successive years. One male may also maintain several display sites within his territory. These male-modified display sites clearly qualify as typical avian courts, and the male's associated advertisement behavior suggests that a polygynous mating system occurs in this species. Males that were known to receive visits by females at their display sites were individuals in adult plumage, that had two pairs of spurs, and that showed a high rate of display site attendance (Davison, 1983a).

In the argus pheasants and peafowl, the largest of the true pheasants, the structural and behavioral trends evident in the peacock-pheasants reach a climax. In the crested argus (*Rheinartia ocellata*) the male lacks definite large and iridescent ocelli, but most of his feathers are covered with a multitude of very small brownish ocellus-like spots. In the great argus (*Argusianus argus*) the secondary feathers are highly lengthened and specialized in pattern to produce a long series of iridescent ocelli organized in a ball-in-socket manner near the feather shaft (Figures 17D, 18C). The adult male:female mass ratio of the latter species is about 1.5:1.

The sexual display of the crested argus (Figure 15D) is apparently rather simple, although males are known to defend and advertise "dancing grounds" on mountain ridges or saddles. These grounds are cleared courts up to about 4 meters in diameter, from which males regularly call over much of the year (Davison, 1978). Similar courts or "dancing sites" are established and advertised by male great argus pheasants; these are also highly dispersed sites averaging several hundred meters apart. They are often placed on hilltops and are kept clean of debris by their attending males. In both species females are probably acoustically attracted to the sites by the loud calling by males; once they enter the display ground the male begins a series of mostly visual signals, including postures that emphasize the ornamental wing and tail feathers (Davison, 1981, 1982; Johnsgard, 1986). In the most elaborate great argus display the male's body is almost fully hidden behind his outstretched, fanned-out wings (Figures 17C, 18A). His feathered ocelli radiate out spectacularly from a focal point near the location of his actual eye, which is barely visible to the courted female and is situated at the

18. Frontal display in the great argus pheasant (*A*). Also shown are male head detail (*B*) and male secondary feather pattern detail (*C*). This drawing has been stylized to aid in visualizing the male's head and body posture during frontal display; the small inset sketch and Figure 17 more accurately depict the male's actual wing and head appearance as seen from the side.

angle of one wing. Even the "highlights" on these eyelike spots are appropriately located, as if they were illuminated from above. Darwin (1871) described these remarkable feathers in detail, and illustrated the graduated sequence by which such remarkable visual signals are progressively formed from nonocelli bars and spots lower on the secondaries.

The displays and social structure of peafowl show some additional interesting developments. The colorful eyelike ocelli (Figure 17B) are located on the highly elongated tail coverts rather than the wings or tail, and are highly iridescent; about 100–175 such ocelli may be present on a single adult male. The adult male:female mass ratio is about 1.5:1. Blue peafowl (*Pavo cristatus*) have traditionally been regarded as having a polygynous, haremlike mating system without territorial boundaries. However, in a feral population studied by Rands, Ridley, and Lelliott (1984), it was found that males defended very small and aggregated display territories, whereas females remained in a flock that ignored territorial boundaries. These authors concluded that "peacocks are lekking birds." Two of the four males that they studied were observed mating; these birds were of intermediate age and were no longer-tailed or larger than the other males, but they did spend more time displaying sexually. A similar social organization and leklike territorial spacing of wild birds in India was also reported by Hillgarth (1984).

The courtship displays of this species were described by Ridley, Lelliott, and Rands (1984), who stated that individual males display from one or two display sites within their small territories. These male displays are mostly frontally oriented to exhibit the ocelli maximally, and may serve to mesmerize the female into immobility. These authors suggested that possibly the larger and more effective the male's ornaments, the better the chance that a female will allow mating. Recently Petrie, Halliday, and Sanders (1991) have indeed proven that those males with the most elaborate trains are most effective in attracting females. On one lek having ten males, over 50 percent of the variance in observed male mating success could be statistically attributed to their relative feather train elaboration, as estimated by the number of ocelli present. Females often rejected potential sexual partners, and in 10 of 11 observed successful copulation sequences the male "chosen" by the female was the one having the highest number of ocelli, at least among the males she had visited. However, there was no significant relationship between the number of ocelli on a male's train and his relative success in male-male agonistic interactions, suggesting that female choice, rather than male competitive dominance, was the critical variable in influencing male copulatory success. The number of ocelli on the male's train may provide a visual index to the male's age, as they increase from the ages of 4 to 12 years, as also do train length and color. Perhaps the relative elaborateness of the train might also provide a convenient visual index to general male vigor and social status. Manning and Hartley (1991) sug-

gested that, since a sample of adult males showed a correlation between train ornamental symmetry and total ocellus number, females might be able to assess a male's
fitness by simply judging the symmetry of his ornamental train during display.

The Social Arena and Lekking Displays of Grouse

If the pheasants are not generally regarded as typical arena and lekking birds the
grouse certainly are; indeed the term lekking was originally applied to the black
grouse (*Tetrao tetrix*), and some purists limit the use of the terms leks and lekking to
the territorial-sexual male assemblages and displays of certain species of grouse.

As with the pheasants, it is worth reviewing the social structures and behaviors of
the nonlekking grouse before addressing the question of typical lek behavior in the
group. This general question was reviewed earlier (Johnsgard, 1983a), and similar
trends to those already described for pheasants are somewhat apparent. In brief, the
characteristics of most nonlekking grouse species include (1) male mating territories
that are simply part of the males' larger overall home ranges, although they may be
only seasonally defended, (2) generally little evidence of sexual dimorphism, dichromatism, or specialized male softparts (comb, esophageal air sacs), (3) no obvious
dominance hierarchy among interacting males, which defend variably sized but typically large home ranges or territories, probably marked by only infrequent boundary
disputes, (4) no clear correlation between the incidence or effectiveness of a male's
territorial defense behavior and his relative mating success, and (5) an individualized
attraction of females to males or vice versa, perhaps facilitated by their overlapping
home ranges. At least in some species (ptarmigans and hazel grouse), individualized
seasonal or even more prolonged monogamous pair bonding may develop, and male
defense of the nesting female or the brood may occur.

Among the lekking grouse, the major characteristics include (1) male display areas
that are not part of a male's normal home range, but instead are visited and intensely
defended for only part of the year, (2) increased sexual dimorphism and dichromatism
plus elaboration of male softpart features that are associated with sexually diethic displays, (3) a dominance relationship established among males, as reflected by their individual ability to attain a territory of high quality (special territorial position and/or
relative size), (4) a probable positive dominance-to-fitness relationship, correlating an
individual male's relative territorial/social dominance with his individual reproductive success, and (5) the attraction of prebreeding females to such relatively dominant
males in order to achieve fertilization, but with no attempts by either sex to establish
individual social attachments (pair bonds) between them. Subsequent nesting by the
females is usually done well away from the nearest lek, and no nest or brood defense
behaviors are ever performed by males.

Some intermediate conditions between these extreme types might be expected to
occur, and the capercaillies seem to provide one such example. Although the territo-

ries of Eurasian capercaillies (*Tetrao urogallus*) are extremely large, they often have their "territory centers" actually located near territorial boundaries, where they may be in visual contact with other males (Müller, 1979). Such strongly defended and advertised areas may not only attract females, but may also place males in directly competitive situations during the female attraction period. Another intermediate condition might exist in the North American blue grouse (*Dendragapus obscurus*). Although male blue grouse territories are usually not contiguous and display sites (which are limited to particular locations within territories) are usually well separated, there have been cases in which four or more males were observed in leklike display groups. The species is considered to be normally polygynous, with the home ranges of females often overlapping those of several males. Males do not defend the female's nest or brood, but occasionally pair bonds are formed that persist until the eggs hatch (Blackford, 1958).

Not only are there intermediate degrees of dispersed display vs. lekking behavior and monogamy vs. polygyny in grouse, but also it is probable that lekking may have evolved several different times in the group, for the conditions that predispose grouse populations to evolve lekking behavior are widespread. One of these predisposing factors might be the evolution of large body size, especially in males, since the larger the adult bird and the more prolonged its individual development, the greater the probabilities of marked individual differences among males in holding territories and in winning battles with rivals. The largest of all grouse, and those having some of the most extreme sexual dimorphism, are the two species of capercaillies. Furthermore, lekking behavior is well suited to and most likely to develop in species adapted to open steppelike environments, which is typical of many rather distantly related species in this group, especially the North American "prairie grouse" of the genus *Tympanuchus*. Such species normally have relatively large home ranges but defend extremely small territories that they can effectively control from a single display site. On the other hand, promiscuous forest-dwelling species tend to defend relatively large woodland territories that may be more or less coextensive with their total home ranges, and that are impossible to control from a single display site.

As in other birds, a major advantage of grouse lekking display is that it allows the females to make rapid and efficient best-mate choices from among the assembled males. Thus females that are fertilized by dominant males on social display grounds are more likely to pass on favorable genes than are those that might be fertilized by lone males displaying away from direct competition with others. For male grouse, the potential reproductive advantages of participating in leks must be balanced against the possible attraction of predators to such assemblages. This is especially true in forested areas or other heavy cover, where the risk of stealthy approaches and attacks by predators may be quite high, especially for peripheral birds. Thus, lekking behavior will often develop in conjunction with a relatively open environmental location and the adoption of appropriate antipredator tactics. Among lekking grouse these tactics include synchronized periods of activity and silence, restriction of most mating activity to crepuscular (dawn and dusk) periods when few strictly diurnal or nocturnal predators are active but when signal transmission is effective, and the evolution of relatively efficient but easily hidden male signals, such as inflatable air sacs and hard-to-locate but far-carrying calls or mechanical sounds.

Lekking displays are therefore most likely to evolve in grouse whenever it is more profitable for a female to seek out a group of displaying males and be fertilized by the most dominant one rather than remain on the territory of a particular male and benefit from whatever resources and protection he might provide her and her brood. For males, it might be better to participate in lek display whenever it is more profitable to compete actively for a favored mating territory, even though this might require a year or two of relatively nonrewarding "apprenticeship" with few or no mating opportunities, than it is to display alone, where there may be a lower probability of attracting the attention of predators but also a far lower chance of attracting and mating with a female. In summary, leklike (mating station) behavior in grouse should potentially evolve whenever the female is fully able to nest and rear her brood without the protection or resources provided by the male (beyond simply fertilization) and additionally when individual differences in male fitness are reflected in differential social or territorial status that females can readily perceive and quickly use when choosing mating partners.

This argument as to the evolution of lekking by grouse is not in direct conflict with, but also does not strongly support, that of Wittenberger (1978). He suggested that the major difference between monogamous and nonmonogamous (potentially lekking) grouse is that females of monogamous species nest and forage within their males' territories, whereas those of promiscuous species do not. Monogamous females might benefit from male vigilance during this critical period, and perhaps are also potentially limited in their overall reproductive productivity by relative food availability in spring. Thus, scarce food resources might favor monogamy in both sexes and the male's defense of a resource-inclusive breeding season territory, whereas more abundant food might favor polygyny or promiscuity and the establishment of short-term mating station territories by males. It seems that food resources will only rarely influence grouse distribution patterns significantly during spring, when green shoots and developing buds or leaves are usually relatively abundant. Potential male participation in nest and brood defense may indeed at times be a significant aspect of the costs and benefits of monogamy versus promiscuity, but its importance may differ considerably at different times and under different ecological conditions. If that is the case, one might expect lekking behavior to vary both spatially and temporally within and between species, rather than showing the species-specific patterns that are typical of all lekking grouse.

In a review of black grouse lekking behavior, de Vos (1979) suggested that lekking in that species might help protect the males from predators (by improved group vigilance), might provide useful information about the location of limited habitat resources (such as localized foraging sites), and could perhaps help protect males from conspecifics who might otherwise territorially influence the population's foraging distribution outside of display areas. For females, lekking behavior might allow them to choose among the best available males (presumably based on their individual phenotypes, their relative territorial positions, or their display activity differences) in a very short time. In de Vos's view, such female mating preferences are not initially responsible for the highly developed social structure of grouse leks. This structuring may instead have resulted from a runaway evolutionary process, in which the above-mentioned advantages associated with male territorial clustering eventually become counterbalanced by the disadvantages of the highly competitive arena system. However, in my view, the capability of females to select among males probably plays a central role in the formation of most grouse leks, in that if females are able to recognize older, more experienced, or otherwise seemingly more fit males, it is to the advantage of younger or less experienced males to establish their own territories as close to such dominant ("hotshot") males as possible. In that way the younger males might be able to divert the attention of a few females attracted to the dominant male's territory long enough to try to mate with them, and would be available to take over the dominant male's territory should he be killed or otherwise be unable to maintain his status.

Oring (1982) recently suggested that lekking behavior probably evolves in birds when one or more of four conditions exist with regard to males: collective male stimulus pooling, information sharing with regard to limited environmental resources, improved predator avoidance through increased group vigilance, and protection against control of limited resources (other than females) resulting from dominance behavior by conspecific males. Oring stated that leks may also evolve when various benefits accrue to females, such as facilitating the assessment of individual male quality, or increased protection from predators during the mating period. Again, I would rank the primary condition determining lek development as the female's potentially active role in mate choice and her associated potential genetic (genotypic) benefits from easily finding and identifying most-fit males, and I would consider only secondarily significant the potential immediate (phenotypic) benefits to males (thus excluding the obvious genotypic benefits accruing to the relatively few successful males).

Stage Displays in Relatively Dispersed Forest Grouse

Like male tragopans and some other pheasants, both dispersed and arena-displaying grouse seem to some degree to restrict their advertising display activities to certain locations having specific environmental conditions. Such favored display sites are likely to be somewhat more secure from predators than others, such as by being unobstructed visually (e.g., on hilltops or in low vegetational cover). The sites may additionally have particular characteristics that may either facilitate the actual performance or influence the broadcasting efficiency of the male's displays, or both.

Several forest-dwelling grouse species fall into this general stage-dependent category, of which the most familiar to North Americans is the ruffed grouse (*Bonasa umbellus*). The male's "drumming" display is well known and has been well studied in many parts of temperate North America. Territorial males tend to select for their display sites those parts of the forest that are sufficiently sparse in ground and shrub vegetation as to provide an unobstructed horizontal view for at least 15 meters in most directions. Good sites are sufficiently screened overhead to help protect the male from aerial predators. The actual display site's physical characteristics may be secondary to these general environmental features, but typically it consists of fallen logs that are large enough to provide fairly level platforms, sufficiently high above the ground for the male to easily observe approaching ground predators or other grouse, and sufficiently rotted to allow the male to dig his claws into the wood while drumming.

The drumming display is a highly stereotyped sequence of wingbeats lasting about 8–11 seconds. It has a rather surprisingly consistent number of about 50 accelerating

19. Aerial and wing-related displays of male forest grouse, including wing-clapping sequence by Franklin's spruce grouse (*A, B*); sharp-winged grouse flutter-jumping (*C*) and in ground display (*D*), plus its wing and attenuated primaries (*E, F*); and drumming sequence of ruffed grouse (*G–I*). For sources see Johnsgard (1983a).

wingbeats, performed as the male sits crosswise on his log perch (Figure 19G–I). The sound, a muffled drumming, is not generated by the male directly striking his wings together but by air compression caused by the strong wing movements (Hjorth, 1970). The resulting low-pitched sound carries well through forest vegetation, but is relatively hard for a human to localize except by perceived changes in sound amplitude as one approaches or moves away from the displaying bird.

In one study (Gullion, 1967) it was found that all males controlling drumming sites were at least 22 months old. Within a resident male's "activity center" an "alternate" drumming male sometimes also was present, and the latter bird might take over a display site if its resident male was killed. Occasionally "activity clusters" of up to eight males might occupy drumming sites in fairly close proximity, this assemblage seemingly representing a kind of exploded lek (Gullion, 1967; Aubin, 1970). However, a significant percentage of the total male population may be nondrummers, including not only sexually immature yearling birds but also some males of older year classes, who may be several years old before they successfully acquire territorial status as drummers.

Several other forest-dwelling grouse species also generate nonvocal acoustic signals during self-advertisement or territorial displays, but not from such specifically definable environmental stages. In addition to its terrestrial "strutting" displays, the Franklin's race of the spruce grouse (*Dendragapus canadensis franklinii*) performs a spectacular aerial wing-clapping display. The clapping is typically performed near the end of an extended display flight during which the male flies through the dark forest with his tail feathers spread and his long, white-tipped tail coverts conspicuously contrasting against them. On reaching a clearing, he makes a single deep wingstroke and then sharply strikes the wings together over his back (Figure 19A). This is quickly followed by a second clap as the bird drops vertically toward the ground and lands (MacDonald, 1968).

In the similarly forest-adapted and closely related sharp-winged grouse (*Dendragapus falcipennis*) of eastern Asia, most of the primary feathers of males are distinctly attenuated toward their tips (Figure 19E, F), which probably produces the strong whistling noises generated during flight or wing flapping. Males perform ground strutting displays that apparently are similar to those of the spruce grouse, but they additionally make very short display jumps, or "flutter-jumps." These displays consist of a preliminary vigorous wing flapping, followed by jumping about a meter into the air while calling loudly and then landing again (Figure 19D). These flutter-jumps may represent a compressed version of the spruce grouse's more extended display flights. Perhaps the mechanical whistling noises resulting from the strong wing flapping also functionally substitute for the spruce grouse's wing clapping as an acoustic component in this display, but so far very little has been written on the sharp-winged grouse's advertising behavior (Johnsgard, 1983a).

It is clear that the grouse of the world provide a remarkable spectrum of mating strategies, ranging from strong monogamy to total promiscuity; of habitat adaptations, ranging from open environments such as arctic or alpine tundra and arid desert grasslands to dense forests; and a signaling variability exploiting visual, vocal, and mechanical generation of display components. They also offer some of the best available information on age-dependent dominance-fitness gradients, on mating behavior site fidelity, and on mate choice behavior among females. Thus, this group deserves special attention.

Arena Displays in the Eurasian Grouse

Among the Old World grouse species, arena-like displays are known to occur in four species, comprising two species pairs. These are the closely related common or Eurasian capercaillie (*Tetrao urogallus*) and the black-billed capercaillie (*T. parvirostris*) of western and eastern Eurasia respectively, plus the Eurasian black grouse (*T. tetrix*) and the Caucasian black grouse (*T. mlokosiewiczi*).

20. Male advertising displays of Eurasian lekking grouse, including black–billed capercaillie in thin–necked upright (*A*) and wide–necked upright (*B*) postures; also Caucasian black grouse flutter–jumping (*C*) and during ground display (*D*). The tails of male Caucasian black grouse (*E*) and Eurasian black grouse (*F*) are also shown. For sources see Johnsgard (1983a).

The two species of capercaillies are in narrow sympatric contact in the general vicinity of Lake Baikal. Males of the black-billed capercaillie (Figure 20A, B) have white on their upper wing coverts and tail coverts and a black bill, whereas the more widely distributed Eurasian capercaillie lacks white on these feather areas but has a whitish bill. In the larger Eurasian capercaillie the average male:female adult mass ratio approaches 2.0:1, with substantial individual variation present, whereas in the black-billed capercaillie the limited available data would suggest a male:female mass ratio of about 1.5:1.

In the black-billed capercaillie males defend small territories in leks that are said to number from 6 to 10 birds, with an approximate average intermale distance of about 50 meters. Male signals include extended series of mechanical-sounding clicking calls, as well as posturing with an extended neck and raised tail. What is so far known of this species' displays suggests that they are quite similar to those of the more widespread Eurasian capercaillie (see below), and hybridization evidently occurs in the limited zone of sympatric contact between them. Indeed, the behavioral differences among males of the two species may primarily have ecological significance rather than reflecting the effects of selection for reproductive isolation as such (Andreev, 1979).

In the case of the black grouse, the two species are well separated geographically from one another. The Eurasian species is widely distributed across much of northern Eurasia, but the Caucasian black grouse occupies only a highly restricted range in the Caucasus Mountains. Males of the two species are also quite similar, but the Caucasian black grouse has more pointed tail feathers that form a V-shaped profile from

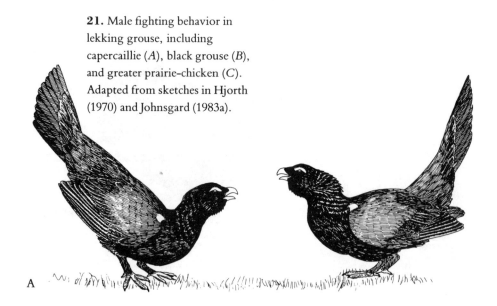

21. Male fighting behavior in lekking grouse, including capercaillie (*A*), black grouse (*B*), and greater prairie-chicken (*C*). Adapted from sketches in Hjorth (1970) and Johnsgard (1983a).

A

above (Figure 20E), whereas in the more widespread Eurasian black grouse these feathers are more truncated and the tail forms a more U-shaped profile (Figure 20F). In both species the males are mostly black, with contrasting white under wing coverts and with crimson eye combs that are engorged during display, but in the Eurasian black grouse the under tail coverts and greater secondary coverts are a contrasting white. In the latter species the male:female adult mass ratio is about 1.4:1, but in the Caucasian species the ratio is evidently nearly equal at about 1.1:1, based on limited available information.

In the Caucasian black grouse typical lekking behavior occurs. The number of males that have been observed at leks is often 10–15, but as many as 30 have been

B

C

22. Male advertising displays of lekking grouse, including greater prairie-chicken (*A*), lesser prairie-chicken (*B*), black grouse (*D, E*), and capercaillie (*F, G*). Also shown (*C*) is the male tracheal and esophageal anatomy of the greater prairie-chicken. Adapted from sketches in Hjorth (1970) and Johnsgard (1983a).

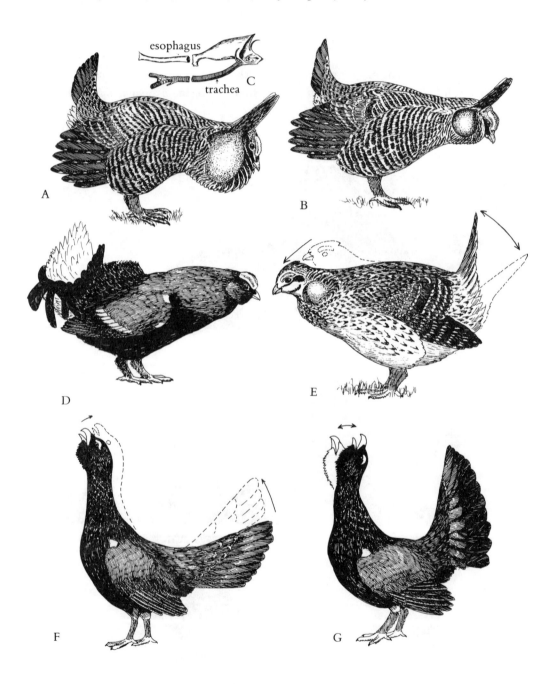

reported. The leks are located on alpine slopes, and probably begin as soon as these sites become snow-free. Territorial advertisement is similar to that of the more common species, but the displays are predominantly aerial, with few if any discernible vocalizations. Instead, flutter-jumping displays (Figure 20C) are common, with whistling noises produced by the wings. Ground postures involve comb engorgement and general feather erection (Figure 20D). No information is yet available on relative individual male dominance and mating success.

By comparison with these two little-studied species, an enormous amount of behavioral information exists for the Eurasian black grouse and Eurasian capercaillie. The capercaillie is of special interest in terms of the evolution of lek behavior, as it occupies a transitional status. Hjorth (1970) and other workers have suggested that it does not qualify as a lek species, in part because it displays in rather heavy woodlands rather than open areas, and also because at least in some areas haremlike groupings of females may associate with males in rather permanent territories (according to observations in Spain and Scotland). However, at least in most regions it does exhibit the three major criteria for lekking species, namely a localized competition by males for territories and mating opportunities, the restriction of the territory's function to that of a mating station, and a strong dominance-to-fitness relationship among the males, in which territorial success is positively related to mating success. This tendency of females to mate with males that are superior fighters has been cited by Moss (1980) as a possible explanation for the very large size of capercaillie males.

Hjorth (1982) judged that the territories of male capercaillies are rather indefinite in size and are somewhat pie-shaped, or radially arranged outwardly without limitation. At the apex of the pie the birds display and defend their mating sites, and the actual defended area of the territory is thus rather small. As in other lek-forming species, the number of males interacting at territorial boundaries probably varies directly with population density, but as many as 10–20 males have been reported present in some parts of Siberia. Small knolls seem to be favored for territories, suggesting that visibility may be important, but a combination of open spaces as well as tall shrubs and very old trees may comprise an important part of the birds' general "stage." Old, open forests with grassy clearings and with piles of stones or rock formations are typical territorial locations, and actual display sites are also often oriented toward the east, where they are illuminated at first sunrise (Johnsgard, 1983a).

For male advertisement, the Eurasian capercaillie stands in a tail-cocked, neck-extended posture (Figure 22F, G). The same or a very similar posture is assumed before soliciting females (Figure 23D). Accompanying this posture is a long series of generally soft and low-pitched calls, including clicking notes, scraping sounds, and (at least in some populations) a rather loud, single "cork" note resembling the sound of a cork being pulled from a bottle. Infrasonic elements have also been reported. As has been observed and confirmed by many observers, nearly all the copulations are performed by a single dominant or alpha male; the author has seen as many as nine

23. Male-to-female displays of grouse, including ruffed grouse (*A*), blue grouse (*B*), greater prairie-chicken (*C*), capercaillie displaying to a soliciting female (*D*), and precopulatory posture of sage grouse (*E*). For sources see Johnsgard (1983a).

females in a single mating cluster, and at least eight of these females were mated during a 90-minute period by the dominant male.

Not only are females attracted to particular dominant males, but fighting behavior among the males (Figure 21 A) is often intense, with severe injuries or even fatalities at times occurring. Even if a dominant male has lost most of his tail feathers through fights, females can still recognize and prefer such alpha-rank birds, and additionally older and more experienced males seem to have better mating strategies than do younger ones. In a German study area, Müller (1979) found that alpha-rank males performed more than 90 percent of the observed copulations, and also maintained the largest territories. Many were fairly old birds; one alpha-rank male was found dead (apparently from old age) when he was over 13 years old. His territory was not taken over by another male until the following spring, indicating that the territorial limits of high-ranking males may be respected by their rivals even during their absence. Mjelstad (1991) has recently reported that the relative intensity of male display provides a visual means of assessing sperm quality in this species, based on studies of captive birds.

The Eurasian black grouse is the classic lek-forming grouse of Europe, and probably more has been written on its lek behavior than on that of any other grouse species. Much of this information has been summarized by de Vos (1979, 1983). As with other lekking grouse, territorial males occupying leks may be categorized in three dominance levels, alpha, beta, and gamma, which are largely age-dependent. De Vos (1983) found that males typically initially established central territories (became alpha-rank) when they were approaching two, three, or four years of age (at about equal probabilities), but about 5 percent initially did so only when almost five years old. Males occupying central territories in their third or fourth spring of life averaged 9.6 copulations per male per season, as compared with 2.3 copulations for second-year males and 1.1 copulations for first-year birds. Of 449 total observed copulations, only 1.6 percent were performed by first-year males. Older males having central territories were responsible for nearly all observed copulations. The maximum number observed for any single male during one season was 20, which was for a male in its third or fourth spring of life. Maximum life duration was judged by de Vos to be 7–8 years for both sexes.

Johnston (1969) suggested that differences in individual male black grouse mating success might be related to their individual variations in lek attendance, general activity and/or courtship intensity levels, differences in their territorial characteristics, or differences in their physical appearances. The most successful males were the ones that displayed most often and were most often involved in territorial disputes (Figure 21B). Most female visits were to males in their third or fourth year of life, and the generally larger territories held by these dominant males gave them increased freedom from harassment by other males when courting females. The appropriate mating strategies of experienced males seemed to be an important part of their attaining the majority of successful copulations. More recently, Alatalo, Höglund, and Lund-berg (1991) reported that those males that were dominant in winter flocks and most successful in winning fights over access to dummy females placed at territorial bound-aries on leks were also the most successful in obtaining actual matings. Females selec-tively mated with those males that were judged (by the authors' criteria) to be the most likely to survive over the next six months, supporting the hypothesis that fe-males selected mates on the basis of male viability traits (rather than on the basis of male "beauty"), thus supporting the "good genes" model of sexual selection.

The primary male black grouse advertisement display (Figure 22D, E) is a tail-cocked, neck-extended and neck-inflated posture, with the comb engorged and the wings slightly lowered. In this "forward" posture the male utters a soft "rookooing" call of complex acoustic characteristics, but of rather low frequency and fairly low amplitude. Postures similar to those of capercaillies and prairie-chickens are assumed during male-male threats and fighting (Figure 21B), but much of the time spent dur-ing agonistic interactions is in the form of ritualized bowing and ground pecking rath-er than actual fighting.

Lekking Displays of North American Grouse

Three or four species of North American grouse regularly perform typical lekking behavior. These include the two forms (often considered subspecies) of pinnated grouse, usually called the greater (*Tympanuchus cupido*) and lesser (*T. pallidocinctus*) prairie-chickens. Two additional subspecies of the greater prairie-chicken also exist (one is now extinct), but these need not be further considered here. The other two lekking species are the sharp-tailed grouse (*T. phasianellus*), and the sage grouse (*Cen-trocercus urophasianus*). The male advertisement displays of the two prairie-chickens are virtually identical except for minor posturing, air sac color, and acoustic differ-ences (Figure 22A, B), and in most respects the male postures and calls of the sharp-tailed grouse also show clear homologies with those of prairie-chickens. Thus, these three general types can be considered as a group, and the detailed differences among them (see Hjorth, 1970; Johnsgard, 1983a) need not be emphasized here. However, the male advertising display patterns and social structuring of the sage grouse are quite different from those of the other species, and will require separate consideration.

In the two prairie-chickens and the sharp-tailed grouse the male:female adult mass ratios average about 1.1–1.2:1, and sexual dichromatism in adult plumages is nearly lacking. In these species the numbers of males participating in single leks varies con-siderably, presumably with population density. Studies of lesser prairie-chickens in Oklahoma suggest an average group of about 14 males, whereas in Nebraska and oth-er locations supporting good populations of greater prairie-chickens and sharp-tailed grouse there is an average of about 10–12 males per lek, but with wide variations. In probably all cases, but certainly at least for greater prairie-chickens and sharp-tailed grouse, the central positions on the lek are consistently held by alpha-level males, and these birds correspondingly dominate the mating opportunities (Lumsden, 1965; Ballard and Robel, 1974).

Male greater and lesser prairie-chickens perform "booming" and "yodeling" displays respectively; the different names refer to their acoustic differences, although their associated tail-cocking and wing-lowering postures are essentially identical (Figure 22A, B). During this display the male's anterior esophagus is briefly filled with air, exposing and expanding bare and brightly colored skin patches at each side of his neck, and a call is uttered as his tail is variously fanned open or snapped shut. The associated vocalization in the lesser prairie-chicken is shorter and higher-pitched than in the greater, but in both cases the collective calls of the males carry considerable distances, and can often be heard more than a kilometer away. Clearly these booming and yodeling displays serve as important male-male dominance and territorial defense signals. These same displays are performed at considerably higher frequencies when females are on the lek. One high-intensity and exclusively male-to-female display used by prairie-chickens as well as sharp-tailed grouse is the prostrate or "nuptial-bow" posture (Figure 23C), which often but by no means invariably precedes copulation. In the normally nonlekking ruffed grouse and blue grouse the high-intensity displays performed by males directly before the female are apparently not structurally different from those performed toward male rivals (Figure 23A, B), perhaps as a reflection of their nonlek social environment.

In the sharp-tailed grouse, the male territorial advertisement display that is homologous with booming in greater prairie-chickens and rookooing in black grouse is the bowing or cooing display. In this display the inflated neck skin is of a different (pink to purple) color than in prairie-chickens (where it is yellow to red), the inflation is not so extreme, and the associated vocalization is a more dovelike cooing. Additionally, cooing may be somewhat more aggressively motivated than is the counterpart display in prairie-chickens, but certainly it is also used in heterosexual situations. Unlike prairie-chickens, male sharp-tailed grouse often "dance" in highly synchronized concert, with outstretched wings and laterally vibrating tail feathers. This display also seems to be more aggressively than sexually motivated, and is often performed in the visual absence of females. When very close to females, males often assume a stationary "posing" posture, or may drop down into a wings-spread posture while facing the female, in a prostrate display that is nearly identical to the nuptial bow of greater prairie-chickens. Not only is this precopulatory posture the same in the two species, but interspecific copulations and hybridization also occur fairly frequently in areas where both species are present.

In a study of the factors affecting male sharp-tailed grouse mating success, Kermott (1982) found that males defending central territories represented only 13 percent of the lekking birds, but performed 93 percent of the observed matings. Such males either gradually attained this territorial status by establishing peripheral territories as yearlings and gradually working inward as older birds died, or more rapidly carved out a central territory by overt aggression. Kermott also concluded that each male is dominant within its own territory, and that the lek should not be considered a dominance hierarchy but rather a true territorial system. I would suggest that grouse leks have characteristics of both, consisting of exclusive-use territories that are hierarchically organized from periphery to center.

In a recent study testing the relative importance of male-male dominance versus female choice in influencing male sharp-tailed grouse mating success, Gratson, Gratson, and Bergerud (1991) found that the most dominant males (those least disrupted during courtship attempts) were *not* the most successful in obtaining copulations. Surprisingly, male mating success and male size were actually somewhat inversely correlated, with smaller males obtaining statistically more matings than larger ones. Centrally located males did obtain more copulations than peripheral ones, and additionally all males tended to move toward the center of leks with increasing age, but perhaps simply as a result of survival rather than through direct competition for interior locations. These authors thus suggested that little intermale competition for spe-

24. Advertisement display of sage grouse, including a strutting sequence lasting 3.7 seconds (*A–G*). Also shown are a front view of maximum air sac inflation (*H*), a rear view (*I*), and agonistic tail rattling (*J*). After sketches in Hjorth (1970) except for *J*, which is from a photo by the author.

A   1.0

B   1.5

C   2.4

D   2.7

E   3.0

F   3.2

cific territorial sites exists in sharp-tailed grouse leks, that dominant males do not always control the most central territories, and that territorial location within the lek is probably not used as such as a specific cue by females in selecting mates. These ideas are at variance with the results of some previous studies of sharp-tailed grouse, and with the general "hotshot" concept of male reproductive success determinants, and need additional confirmation.

In another recent study on mating behavior in sharp-tailed grouse, Landell (1989) reported that territorial location (central males favored), size (males with shorter wing lengths favored), and agonistic behavior (males with higher rates of agonistic interactions favored) were all correlated with relative male mating success, but that males' ages and relative display rates were not. Female aggregation tendencies, probable copying behavior, and apparent male sampling behavior all seemingly influenced choice of mates by females, but female-female aggressive interactions did not.

The sage grouse represents the most highly organized of all species of lekking grouse, at least on the basis of the usual large size of the leks (up to several hundred

H
3.2

G
3.7

I

J

males reported historically, but now averaging about 50 males in good habitats) and the correspondingly small average size of the individual territories held by participating males (averaging about 10 meters in diameter). Adult males differ considerably from females in plumage pattern, color, and body mass. Thus, the mass ratio of adult males to females in this species averages about 1.55:1, but males exhibit considerable individual mass variability. This mean dimorphism ratio is somewhat less unbalanced than that of the even larger Eurasian capercaillie, which is only marginally leklike in its male territorial spacing behavior. However, the remarkably large body size in male capercaillies may be partly related to ecological adaptations or to behavioral factors other than lek-related male competition, such as their very high male-male aggression levels (Moss, 1980; Johnsgard, 1983a).

The "strutting" display of the male sage grouse serves not only as a male-male territorial challenge but as a male-female sexual signal (as is also true of capercaillies). It consists of a remarkably complex sequence of stereotyped postures, movements, and associated mechanical as well as vocal sounds, which have been described in detail by Hjorth (1970). The most interesting visual feature of this display sequence is a series of rather brief but increasingly stronger exposures and inflations of the frontally situated olive-colored esophageal air sacs (Figure 24A–H). Most of the sounds associated with strutting are mechanically produced, either by scraping the feathers of the anterior wing coverts over the bristly white breast feathers, or the plopping sounds produced by the rapid deflation of these air sacs. Sage grouse have very few other territorial or sexual signals (flutter-jumps are lacking, as apparently also are unique precopulatory displays or specific calls or postures directed only toward females). Overt fighting between territorial males is common, as are aggressive tail cocking and tail rattling while the birds are laterally oriented toward one another (Figure 24J). Such tail rattling often precedes fights that consist of strong wing beating against the opponent.

Females seem able to locate alpha-level individuals unerringly from among the sometimes rather vast assemblage of displaying males. They quickly gravitate to such males, often forming mating clusters of hens waiting to be mated. The large size of the males seemingly makes copulation attempts rather awkward and difficult (Figure 23E). Most studies of sage grouse have indicated that only a few alpha-level males perform the majority of copulations. Scott (1942) noted that only a few "master cocks" performed 74 percent of 174 observed copulations, and Hartzler and Jenni (1988) estimated that only 20 percent of the two-year-old and older males performed 80 percent of all 515 copulations over a three-year period. Lumsden (1968) found that only two alpha-rank males accounted for more than half of the 51 copulations he observed at one lek. Wiley (1978) also reported that two males accounted for 86 percent of 42 copulations seen at one lek. However, he suggested that the social interactions of the females, such as their collective gravitation toward the center of the lek, and not merely the females' individual responses to the males' displays, might account for most if not all of their apparently selective breeding behavior. Thus in Wiley's view, age, with its concomitant shifting of older territorial males toward the mating center, rather than "beauty," might provide the key to reproductive success among individual male sage grouse. However, Hartzler and Jenni (1988) questioned whether a centripetal movement toward the lek center existed in their lek, and also rejected the idea that the males are organized hierarchically on the lek.

Lumsden (1968) observed that female sage grouse gathered around specific males, rather than congregating at particular mating spots in the lek, and Hartzler and Jenni (1988) believed that females may be attracted to the most "conspicuous" males, namely those displaying more frequently than others when females are nearby. Gibson and Bradbury (1985) similarly reported that male sage grouse mating success is strongly correlated with individual differences in male lek attendance and with the display rate and certain acoustic characteristics of their courtship displays, suggesting that these behavioral traits may provide the cues by which females choose their mates. Additional observations by Gibson, Bradbury, and Vehrencamp (1992) on two leks over a four-year period suggest that female choice of mates is based on individual male vocal display traits, the location at which the female mated during the past year, and copying the choice of other females.

Recently Boyce (1990) has obtained some evidence that female sage grouse can recognize and thus avoid mating with parasite-infected males, which also suggests that female choice of specific mating partners may be an important element in male mating success. However, Gibson (1990) found that blood parasite load did not measurably affect male display performance or relative male mating success.

Most recently, Gibson (1992) supported the view that the responses of females to males through female choice may influence male lek-clustering behavior. Specifically, he suggested that "temporal spillover" (fidelity of females returning to previous-year mating sites) rather than "spatial spillover" (reproductive advantages to neighboring males of being territorially situated near "hotshot" males) may cause male aggregations at leks. Supporting this position, he reported that territories vacated by the most successful males served as foci for male territories the following year, and furthermore found that over a seven-year study period the fidelity of males to previous territories increased with their own prior mating success, but not with that of neighbors. However, Hartzler and Jenni (1988) found no evidence that females select mates based on their occupation of a traditionally successful territory, and said furthermore that among males there is no more competition for the previous year's successful territorial sites than for any other site on the lek. The costs to females in making active mate choice (increased travel, increased predation risk) are probably very low (Gibson and Bachman, 1992).

25. Head (*A*) and strutting posture (*B*) of male wild turkey, as compared with strutting posture (*C*) and head (*D*) of ocellated turkey. After various photographic sources.

The Social Structure and Sexual Behavior of Turkeys

There are only two species of turkeys in the world, both native to northern Central America and the southern United States. The common wild turkey (*Meleagris gallopavo*) has long been domesticated and has been extensively modified in appearance as a result. However, the ocellated turkey (*Meleagris ocellata*) has never been domesticated and its behavior in the wild is still only quite poorly studied. The two species are rather surprisingly different in appearance, especially in the case of adult males (Figure 25), but they may be more closely related than is suggested by these external and sex-related traits. Adult ocellated turkey males weigh about 5 kilograms and females about 3, representing a male:female mass ratio of about 1.7:1. In the somewhat larger common wild turkey males typically weigh about 7.4 kilograms and females about 4.2, representing a similar male:female mass ratio of about 1.8:1. There is some question as to whether turkeys are more closely related to the grouse (as might be suggested on zoogeographic grounds, as well as by similarities in their sexual behavior) or to the Old World pheasants (as is generally believed to be the case, from various lines of evidence), but that is not of direct importance in terms of the present discussion.

The social and sexual behavior of the ocellated turkey can be summarized fairly easily, based on studies of wild birds by Steadman, Stull, and Eaton (1979) in Guatemala. There the birds maintain fall and winter flocks of up to 16 birds, each with up to three adult males plus varying numbers of yearling males and females. The flocks break up in spring, when the adult males begin to strut and gobble, each trying to attract its own group of females and yearling males. A gobbling and strutting male (Figure 25D) defends an area near the females' roosts for a few hours each day, but does not maintain a resource-inclusive territory. The gobbling period lasts about three months, reaching a peak in May, and display is most intense in early morning and again just before dark. In a flock having three males, one male was evidently dominant, not only gobbling much more often than the others but also being the only one observed to copulate. In another flock of three adult males, two of the males were seen copulating, suggesting that in this species a single male may be unable to maintain sexual dominance through the entire breeding season. Unlike in the common wild turkey, the yearling ocellated turkey males remained with the females during the mating period, perhaps because they were considerably smaller than the adult males and thus posed no competitive threat to them. In short, it would appear that this species is polygynous, with adult males competitively trying to acquire harems during the breeding season but not maintaining separate mating territories. Breeding flocks sometimes have more than one breeding male present, and thus dynamic rather than static male dominance relationships may be present.

The common wild turkey has been far more fully studied than the ocellated turkey, and some frequently cited observations are those of Watts (1968) and Watts and Stokes

(1971), based on studies of nearly 600 marked wild birds in Texas. There, winter flocks were composed of birds that recognized one another individually. Winter flocks of males consisted of apparent "sibling" groups, whereas most summer flocks were composed of two of three females and their broods. Sibling groups of juvenile males were gradually assimilated into adult male flocks; thus male winter flocks typically consisted of 10–25 birds, representing various age classes or "brotherhoods" of sibling groups. Within each sibling group a permanent peck order existed. However, such sibling groups also typically fought other such groups as a concerted unit, thereby establishing an age-related relative overall dominance hierarchy among males. These male-only flocks began to break up in February, with the various age classes of male sibling groups displaying daily on traditional display grounds near female roosts. During strutting (Figure 25B), the most dominant sibling group of males mixed freely with the females, but only the alpha male of the dominant sibling group actually copulated with females during this period. Of 170 males in four display groups, no more than six males were responsible for all the observed copulations. The aggregation of the sexually active males on regular display areas (but not limiting their display activities to specific territories nor to individual display sites) and the restriction of matings to a very few alpha-level males suggest a leklike social behavior not very different from that of sage grouse or other lekking grouse. However, the maintenance of sibling male social bonds would seem to be unique, and the mobility of the displaying birds is seemingly analogous to that of duck social displays on aquatic arenas.

In a critical review of these studies, Balph, Innis, and Balph (1980) questioned whether the "sibling groups" as described by Watts actually contained only brothers, and instead suggested that kin selection could produce cooperative matings not only among brothers but also among more distantly related individuals. However, they also strongly doubted the need to invoke kin selection at all for the observed restricted mating activities, suggesting that in this specific flock a temporary unusually high density may have affected normal dominance relationships, and furthermore noted that permanent year-round flocking of adult turkeys could generally promote stable social relationships and reduce aggression within such groups.

5

Bustards: Feathered Balloons and Aerial Rockets

The bustards are a group of 22 generally arid-adapted Old World species that range in size from relatively small birds about the size of partridges (or about 200 grams) to very large species whose males have been reputed to be among the heaviest of all flying birds (over 20 kilograms). Their sexual dimorphism ratios, based on linear wing, tarsus, and culmen measurements, range from less than 1:1 in the case of some small species in which the males perform aerial displays, for which acrobatic abilities have perhaps selected for small male size, to as high as 1.3–1.4:1 in the case of the largest species, the males of which only perform ground displays. In these latter species the adult male:female adult mass ratios sometimes reach 2:1 in individual cases, but it is typical of these especially large species for a considerable variation in adult male weights to exist, as noted earlier for musk ducks and capercaillies. Presumably these variations in adult male mass are largely age-related, and they may play important roles in affecting individual male dominance characteristics or individual male abilities to attract females (Johnsgard, 1991).

The mating systems of bustards are not yet entirely clear, but at least the majority of species appear to be nonmonogamous. Although monogamous mating has been suggested for several species of the mainly African genus *Eupodotis,* it has not been proven as the typical pattern for any. The newly hatched young of bustards are precocial, but they grow very slowly, and maternal care and actual feeding of the young may persist for long periods. The male is not yet known to play an active role in nest site selection or in the protection of the nest, incubating female, or brood in any bustard species. However, in some seemingly monogamous species the male has been reported to warn its apparent mate of approaching danger, and possible instances of nest site guarding have been suggested in some apparently nonmonogamous species.

Monogamy is most likely to be present in relatively small species such as the little brown bustard (*Eupodotis humilis*), in which the presence of the male may be needed to help protect the female and her nest or brood during the breeding season. Sexual dichromatism is seemingly virtually lacking in this species, and so too is sexual

26. Male kori bustard in balloon
display near female (*A*, after
photo by author). Also shown
are a male trachea and esophagus
of Australian bustard (*B*) and a
male trachea of the kori bustard
(*C*). For sources see Johnsgard
(1991).

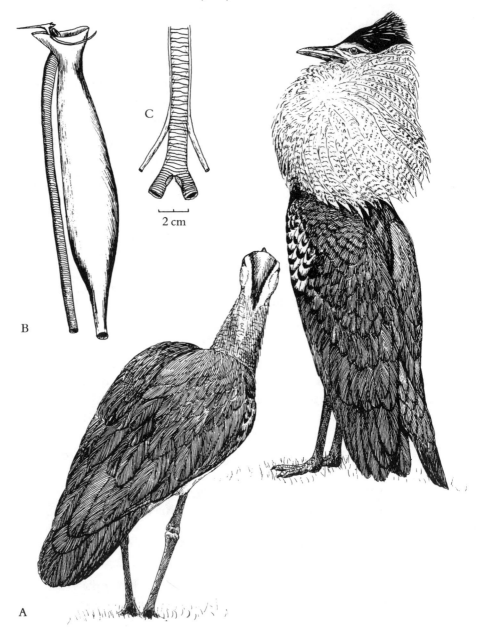

2 cm

dimorphism, judging from the limited available data on adult measurements. Other somewhat larger species of this genus have at times also been assumed to be monogamous, based on the presence of duetting calls of seemingly paired adults and the apparent absence of special display-related structures such as an inflatable crop or esophagus in males.

Assuming that at least the majority of bustards are potentially if not regularly polygynous or promiscuous, a second question is whether males maintain resource-inclusive territories or instead only defend mating station territories. The latter seems to be the case for the largest species of bustards, namely those in the genera *Otis* and *Ardeotis*. However, some of the smaller *Eupodotis* species perhaps maintain permanent if ill-defined resource-based territories within a larger home range. Some of these apparently territorial species (such as *vigorsii* and *rueppellii*) live in nearly vegetationless desertlike areas where the limited food supplies may require maintenance of year-round resource-inclusive territories. Groups of bustards residing within such areas may represent a pair and their immature offspring.

To summarize probable mating systems in the bustard family based on the available information recently at hand (Johnsgard, 1991), it appears that five species are possibly monogamous, two species may have dispersed (noncontiguous or isolated) male territories, males of eight species have what have been described as probable semidispersed or exploded lek distributions, one species is promiscuous and is "usually leklike," and one species exhibits typical lekking or arena behavior ("clustered male territories"). The last two mating types are of special concern in terms of the evolution and mechanics of arena behavior in bustards, but representatives of the other types will be discussed in a more general way.

As noted earlier, some species such as the white-bellied (*Eupodotis senegalensis*) and black-throated (*E. vigorsii*) bustards possibly maintain monogamous pair bonds, at least in captivity. In these two species the esophagus of the male is apparently not inflated during sexual display, although the male's crown and throat feathers may be erected to form a ruff, and no aerial display flights by males have yet been documented. A kind of "group territoriality" has been described for the white-bellied bustard, during which single birds or groups utter loud and deep calls that are often answered by another group (Kemp and Tarboton, 1976). The functions of such interactive calling can barely be guessed at, but conceivably these calls may facilitate spacing of adjoining territorial pairs or groups, as well as potentially serving as male advertising for mates.

Two species of bustards are believed to have well-dispersed or isolated male territories. Both have well-developed sexual dichromatism and moderate sexual dimorphism in mass (about 1.1–1.5:1). One of these is the desert-adapted houbara (*Chlamydotis undulata*) of northern Africa, and the other the grassland-adapted little black bustard (*Eupodotis afra*) of southern Africa. In the little black bustard there is no evidence of permanent pair bonding, and the females may perhaps wander from one male's territory to another until they are eventually fertilized. Males advertise their territories with loud, raucous calls and with short but conspicuous "aerial rocket" displays, during which they fly upward while calling loudly, circle while performing deep wingstrokes, and then descend in a fluttery glide. At least at times the males' territories are evidently rather small and probably contiguous rather than large and dispersed, since as many as five males have been observed to call and circle with a female if she is flushed (Kemp and Tarboton, 1976).

In the houbara, polygyny or promiscuity probably prevails. Males evidently display solitarily in well-separated display area territories that are perhaps barely in visual contact with one another, the males ranging from about 0.5 to 1.0 kilometers apart. No male aerial displays are definitely known to occur in this medium-sized species; the absence of concealing ground vegetation probably makes aerial displays unnecessary. Instead, the territorial males perform a number of terrestrial strutting or running postures, using conspicuous feather erection of their ornamental crown and neck feathers (Figure 29A, B). Specific male display vocalizations during advertising behavior have not yet been been reported, but the male's elongated white neck feathers and his white under tail coverts probably provide highly effective long-distance visual signals in this species' open desert habitats.

Dispersed Lek Displays among Bustards

As noted above, eight species of bustards have what appear to represent exploded lek or semidispersed breeding distributions. These include at least three of the four very large *Ardeotis* species, at least one moderately large species of *Neotis,* and several considerably smaller species of *Eupodotis.* The four *Ardeotis* species of Africa, India, and Australia are of special behavioral interest because of their long-distance visual and acoustic signaling "balloon displays," which are highly effective in their relatively open grassland or semidesert habitats. Similar inflated-neck terrestrial balloon displays occur in males of the comparably sized Denham's bustard (*Neotis denhami*) of eastern Africa, which additionally has substantial sexual dimorphism in adult body mass. The several smaller African and Indian *Eupodotis* species that seem to have exploded lek male spacing patterns often show considerable sexual dichromatism but little or no sexual dimorphism (possibly even slightly reversed dimorphism in a few). These species also exhibit some remarkable but different adaptations for both aerial and terrestrial display in generally grassy or grass and brush habitats.

In all of the *Ardeotis* bustards there is very substantial sexual dimorphism among adults, although relatively little sexual dichromatism is present. Not many weight samples exist for adults of most species, but in the great Indian bustard (*A. nigriceps*) males have reportedly weighed up to about 18 kilograms in historic times. More recent figures suggest that 14–15 kilograms is now the typical maximum, and an adult male:female mass ratio of about 2:1 may also be typical. In the kori bustard (*A. kori*)

27. Advertisement displays of male bustards, including kori (*A*), Arabian (*B*), great Indian (*C*), Australian (*D*), black-bellied (*E*), Denham's (*F*), and rufous-crested (*G*). For sources see Johnsgard (1991).

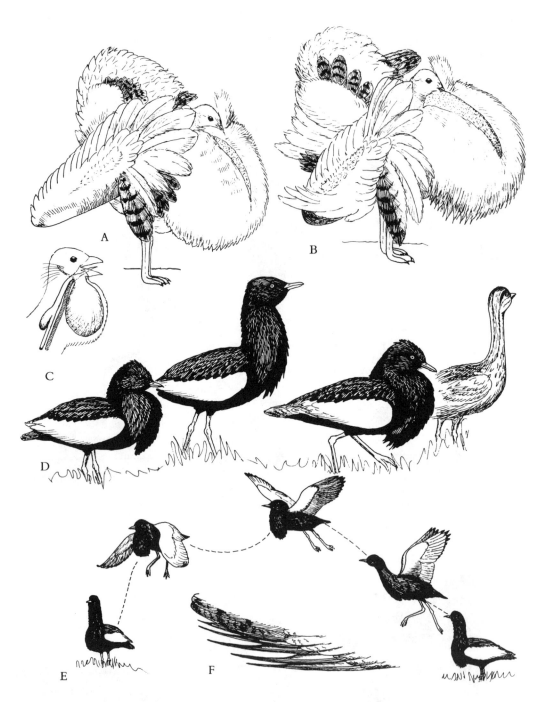

28. Male displays in lekking bustards, including balloon display of great bustard (*A, B*) with an illustration of its inflated air sac (*C*), and terrestrial pumping display (*D*) and aerial rocket display (*E*) of Bengal florican. Also shown are the attenuated outer primaries of a male lesser florican (*F*). For sources see Johnsgard (1991).

males have reportedly attained weights of up to 19 kilograms, but probably 10–13 kilograms is more typical, along with male:female mass ratios of about 2:1. Very few weight data are available for the similar but somewhat smaller Arabian bustard (*A. arabs*), but males have reportedly weighed up to 10 kilograms. In the Australian bustard (*A. australis*) old males may weigh up to 14.5 kilograms and females up to about 6.4 kilograms, with an adult male:female mass ratio of about 1.3–1.4:1 probably fairly typical. In at least the Australian bustard, males apparently continue to increase in adult weight with increasing age, and the absence of extremely heavy males in present-day populations of *Ardeotis* bustards, as well as in great bustards (*Otis tarda*), may simply reflect the fact that under present conditions few if any males survive long enough to attain their maximum potential weights.

These enormous male bustards may be seen for at least a kilometer under normal viewing conditions, and their visibility is supplemented by the erection and inflation of white to grayish white neck feathers during display. In at least two and probably all species of *Ardeotis* the male's esophagus is inflatable, producing a distinctive "balloon display." During this display the male additionally produces a low, booming vocalization that is not of great amplitude but that can be readily heard for very considerable distances. In the kori bustard, the displaying male raises his tail to an erect position prior to display, exposing the whitish under tail coverts, and slightly droops his folded wings, while holding his neck feathers and crest erected (Figure 27A). He then lowers his tail so that it is in line with his body, and begins to produce a series of progressively more rapid booming notes with the neck maximally inflated (Figure 26). He ends each call series with a slightly convulsive jerk and by opening the bill briefly, as if expelling some excess air. Judging from personal observations of captive birds, female kori bustards are attracted by this display, but at the same time they keep a rather wary distance from the much larger male (Figure 26A). The sight of a nearby female causes the male to display with even greater intensity, uttering his call sequences at regular intervals of about 30 seconds.

A very similar sequence of posturing and low-pitched calls occurs in the other *Ardeotis* species, judging from currently available information. In the great Indian bustard the male's display (Figure 27C) can be both heard and seen from a kilometer or more away, and the interval between calls in that species averages about 14 seconds. Total individual display periods may range from a few minutes up to about three hours in length; display is most intense during early morning and evening, especially just after daybreak. The seasonal peak of display occurs early in the full monsoon season, which immediately precedes the main breeding period. In one enclosed study area of 325 hectares, nests of 12 females were found during one year and 9 in another, all of which apparently had been fertilized by the single dominant male whose display site was located within this enclosure. Other males are not allowed near the display sites of such dominant males, and when one bird tried to establish a display site about a kilometer away from a dominant male it was unsuccessful. Male territorial limits apparently extend as far as the resident male is able to see intruders (Ali and Rahmani, 1984; Manakadam and Rahmani, 1986).

In the closely related Australian bustard, studies of captive birds (Fitzherbert, 1978) have provided additional useful information on social and sexual behavior. Male displays are usually performed at a very small display site of about a square meter, on which the male stands. The balloon display (Figure 27D) posture of this species is nearly identical to that of the great Indian bustard, and the call is a series of guttural notes lasting about 2.5 seconds. When displaying near a female, the male performs these sequences at intervals of only about 7 seconds, as compared with average 12-second intervals during more passive display periods. The male's call is audible for distances of up to at least 300 meters, and his white, pendulous throat sac may be seen from at least a kilometer away. Although some display may be performed in birds as young as two years, males may perhaps not become fully sexually mature until they are about five years old. Males also increase in weight with advancing age, to an undetermined limit. However, major seasonal variations in body mass also occur, with peak seasonal weights (up to 50 percent above nonbreeding seasonal weights) attained at about the time sexual display begins. During the peak period of displays and egg laying, adult males may be engaged in partial or intense display activities for as much as 95 percent of the daylight hours, with their foraging opportunities consequently greatly reduced. Females are evidently promiscuous and are reportedly attracted to the most dominant available male (Fitzherbert, 1978; Johnsgard, 1991).

In the Denham's bustard males weigh up to about 14 kilograms and average nearly twice as heavy as females, making this the largest of the *Neotis* species and comparable in mass to the *Ardeotis* bustards. The males occupy individual territories well separated by distances of about 700–2,000 meters, but these still seem to be organized into a generally leklike clustering (Tarboton, 1989). The males display to one another and to females with inflated-neck balloon displays (Figure 27F) and with associated booming calls that are audible for considerable distances. Their posturing is visible as far as two kilometers away. Males appear to play no further role in reproduction after fertilization has occurred.

Probably many of the smaller African *Eupodotis* species show comparable exploded lek characteristics, although good comparative data are still largely lacking. Among these are the African black-bellied bustard (*E. melanogaster*) and the rufous-crested bustard (*E. ruficristata*), both medium-sized to very small bustards of grassland habitats. In both of these species males lack inflatable necks but have mostly black underparts and black-and-white wing covert patterns that probably serve as important visual signals during territorial display flights and terrestrial displays. In the black-bellied bustard males weigh about 2–3 kilograms and average only slightly more than females. The birds display in well-defined territories, using both terrestrial "strutting" displays with accompanying moaning or popping calls (Figure 27E) and, less frequently, conspicuous and rather extended display flights above the territory that seem to exhibit the bird's black underparts and contrasting white underwing patterning.

The rufous-crested bustard is even smaller, with adult weights typically of about 500–800 grams in both sexes. Like the black-bellied bustard it exhibits almost no sexual dimorphism, although sexual dichromatism is substantial. In this savanna-adapted species male ground display consists of tail cocking, crest erection, and wing drooping while calling (Figure 27G). The relatively more conspicuous aerial displays of this and several other small African *Eupodotis* bustards consist of short "aerial rocket" flights that are usually of high trajectory but cover little actual horizontal distance. Such flights apparently serve to raise the male above the level of the surrounding vegetation, and briefly expose him to the view of other males or nearby females.

The two Indian species of *Eupodotis,* which are often called "floricans" and are at times separated generically from their African counterparts, provide good examples of fairly well-studied species that also seem to be distributed in exploded leks. The average distances between territorial males is correspondingly far less in these very small (200–350 grams) grassland-adapted bustards than in the much larger exploded lek *Ardeotis* species. Both species of floricans exhibit marked sexual dichromatism. However, males appear to average slightly smaller in body dimensions and body mass than females, perhaps because of the apparent importance of the males' aerial displaying abilities.

In the Bengal florican (*E. bengalensis*) adult males are entirely black, except for white wings, and females are a concealing brown. In this species the territorial males are well separated from one another, usually being out of visual and auditory contact and with at least 500 meters between their display sites. Male advertisement behavior is confined to early morning and evening hours, with most displays occurring under clear skies with little or no wind. In addition to ground displays in which a male closely follows a female while performing head-pumping movements (Figure 28D), males regularly perform short display flights comparable to the aerial rocket flights of the smaller African bustards. During such flights, which last 6–9 seconds, the male may cover 20–40 meters and reach an altitude of 3–4 meters. Whistling notes are uttered at the apex of the flight, and the last phase consists of a nearly vertical drop back to earth (Figure 28E). The male may then walk back to his starting point, but there is no special display site, and successive jumps may be performed from various parts of the male's territory (Inskipp and Inskipp, 1983; Narayan and Rosalind, 1988).

In the lesser florican (*E. indica*), breeding males are black only on the head, neck, and underparts and have less white on their wings. However, the outer primaries of the lesser florican are more highly specialized for sound production by their extreme attenuation (Figure 28F), and the male also has six elongated and ornamental cheek plumes that are strongly erected during display. In this species males often have their display sites on areas of open and raised ground that offer excellent visibility, and the males tend to confine their displays to specific sites ranging in size from about a meter square to about 15 by 15 meters. Distances between adjoining male territorial sites ranged from 275 to 450 meters in one study, and in good habitats 10–15 males may be present in areas of about 100 hectares. In one study area 8–9 males displayed on an area of 350 hectares, with their major displays being short rocket flights above the low vegetation. A loud rattling sound accompanies the jump, perhaps produced by the attenuated primaries and audible for nearly half a kilometer, and a froglike croak is also uttered at the zenith of the flight. Displays are evidently performed much more frequently in this species than in the Bengal florican, with display leaps occurring about once per minute during active periods and up to 400 or more performed in a single day. Peak activity occurs during early morning and evening hours, and especially during cool or cloudy weather. Presumably females are attracted to such displaying males and copulation probably occurs on the male's territory. Male advertisement behavior begins with the onset of summer rains and associated grass growth, and lasts about three months. Ridges are evidently first taken by territorial males, and later-arriving males probably take up nearby positions that are as near as possible to these prime sites. Fighting between males is common during this period, but it is still not known whether a clear dominance hierarchy exists among the territory holders.

It is not behaviorally far from the condition just described for *Ardeotis* and *Eupodotis* to that which characterizes the little bustard (*Tetrax tetrax*), namely a fairly definite, if still somewhat dispersed, lek arrangement. Territorial males hold breeding season territories averaging about 1–3 hectares in area, but which are clustered into recognizable display centers. Females visit such display centers for copulation but nest independently of the males and their territories. Schulz (1985, 1986) described this breeding system as male dominance polygyny, with an intermediate level of male dispersion. He noted that the little bustard has some similarities to the capercaillie in that males of both species display in somewhat clustered display centers. Adult male little bustards lack an inflatable esophagus but develop a fleshy neck enlargement, and feather erection maximizes the apparent size of the male's neck even more (Figure 29C). Adults exhibit considerable sexual dichromatism in plumage, and adult males are apparently slightly larger than females, although few comparative weight data yet exist for breeding condition birds.

Breeding season territories of male little bustards are located in areas having open, panoramic views and with only low vegetation. The males vigorously defend and visually advertise these territories, in part by erection of their strongly contrasting neck feathers. Male calling, together with wing-generated whistles produced by the vibrations of specialized pointed and shortened primaries, evidently serve primarily for territorial proclamation. Nonvocal sounds are especially commonly produced by males during low light levels, including intense foot stamping. Territorial wing beating ("wing-flashing") also occurs while the bird remains on the ground. Males addi-

29. Male displays of smaller bustards, including houbara running with erected neck feathers (*A, B*) and little bustard, showing neck feather erection (*C*), precopulatory display to a dummy (mounted) female (*D*), display flight (*E*), and display leap (*F*). The attenuated seventh primary is also illustrated. For sources see Johnsgard (1991).

tionally have a flutter-jumping display ("wing-flash-and-leap") during which they jump vertically upward with deep wingstrokes, exhibiting their highly contrasting wing patterns (Figure 29F). This display lasts only about a second, and the bird raises itself only slightly more than half a meter above the ground. Such visual signals nevertheless carry great distances in open country and evidently serve to attract females,

30. Male advertisement behavior of great bustard, including normal posture (*A*), intermediate stages of display (*B, C*), and maximum posturing (*D*). After a photo by W. Gewalt and sketches in Hidalgo and Carranza (1991).

and thus are probably functionally analogous to the aerial rocket displays of *Eupodotis* species. There are also more extended display flights (Figure 29E) that are performed by males within their territories, which are characterized by their very shallow wing-beats as well as gliding phases. When females are within the territory of a displaying male they are followed closely by males (Figure 29D), who make repeated copulation attempts.

The Great Bustard and Its Flexible Mating Strategies

The last species of bustard that needs individual consideration, the great bustard, is perhaps the most interesting of all from an ethological standpoint. Not only are its general behavior, biology, and ecology perhaps the best studied of all bustards', but its mating system may also be the most flexible of any known bustard species. Mating patterns range under differing conditions or at different times from seemingly monogamous pair bonding with something approaching resource-based territories, to polygynous or promiscuous forms that essentially represent typical lek systems. It has recently been suggested (Carranza, Hidalgo, and Ena, 1989) that this remarkable flexibility may be related to the fluctuating and unstable habitats in the species' present-day range, with varied ecological conditions prevailing between different areas and during different years. Thus, variations in mating systems may exist not only between different populations, but also in the same population during successive years.

The great bustard exhibits only slight sexual dichromatism but substantial sexual dimorphism. This dimorphism extends to the esophageal structure, which in adult males is highly inflated during display (Figure 28C). Historically, males (presumably very old ones) reportedly weighed as much as 21 kilograms; adult males in present-day populations may weigh up to about 16 kilograms. Wild adult males have recently averaged about 9 kilograms, whereas adult females average about 4.5 kilograms, which represents a male:female mass ratio of about 2.0:1. Together with the *Ardeotis* bustards these are the largest of all bustards, and adult males are among the heaviest of flying birds. Males apparently mature at 5–6 years and females at 3–4 years. The adult or "operational" sex ratio currently favors females in wild populations, but legal or illegal shooting of adult males probably has greatly influenced this particular trait, and perhaps also accounts for the failure of any present-day males to reach the extreme weights that were apparently fairly common about a century ago.

In field studies of nine subpopulations of great bustards in Hungary (Sterbetz, 1981), individual variations in pair-bonding tendencies were very evident. In two of the nine subpopulations the birds were seemingly monogamous. Males followed the fertilized females to their nesting sites and apparently stood guard nearby, at times remaining near the female until the brood-rearing stage. However, in most of the subpopulations the mating patterns ranged through harem polygyny to complete promiscuity.

In the last situation, leklike clusters of the sexually active males formed, which Sterbetz referred to as "rutting packs." Sexual segregation is typical of great bustards throughout the year, with the sexually immature males forming smaller foraging groups that remain near or within the communal display areas of the displaying adult males. The adult males of the nonmonogamous populations usually space themselves apart at distances of 50 meters or more and display communally. However, unlike in typical leks, definite individual territorial boundaries are not maintained within this display ground, and instead the males seem to move about and display at random at various display stations. Females may also seemingly mate at random with males they encounter in such situations, but more probably they selectively mate with particular males based on individual variations in their appearance or display behavior.

In southern Portugal a duality of mating strategies has been observed similar to that reported from Hungary. Most males exhibited nonterritorial but otherwise leklike behavior, whereas a few exhibited definite territorial defense and a haremlike form of polygyny (Cramp and Simmons, 1980). In the latter situation the adult females established a social dominance ranking, with the most dominant female of a harem sometimes chasing those females that were receiving attention away from the dominant male. The possible reproductive advantages and disadvantages of these two simultaneously operating mating systems remain to be explored; it has been suggested that the younger males may be prone to attempt monogamy, whereas older and more experienced males might be more likely to compete for the greater reproductive payoffs associated with polygynous or promiscuous mating strategies (Johnsgard, 1991).

In southern Spain, the great bustard was intensively studied in two successive years and among two different populations about 275 kilometers apart (Carranza, Hidalgo, and Ena, 1989). In one of these areas, the mating system seemed to consist predominantly of polygyny, based either on resource (nesting habitat) defense or on female defense in the form of harem acquisition. In that area, 17 of 21 nests were found in areas used by a single territorial male. In the other area, where average female flock size was lower but more females were present per adult male (operational sex ratio of 1.27 females per male), the prevalent mating system was based on individual male dominance in a lek-type competitive situation, but without apparent territorial boundaries. However, solitarily displaying males could be found in strategic zones within the general area of male display, forming what could be regarded as an exploded lek.

Regardless of the kind of mating strategy that is locally employed, the male advertising behavior of the great bustard during the breeding season is consistently essentially the same. It has been described in many places (Cramp and Simmons, 1980; Johnsgard, 1991), and only the broad outlines need be mentioned here. Not surprisingly in view of the large size of males, aerial displays are wholly lacking. The entire male advertising display is one that has been evolved to achieve long-distance effectiveness (the displaying birds are visible over several kilometers of distance), as well as effectiveness in low light intensities (extensive white plumage exposure). Interestingly, the display is essentially silent, although a low and drawn-out groaning call is produced by displaying males in intensely hostile situations. During the male's balloon display (Figures 28, 30) the extreme tail cocking, neck inflation, and erection and even reversal of many of the white under wing coverts cause the bird to be visually transformed into what appears to be a gigantic white powder puff, so that it becomes hard to separate visually the bird's front from its rear. This astonishing posture may be held for as long as about eight minutes, but usually lasts about two. During full display the bird stands rather motionless, but between displays it may move about somewhat, often while remaining in partial display posture.

In common with other bustards, there are no apparent differences between the hostile displays directed toward other males and the sexual displays directed toward females. When females appear and approach a displaying male the intensity of his display increases, but its form remains the same, just as in sage grouse. Unlike in lek-forming grouse, most copulations apparently occur at the edges of the collective display ground rather than at its center, but in both cases copulations probably involve the most experienced and most socially dominant males (Cramp and Simmons, 1980).

To summarize, the bustards appear to show all gradations from monogamous to promiscuous mating systems, just as in grouse and waterfowl, and in the case of the great bustard these gradations are seemingly present within a single species. Such variations in mating strategies suggest that shifts from one type to another might occur fairly easily under varied social or ecological conditions. These might include differences in tertiary sex ratios, varied amounts of female clustering, differences in local distributions of important ecological resources including foraging or nesting habitats, and perhaps variations in overall population density in different localities or at different times.

6

Sandpipers, Snipes, and Ruffs: Rendezvous at the Lek

The very large assemblage of shorebirds of the world exhibits a considerable diversity of mating strategies (Johnsgard, 1981), but probably the most common pattern is that of seasonal or at least single-brood monogamy. More prolonged or indefinite monogamous matings also occur in a few species that have strong nest site fidelity or exhibit little migratory tendencies. Monogamous matings are perhaps especially favored in those generally small and vulnerable birds where the presence of the male for nest defense, helping to brood the young, and similar activities makes biparental rather than uniparental care advantageous. Opportunities for arena behavior, conversely, are limited to the relatively few shorebirds that exhibit uniparental brood care. Among the Eurasian and North American shorebirds, promiscuous or polygynous matings probably are typical of the European and American woodcocks (*Scolopax rusticola* and *S. minor*), white-rumped sandpiper (*Calidris fuscicollis*), pectoral sandpiper (*C. melanotos*), curlew sandpiper (*C. ferruginea*), buff-breasted sandpiper (*Tringites subruficollis*), great snipe (*Gallinago media*), and ruff (*Philomachus pugnax*). Polygynous matings may also at least occasionally occur in a few other rather poorly studied Asian sandpiper species, such as the Temminck's stint (*C. temminckii*) and sharp-tailed sandpiper (*C. acuminata*), but too little is yet known of such species to comment on them. It is also probable that some sanderling (*C. alba*) populations in the Canadian arctic exhibit serial polygyny, whereas others in Greenland may be monogamous (Pitelka, Holmes, and MacLean, 1974; Johnsgard, 1981).

Excluding the great snipe and woodcocks, adult males of these promiscuous or polygynous species tend to average larger than adult females. The increasingly dimorphic adult mass ratios among them are as follows: curlew sandpiper 1:1, sharp-tailed sandpiper 1.1:1, white-rumped sandpiper 1.15:1, buff-breasted sandpiper 1.3:1, pectoral sandpiper 1.4:1, and ruff 1.7:1 (Johnsgard, 1981). The ruff, which has the highest degree of sexual dimorphism yet reported for any polygynous shorebird, also exhibits a high level of sexual dichromatism, which is otherwise very slight or virtually absent among these six species.

Reverse sexual dimorphism also occurs in a few shorebird species. Thus, adult females of the great snipe are slightly larger than males, with a mean male:female adult mass ratio of 0.86:1; adult males also exhibit more white on their outer tail feathers (Höglund, Kålås, and Løfaldi, 1990). A varied amount of reverse sexual dimorphism similarly exists in woodcocks, the American woodcock having an average adult male:female mass ratio of 0.77:1 and the European woodcock a nearly even ratio of approximately 0.95:1. In the case of the woodcocks there is essentially no sexual dichromatism, although in the American woodcock males exhibit outer primaries that are specialized for aerial sound production.

In neither of these hard-to-observe woodcock species has anything approaching typical avian arena behavior been described, but a brief summary of their unusual mating strategies may nevertheless be warranted. With regard to the European woodcock, it is generally believed that adult males are territorial during the breeding season, with aerial patrolling (called "roding") regularly performed each dawn and dusk along particular flight lanes, and with such territories probably maintained by aggressive interactions. If territorial size can be estimated by the area traversed by such roding flights, it must range from as little as about 6 hectares (15 acres) to as much as about 135 hectares (333 acres). However, at least some common use of flight lanes may occur among different roding males, and radio-tagged males have been observed to use common foraging areas (Hirons, 1980). Thus spatially defined, exclusive-use male territories may be poorly developed or even nonexistent and certainly do not include all the land area below the roding males. During a single breeding season a territorial male may mate successively with as many as four females, who apparently attract the attention of roding males and induce them to land and court them. Such "pair bonding" probably lasts only a few days, or presumably just long enough to insure the female's fertilization. The male probably remains with each mating hen until her clutch is complete, then resumes his display flights. Nests are typically scattered but may at times be somewhat clustered, suggesting that nesting females are nonterritorial and also are not dependent upon direct mate protection. There is no evidence that the male participates in nest defense or brood care (Hirons, 1980; Cramp and Simmons, 1983).

The aerial male displays of the American woodcock are quite different from those of the European species (Hirons and Owens, 1982). Rather than covering a considerable area during display, male American woodcocks perform almost vertical display flights over territorially defended "singing grounds" that may range in size from about 1.8 to 3.2 hectares (4.4–7.8 acres) (Weeks, 1969). The aerial acrobatics associated with these display flights probably attract reproductively ready females, and perhaps have resulted in sexual selection favoring small male size and increased aerial maneuverability rather than increased male size and strength. By comparison, the more normal mode of flying during territorial roding in the approximately twice as heavy European woodcock probably places less importance on maneuverability in attracting females, and a larger body size may be desirable during hostile male-male interactions.

The American woodcock is currently believed to be essentially polygynous, but too little is known of male spacing patterns and reproductive success in either woodcock species to judge the degree of male competition for mates, or whether any individual differences in male mating success can be related to corresponding differences in male territorial quality or advertising display efficiency. The woodcocks therefore do not appear to provide any clear model illustrating the incipient pattern or evolutionary direction of arena behavior in shorebirds. Perhaps they represent the first of the three basic variations on polygyny proposed by Emlen and Oring (1977), namely polygyny controlled by mate defense or by resource-based territoriality, assuming that fertilized females actually do subsequently nest within the defended territories of "their" males, or are at least protected to some degree by them during critical breeding periods.

Evolutionary Routes to Arena Behavior
in Polygynous Shorebirds

In the polygynous or promiscuous pectoral, white-rumped, and curlew sandpipers, differential male reproductive success is not obviously controlled by resource-based or mate defense territoriality. In all of these tundra-breeding species the females apparently do not usually place their nests within male territories, or even in areas heavily used by males, and thus they do not seem to depend on male territorial resources or on male defense of themselves, their eggs, or their chicks. However, male pectoral sandpipers have occasionally been seen associating with females during the incubation period, and possible male participation in brood defense has been reported, so some temporary pair bonding may perhaps exist in that species; and the same applies to white-rumped sandpipers.

One possible shorebird mating strategy lies in increased individual male control of reproductive access to breeding females, whether through rapid sequential attraction and short-term pair bonding (perhaps the usual situation in pectoral sandpipers) or through simultaneous harem accumulation among relatively dispersed or somewhat clumped males (as may occur in curlew sandpipers). In the still little-studied curlew sandpiper males may maintain definite adjacent territories in some low-density years, whereas in other years of higher density the birds appear to be more mobile, with centers of courtship and territorial activity shifting from day to day. Similarly, in the simultaneously or sequentially polygynous white-rumped sandpiper the population density may also influence spacing patterns, but the males tend to show clumped territorial distributions even in years of higher density (Pitelka, Holmes, and MacLean, 1974).

Another potential strategy lies in differential male dominance and female attraction abilities among highly clumped, leklike aggregations of males during the breeding season (Emlen and Oring, 1977). Examples of this strategy occur in several shorebird species, such as the buff-breasted sandpiper (which appears to vary between typical lekking behavior and resource defense polygyny), and in the classic lekking behavior of the great snipe and ruff.

The breeding behavior of the pectoral sandpiper is somewhat better understood than is that of either the white-rumped or the curlew sandpiper. This species is also of

31. Display behavior of male pectoral sandpiper, including parallel march (*A*), breast inflation (*B*), hoot display flight (*C*), posthoot soaring (*D*), and precopulatory approach (*E*). After photos by Myers (1983).

special interest inasmuch as it shows an unusual degree of adult sexual dimorphism in relationship to its nonlekking mating behavior. The adult breeding male can greatly expand its neck and chest during advertisement display, apparently through the inflation of its esophagus, which is visually supplemented by fatty neck tissue and the erection of its ventral neck and breast feathers (Figure 31B). This not only serves to increase the bird's apparent size but probably also affects the pitch of the male's associated hooting call. This far-carrying call has a fundamental frequency of only about 400–500 hertz, which is similar in pitch to that of a booming greater prairie-chicken or a kori bustard during its balloon display (Johnsgard, 1981, 1983a).

The pectoral sandpiper's remarkable advertising display has been well described by Myers (1983). Territorial males are easily identified by their hanging feathered "bibs" (Figure 31B), which balloon out even larger when the male takes off and performs a display flight above his territory. During this flight he utters a prolonged series of hooting notes at the rate of 2–3 per second, which lasts 10–15 seconds. The neck sac is "pumped" up and down in synchrony with the hoots (Figure 31C). Males often fly just barely above the females standing on the tundra. After passing beyond a female, the male may begin a series of rapid alternating flutters and soaring glides (Figure 31D), never rising more than a few meters above ground level. He may thus fly in a broad arc of up to about 100 meters in diameter before coming back and landing at the point where he first took off. Males are intolerant of other males on their territory, and those having neighboring territories often perform long border encounters, walking parallel to one another along these boundaries with lowered wings and outstretched necks (Figure 31A). However, when approaching a receptive female, the male utters an excited but muted version of his hooting call, along with a continuous sequence of growling and squawking notes, while simultaneously cocking his tail, expanding his bib, and drooping his wings. The final precopulatory display is a series of frenzied, wheezing calls, uttered as the male maneuvers behind the female and spreads his wings in preparation for mounting (Figure 31E). Males mate promiscuously, and will display toward any female that lands on their territory or even flies by overhead.

Each male defends a breeding territory of coastal tundra that ranges from about 1 to 10 hectares (2.5–25 acres) in extent. These territories doubtless provide important food resources for both males and females. Territories are well scattered over the tundra, with little indication of male clustering. Evidently each breeding female spends much of her time within a single male's territory, but she sometimes strays to other areas too, and perhaps may even be fertilized by more than one male. The male's interest in a particular female apparently terminates when her clutch is completed, although it has been suggested that males might at times perform nest defense behavior. However, all the males will have abandoned their breeding grounds by the time the chicks hatch, so any male defense of the female or eggs must be confined to the prehatching phases of the nesting cycle (Myers, 1983).

Myers (1983) suggested that the pectoral sandpiper falls near the lekking buff-breasted sandpiper and the ruff in its polygynous breeding behavior. However, the considerable territorial spacing and the defense of fairly large and exclusive-use male territories also affiliate it with the more typically monogamous species of tundra-nesting shorebirds. It is perhaps possible to say that, if a first step toward arena behavior occurs in the pectoral sandpiper, the buff-breasted sandpiper represents the second definite step.

The Buff-breasted Sandpiper and Its Variable Mating System

The buff-breasted sandpiper is a distinctively colored arctic-breeding shorebird; both sexes exhibit a mostly pale cinnamon-colored plumage during the entire year, in which the tips of the flight feathers are uniquely patterned with a fine black marbling

32. Display behavior of buff-breasted sandpiper, including single wing-up male displays (*A, B*), posture of male when leading female to copulation site (*C*), double-wing embrace by male toward two receptive females (*D*), and double-wing embrace by male (*E*). After photos by Myers (1979).

that is more evident in males than in females. Males are considerably larger than fe-males (averaging about 70 vs. 53 grams in the breeding season) but otherwise are very similar to them. The birds breed in high-arctic upland tundra, especially along grassy ridgetops or other areas having a rather scanty vegetational cover.

Some of the best available observations for the sexual behavior of this species are those of Pitelka, Holmes, and MacLean (1974) and Myers (1979, 1989). These have recently been supplemented by somewhat divergent observations and conclusions about the species' territorial spacing pattern and mating system by Cartar and Lyon (1988). Probably very soon after the males arrive on their breeding grounds they establish territories in areas of dry and flat tundra, especially on sites having short sedge- and grass-dominated vegetation. Males typically gather in groups of 2–10 birds, each of which defends a territory some 10–50 meters in diameter, a spacing pattern described as exploded leks by Pitelka, Holmes, and MacLean (1974). Myers (1979) judged the male territories to be considerably larger, of from less than 1 to about 4 hectares (or up to about 10 acres) in area. The males' display sites consist of well-drained grassy areas having closely spaced sedge tussocks about 20 centimeters high, from which resident males chase intruding males. The birds seem to concen-trate their display activity during times when light levels are low, during which the white, mirrorlike undersides of the wings become effective and probably very important long-distance visual signals. Males often direct their displays toward any moving object, including humans. Their primary advertisement displays consist of the wing-waving or single wing-up display (Figure 32A, B) and the double-wing embrace (Figure 32 D, E).

Advertisement display often begins with single wing waving, which is generally directed toward a female in view on the ground or in the air. Almost no sounds are made by such displaying males, although at times they will perform short, vertical flutter-jumps that may perhaps produce some associated wing noises. When one or more females have been thus attracted to a male's territory, he attempts to lead them toward his copulation site by cocking his tail and ruffling his back feathers while walking away (Figure 32C). Up to as many as six females may be simultaneously attracted to such a displaying male. As soon as a female has been attracted the male will hunch over, ruffling his back feathers and marching in place at a faster tempo. He will then suddenly rear back and open both wings, with his bill tilted variably up-ward. At this time he also vocalizes with a repeated *tic* call, synchronized with his footsteps. Females are strongly attracted to this "double-wing embrace," and several may gather within the arc of his outstretched wings, seemingly closely inspecting his marbled underwing feathers (Figure 32D). As a final precopulatory step a female will become receptive by spreading her own wings. At this moment other rival males from adjacent territories may rush in and try to interfere with the copulation attempt.

Unfortunately there is still no information on possible individual variations in male mating success.

In such a leklike situation the females may be expected to nest well outside the area occupied by the displaying males, and that general trend seems to be documented for some breeding areas. However, Cartar and Lyon (1988) found that all of six nests that they found on Jenny Lind Island during the major period of display were within the territories of a displaying male. Not all territories were fully plotted on their sketch map, but these ranged up to about 200 meters in diameter (or roughly 4 hectares in area), and had up to two nests present in a single territory. Three of the territories were contiguous, but clumping was not strongly evident. These authors concluded that at least on Jenny Lind Island the birds did not lek, but instead defended typical nesting territories and probably practiced resource defense or mate defense polygyny.

The Great Snipe and Its Lekking Behavior

Of all the 16 species of snipes in the world, the great snipe is the only one known to exhibit lekking behavior. Such behavior is certainly not to be expected in this group, since in general snipe are believed to be monogamous. The pair bond probably persists through the chick-rearing period, although in some snipes brood care is evidently only performed by the female (Johnsgard, 1981). The great snipe differs only slightly from the other snipes in appearance, and the sexes are nearly identical in body measurements and adult plumages. However, breeding females are probably about 10 percent heavier than males (Avery and Sherwood, 1982), and they also have slightly larger linear measurements (Lemnell, 1978; Höglund, Kålås, and Løfaldi, 1990).

Male great snipe establish very small, contiguous territories in leklike assemblages during their mating season, which they occupy and defend on a nightly basis. Of the territory sizes that have been fairly closely estimated, the observed range is from 20 to 120 square meters, with the larger estimates perhaps being somewhat more reliable. A few locations within each male's territory are preferred display sites; these are often mounds of sphagnum moss that are slightly elevated above the surrounding land and provide generally good visibility. In one recent study (Avery and Sherwood, 1982) from 12 to 16 males were regularly present on a lek, in another 6–12 males were present (Lemnell, 1978), and in a third 6–8 males were present (Höglund and Robertson, 1990). There are historic records of great snipe leks containing as many as 50–60 males, but 10 males now seems to be close to the mean in Scandinavia, with a maximum of about 15–20 present at the peak of the display season.

There is a bimodal temporal pattern of display activity in Scandinavia, with peaks occurring about an hour before midnight and again about three hours after midnight (Lemnell, 1978). Intense fighting over territories occurs early in the season. It is known that birds holding central territories are generally older than peripheral males. Additionally, centrally located males have a larger average number of white tail feathers (which are important display features) than do peripheral ones. This visual distinction is possibly a reflection of age-related male plumage variation (Höglund and Lundberg, 1987).

Between active displays, males may assume a "normal" oblique posture (Figure 33A), which also serves as a preliminary ("phase 1") posture to the drumming sequence. When in an aggressive situation, the posture assumed by males is much more horizontally oriented (Figure 33B), with the bill directed toward the opponent. Two males at the edges of their contiguous territories may run in parallel along their common territorial boundaries (Figure 33H). One may also try to intimidate the other by performing wing-shivering and tail-rattling displays (Figure 33G), especially when the two have come within a meter of one another. When the two are even closer they are likely to orient themselves in parallel, perhaps treading back and forth with each

33. Lek displays of male great snipe, including normal posture (*A*), horizontal posture (*B*), high-crested posture at start of drumming phase (*C*), middle of drumming phase (*D*), tail-raising phase (*E*), wing flap phase (*F*), tail rattling and wing shivering (*G*), parallel running (*H*), and detail of tail (*I*). After drawings and photos in Lemnell (1978) and (*E, F*) cine film analysis.

bird continually changing the positions of its body and head. During such encounters the white outermost rectrices may be exposed through tail fanning or tail turning. Conspicuous preening and even pseudosleeping postures may also occur during such intimidation situations. Fighting is done with bill-thrusting or bill-sparring movements, but few such attacks seem to hurt either of the opponents. Leaping attacks involve both the bill and the feet and are the most intense of the various agonistic encounters, which at times last for several minutes (Lemnell, 1978).

The primary advertising display of the male great snipe is called "drumming." Some aerial display activity also occurs, namely noisy flutter-jumps up to about a meter high and some more extended flights, but these seem to be of reduced significance as primary male signaling behaviors. All of these displays seem to have the same function, namely announcement of the male's territorial position both to rival males and possibly to receptive females.

The male's drumming or self-advertisement display has been described by various authorities (Ferdinand, 1966; Lemnell, 1978; Avery and Sherwood, 1982). Lemnell divided the display sequence into four phases, of which the first corresponds to the normal oblique standing posture (Figure 33A) mentioned earlier. In this posture the birds begin to utter a songlike twittering (or "bibbling") of about eight notes per second, the notes having a fundamental frequency of about 3 kHz and being audible to humans for up to about 100 meters under favorable conditions. A low-frequency motorlike sound of uncertain origin and having a frequency of about 100 Hz occurs alternately with these vocalizations. In the second phase the male stretches his neck and raises his bill to the horizontal or slightly above. The bibbling notes continue, with their frequency and amplitude increasing and their cadence becoming faster (about 11 per second). The low-frequency sound decreases in intensity during this phase.

The third or peak drumming phase is initiated by a high-crested posture (Figure 33C). During this phase, which lasts nearly two seconds, the bibbling becomes a loud metallic clicking, reaching a maximum fundamental frequency (to 6 kHz) and loudness (audible for about 200 meters) in the middle of this phase (Figure 33D), while the low-frequency motor sounds also become very loud. Near the end of this phase the tail is suddenly raised, strongly tilted laterally, and widely spread, which briefly exposes the white outer rectrices (Figure 33E). This is followed immediately by a sudden single wing flap (Figure 33F), which very briefly flashes the zebralike axillaries and patterned wing undersides. The tail is then quickly lowered and closed. During the final phase, which lasts about a second, the male's neck is stretched upward and his bill is widely opened and raised. The high-frequency clicking vocalization becomes a more continuous, frequency-modulated "whizzing" sound that gradually but irregularly descends in both frequency and amplitude. After the whizzing is completed the male gradually returns to his normal oblique position, which is attained about five seconds after the start of his display (Lemnell, 1978).

Females typically visit the lek around midnight and are quite secretive, making close observations almost impossible. However, it is known that territorial males shift to display sites as close as possible to any visiting female, and they actively perform flutter-leaping and drumming. Males may also wing-tremble, tail-flick, and crouch before a nearby female. Precopulatory squatting by females has been observed, but little detailed information on copulatory behavior yet exists. Interference by other males may sometimes prevent completed copulations, although treading lasts only a few seconds (Lemnell, 1978).

Recently evidence has developed favoring the idea that classic sexual selection occurs in this species, even though it is essentially sexually monochromatic and exhibits slightly reversed sexual dimorphism. Höglund and Lundberg (1987) initially reported that individual male mating success is strongly skewed, with the most successful male in each of two leks studied obtaining about 30 percent of the observed matings (totaling 53), whereas five other males on both leks obtained no matings at all. The most important variables determining a male's success seem to be his display frequency and the nearness of his territory to the center of the lek. Other male variables, such as the number of white tail feathers, size and mass criteria, fighting frequency, song characteristics, and the incidence of territorial attendance, were not found to be statistically significant. A study of centrally located males showed that they differed from peripheral ones in having higher overall levels of nightly lek attendance, and in being older and correspondingly having more white on their outer tail feathers. Only 2 out of 15 central males (13 percent) disappeared from their leks during a single display season, but 9 of 33 peripheral males (27 percent) disappeared during that same period, suggesting that a higher level of lek fidelity exists in central males than in peripheral ones. Territorial males also weighed about 3–6 grams more on average than did non-territorial males.

More recently, Höglund, Eriksson, and Lindell (1990) have confirmed that adult male great snipe have more white on their tail feathers than do females, and that the amount of white present in individual males increases with age, at least through their second year. Males with experimentally enlarged white areas on their tails did not seem to benefit in male-male competitive territorial encounters. On the other hand, females seemed to use the amount of white present as a cue in choosing mates, since 10 of 17 observed copulations were made by a male with an experimentally enlarged white tail pattern. These authors concluded that females decide on a specific mating partner based on several criteria. One of these criteria is the amount of white in the tail, which is a visual signal that probably works best at rather short distances under nocturnal conditions (Figure 33I).

Höglund and Robertson have recently (1990) tried to determine how females selectively choose centrally located males for mating. Females may simply choose them on the basis of their lek position, which reflects their correlated dominance-related ability to hold such a territory (the position hypothesis), or because females can recognize and are attracted to a specific vigorous and/or unusually attractive male (the attractiveness hypothesis). Under the latter hypothesis, a dominant male's interior territorial position might thus be simply a result of less dominant males gathering around him and attempting to intercept females that may be attracted to his individual appearance or behavior. By experimentally removing the dominant male, Höglund and Robertson determined that the vacated territories of dominant birds were left unoccupied for a time, but when less dominant males were removed these vacated territories were immediately occupied by new males. Additionally, dominant males ignored a taxidermically mounted male dummy, whereas less dominant males approached and often attacked the dummy bird. Finally, on a small lek consisting of only about seven males, where no effects were evident of central versus peripheral position, individual male mating success was found to be correlated with relative male-male dominance and male display rate. All these results favor the attractiveness hypothesis and are counter to the position hypothesis as a mate choice mechanism among females. Höglund and Robertson furthermore suggested that great snipe leks probably initially develop because some males, being unable to attract females on their own, tend to associate with more attractive males. This explanation is similar to the general "hotshot" hypothesis but emphasizes that female mating preferences for specific males, rather than merely reflecting male-male dominance relationships, may be especially important in the early development of lekking behavior.

The Ruff, Its Arena, and Its Polymorphic Males

Of all the arena birds of the world, the ruff exhibits not only strong sexual dichromatism and dimorphism but also the most remarkable male plumage "polymorphism" (=polychromatism) of any arena bird species. Indeed, in no other avian species is more variability present in the appearance of breeding males than occurs in the ruff. In nuptial plumage the male's ornamental feathers of the neck ("ruff") and crown ("head tuft") range from nearly perfectly white to almost totally black. There is a continuously intervening range of grays, browns, reddish browns, and beige tones and patterns between these extreme phenotypes (Figure 34). Hogan-Warburg (1966) determined that breeding males may vary in (1) the basic and secondary colors of the ruff and head tufts, (2) the specific feather patterns of these areas, (3) the color and pattern contrasts between these two areas, and, to a lesser extent, (4) the color of the yellow to red skin wattles of the face and (5) the wing covert colors. The length and brilliance of the nuptial feathers, the size of the facial wattles, and the color of the bill and legs are all directly influenced by age. However, individual males maintain their nuptial plumage colors and patterns throughout their lives, and thus these features are genetically controlled. The males' varied plumage condition thus appears to be a case of dynamic, apparently genetically balanced polymorphism.

Leks of this species typically are located in moist, grassy meadows that are occupied by about 5–20 males, or rarely up to as many as 50. Leks designated by Hogan-Warburg as "small" had 3–8 males; those with 15 or more were considered "large." Leks are active in northern Europe from April to June, and occur in traditional sites that may persist for decades, perhaps at times even as long as a century. A typical lek is only about 20 square meters in area, and in favorable habitats separate leks may be spaced in clusters that are separated by only about 50 to a few hundred meters. Such clustered groups of leks in turn are often separated by several kilometers from other

34. Comparison of female ruff ("reeve") (*A*) with polychromatic males of white (*B*), intermediate (*C*), and dark (*D*) morphs. After photos by author.

such groups, depending upon habitat. Individual male display sites ("residential areas" of Hogan-Warburg) are typically only about 30 centimeters in diameter, and the distances ("interresidential areas") between separate display sites average 1–1.5 meters. Each display site is occupied by a separate male. These display sites or residences are unlike typical closely contiguous territories in that the interresidential areas are not contested, although the display sites (residences) themselves are strongly defended.

The displaying birds normally arrive at the lek each day before dawn, with a peak in morning display occurring shortly after sunrise. There is a second late afternoon period of activity as well. Females visit various leks, either singly or in small groups, but their visits are of relatively short duration, with a peak in visiting shortly after sunrise. Females may leave the lek without copulating, or they may choose to mate with a particular male. Some males are chosen more frequently than others for copulation, which occurs directly on a residence site.

Hogan-Warburg recognized several categories of males based on their territorial status. Two subgroups of residence-holding or "independent" males comprised about 60 percent of her total observed male population. These include "resident" males, which have established regular display sites (residences) on the lek, and "marginal" males, which are potential possessors of residences that at any given time occupy only peripheral display locations on the lek. A second major category consists of "satellite" males, which never acquire permanent resident status and which represent a minority of the total male population. Among satellite males a further if less clear-cut distinction may be made between "central" satellite males (those that visit residences more frequently and are more often tolerated by residence holders) and "peripheral" satellite males (the more mobile and less well tolerated individuals). Transformations of males from satellite to independent status or vice versa were not observed by Hogan-Warburg, although within these broad categories their status was subject to ontogenetic change. Thus, young independent males may be initially only marginal in status, but later attain the status of resident male when fully mature. Similarly, young satellite males may shift with age from peripheral to central status. However, males of either category may eventually decline in status when too old to maintain their higher rank (Hogan-Warburg, 1966).

A recent study by Lank and Smith (1987) indicates that a very substantial proportion of male ruffs (they estimated 88 percent) may never try to occupy leks. These males spend as much time as do lekking males in courting females, but use one of two alternate strategies. The first strategy is to directly pursue and court females (the "following" strategy), and the other is to wait for females to appear at resource-rich sites, with temporary leks being established at such locations ("intercepting" strategy). However, males displaying on permanent leks mated at significantly higher rates than did "followers," and at slightly higher rates than did "interceptors." Displays of males using these nonlekking mating strategies were apparently similar or identical to those of lekking males, and about 90 percent of all social display in the population under study was estimated to occur off regular leks. Probably much of this off-lek display is performed by marginal or satellite males that also spend varied amounts of time on established leks. The authors indeed suggested that satellite behavior in ruffs may have arisen as an outgrowth of following behavior, during which these mobile males began to follow females to established leks and continued their displays there.

In summary, it would appear that as many six recognizable types of male behavior patterns and mating strategies may exist in ruffs:

> Lek-centered male types
>> Independent males
>>> Resident males
>>> Marginal males
>> Satellite males
>>> Central satellite males
>>> Peripheral satellite males
> Off-lek male strategies
>> Followers
>> Interceptors

Female preference for mating with particular males is most evident on large leks. In the case of multiple occupancy of residence sites (which are sites held by a resident male plus a variably tolerated satellite male), competitive behavior among the inter-acting males as well as female choice determines the actual mating partner. Although both resident and satellite males may succeed in copulation, marginal males succeed only very rarely (Hogan–Warburg, 1966).

Hogan–Warburg argued that this unusual variation in male status is an example of balanced behavioral polymorphism, which she linked to the apparent survival value of having some satellite males present on a lek. Thus, although they are not residence holders, the presence of satellite males increases the overall chances for male copula-tory success on small leks (presumably by increasing the visibility of such small as-semblages and thereby attracting females to such leks). They promote furthermore the establishment of new leks and the maintenance of several leks in a single area. The proportion of copulations obtained by satellite males varies from lek to lek; in large leks the satellite males have been found to obtain relatively few copulations, since the resident males are highly intolerant of them. However, in small leks where they are more readily tolerated they may perform a significantly larger proportion of copula-tions than would be expected by chance.

Male plumage polychromatism may also be selected for under this mating system. The marked nuptial plumage diversity of independent males may help facilitate their individual recognition by both females and other males. Contrariwise, the relatively

35. Display behavior of male ruff, including sleeked upright posture (*A*), wing fluttering (*B*), strutting (*C*), oblique posture (*D*), front (*E*) and side (*F*) views of squatting male, and copulation (*G*). After photos by author except for *B* and *C*, which are based on sketches in Hogan-Warburg (1966).

reduced plumage diversity among satellite males (which tend to have predominantly white nuptial plumes) may serve primarily to identify and help visually distinguish the satellite males as such, both for females and also for other males. Hogan-Warburg suggested that the extremely tightly packed spacing pattern of ruff leks may make male plumage diversity and corresponding ease of individual male identification more important for ruffs than for other avian lekking species. Furthermore, in contrast to many lekking grouse and bird-of-paradise species, the ruff constitutes a monotypic genus. Thus, there are no opportunities for interspecific hybridization resulting from female "mistakes" in selecting an inappropriately plumaged male partner.

Although no specific male plumage color or pattern is obviously preferred, females consistently choose to mate with males having an optimal development of the nuptial plumage, whether they mate with satellite or with resident males. On large leks, attractive resident males holding unshared residences are generally most successful. At least on one large lek, a centrally located male was preferred for mating, but at another two males occupying edge positions were more successful. Selection of males by females is apparently also partly influenced by the locations of females already assembled at particular residences on the lek. Significantly, the most successful males on both large and small leks were those that tended to rise up from their immobile "squat" display (Figure 35E) and perform various other displays. These include female-oriented tail swaying, and especially such clearly agonistic displays as the "low-horizontal" (similar to the horizontal threat display of the great snipe) or the "spread-tail-forward" (a male directs his bill toward a male opponent while partially crouching and spreading his tail). Both of these postures tend to hold other potentially interfering males at bay during copulation attempts.

On small leks, precopulatory crouching by females often occurred only immediately after a courting male had returned to his residence following a charge toward or attack on other males. A quantitative correlation between the results of such agonistic male-male encounters and relative male mating success was not attempted by Hogan-Warburg. However, relative male fighting success might also be expected to strongly influence male attractiveness and associated male mating success, since many attempted copulations are interrupted by the presence of competing marginal and satellite males. The importance of male fighting success in influencing mating success might also help explain the substantial (1.7:1) male:female mass dimorphism ratio, which is the most unbalanced of all shorebird species'.

Male displays of ruffs are quite numerous and diverse, as described by Hogan-Warburg (1966), but most can be considered as hostile in motivation. The normal oblique posture (Figure 35D) is assumed by resident males when not directly engaged in social interactions. Marginal males are more likely to adopt an upright posture, such as the "sleeked upright" (Figure 35A), which is an alert posture that is also assumed by resident males on a partially occupied lek. Various "forward" displays, with bill pointing, forward bill thrusting, and successive levels of threat, charge, attack, and overt fighting, are commonly performed by rival resident males when females are present on the lek. Contrariwise, bill thrusting toward the ground, turning away, and retreating provide submissive signals. Wing fluttering (Figure 35B) may be performed by either resident or satellite males upon seeing approaching conspecifics in flight. Wing fluttering and similar flutter-jumping and air-hovering expose contrasting white underwing patterns that are visible for great distances. These conspicuous wing movements may be combined with tail trembling and tail lifting in some individuals.

Male behavior in the presence of females on the lek is quite different. The male faces a landing female in a somewhat crouched posture, with his tail cocked, his breast touching the ground, and his bill pointed toward her. The male then quickly sinks into a squatting position and remains motionless and silent before her, his bill pointing strongly downward. The squatting bird is typically positioned so that his bill is directly oriented toward the female, or he may be more laterally situated so that the female remains visible with one eye (Figure 35E). When a resident and satellite male share a display site the satellite males squats farther from the female than does the resident bird. At various times the satellite male may rise from his squatting posture and slowly approach the female in a copulation-inviting "strutting" posture (Figure 35C). This usually occurs when the resident male has briefly risen and left his display site to threaten or attack another nearby male. Females typically retreat from such strutting birds, but at times they may stop and crouch in a precopulatory soliciting posture. When the female thus crouches, the resident male often attacks any nearby satellite male. Or he too may quickly adopt a precopulatory display (a half-squatting, tail-cocked posture facing the female) and then quickly try to copulate with the crouching female (Figure 35G).

Van Rhijn has expanded the studies of ruff behavior begun by Hogan-Warburg. He has suggested (1973) that the independent or satellite status of males depends on both genetic and environmental factors; the positive correlation between varied male mating strategies and plumage categories was offered as evidence for this position. He also speculated that independent and satellite males may be homozygous with respect to alternate alleles of a single gene regulating these traits. Furthermore, in his view heterozygous birds might be those having atypical male plumages, and the maintenance of male plumage and behavioral polymorphism may be a consequence of the "superiority of heterozygotes," namely those males having atypical nuptial plumages. It seems highly unlikely that such extreme present-day male plumage variability (with nearly 50 male plumage phenotypes known, including 12 basic color types and four basic pattern types) and associated male behavioral variability could possibly be controlled by a single gene. Van Rhijn later (1991) noted that an earlier plumage analysis by the Russian geneticist L. V. Ferri had suggested that about seven gene locations would be needed to explain male plumage variability alone, and so van Rhijn's simpler explanation does not seem to be adequate to explain either the behavioral complexity of the varied male mating strategies or the extreme breeding plumage variability in male ruffs.

Höglund and Lundberg (1989) recently reported that dark-morph males average slightly larger in various morphometric measurements than do white-morph ones. This observation would predict that the darker birds should have a higher overall rate of copulatory success than white ones in competitive situations, and that perhaps other environmental factors may be significant in maintaining the present degree of male plumage polymorphism. However, Hill (1991) has analyzed various components of individual male mating success, finding that success was unrelated to the darkness of the nuptial plumage of independent males, the male's territorial location, or his rate of courtship displays. Contrariwise, positive correlates of mating success included a high frequency of visits by satellite males on an independent male's territory, a consistency of lek attendance, and low rates of aggressive behavior. Hill's conclusions, however, are not based on large sample sizes of individually marked males, and have not yet found widespread acceptance.

In a related question, Lank and Smith (1992) have recently confirmed that wild female ruffs are attracted to the larger of two adjacent male groups for lekking activity, to the extent that males in the larger group had significantly higher rates of female visitation. This result of course supports the general theory of female choice as a significant driving force in male lekking behavior.

Van Rhijn (1985, 1991) has recently offered a hypothetic scenario to explain the social evolution in ruffs as well as other northern hemisphere sandpipers, but these ideas are clearly impossible to test and seem rather unlikely. He has also suggested that the behavioral dimorphism of male ruffs may have resulted from satellite males tending to assume a strategy of mimicking females at leks, whereas other more intruding males may prefer to stake out residences on leks containing other males that tend to resemble themselves. There seems to be no direct evidence for either of his hypotheses. Hogan-Warburg (1993) has already criticized these ideas and has suggested that females choose individual mates (resident or satellite) on the basis of behavioral differences among them. These differences especially involve the ability of a male to interrupt his squat display during female visits, and also on the differential tolerance of resident males toward intruding satellite males, with the more intolerant males being favored.

7

A Potpourri of Kakapos, Hummingbirds, and Lyrebirds

One of the rarest birds in the world is the New Zealand kakapo (*Strigops habroptilus*), an enormous mossy-green-colored parrot that is one of only two nocturnal parrots as well as the only flightless parrot in the world. It is also the only parrot in the world that performs lekking behavior, an activity that is highly unexpected within a group of generally extremely monogamous birds. Among the more than 300 parrot species, the only other known to be nonmonogamous is the New Zealand kea (*Nestor notabilis*), which is apparently polygynous but is a harem-forming rather than an arena species. Both of these parrots are large, but the kakapo is the heaviest of all the world's parrots, with males averaging about 2 kilograms and females about 1.3 kilograms (Merton, Morris, and Atkinson, 1984), representing an adult male:female mass ratio of 1.6:1 The sexes are virtually identical in adult plumage.

Information on the kakapo's social behavior is extremely limited. This is partly because of the birds' nocturnal activities and the difficult, almost inaccessible terrain in which they live, but especially because of their present extreme rarity. Until the present century the birds were fairly widespread throughout New Zealand, but they disappeared from North Island and most of South Island during this century. On the latter they were apparently confined to the Milford District of Fjordland. Some of these birds (five males) were transferred by conservationists to Maud Island in Marlborough Sounds in an effort to save the species. By the late 1980s this population had apparently entirely disappeared. However, in 1977 a small previously undetected population (originally optimistically estimated to be about 200 birds) was discovered in southeastern Stewart Island. At the time of its discovery this population was already declining as a result of predation by cats and perhaps other predators, and by 1982 it was believed to number fewer than 100 individuals. Some of these birds (13 males, 9 females) were subsequently captured and moved to Little Barrier Island, and two Stewart Island females were also released on Maud Island (Forshaw, 1989). Between 1987 and 1989, 20 male and 9 female kakapos were additionally transferred to Codfish Island. The total world kakapo population was estimated at about 55 individuals (mostly males) in 1991 (Powlesland et al., 1992), nearly all of which are in predator-free environments. Thus, there is still a slim hope that the species may survive into the next century, although the odds against it presently would seem to be considerable.

Most of what is presently known of the kakapo's social behavior has been provided by Merton, Morris, and Atkinson (1984), which has been supplemented by the more popular accounts of Cemmick and Veitch (1987) and Morris and Smith (1988). These accounts relate only to male behavior patterns; visits by females are apparently brief and infrequent, and copulation has not yet been observed, although it is believed to occur near the display sites or "bowls" that are constructed by males. Observations at two nests indicate that no male involvement occurs at the nesting stage of the breeding cycle.

Kakapos of both sexes are solitary birds, and, although virtually flightless (they are able to glide for short distances), individuals may occupy steeply sloping territories of up to about 50 hectares. Home ranges of females overlap those of males but are usually located some distances from the males' display sites. The birds forage on a wide array of plant materials but few if any animal materials. During the breeding season each male establishes a localized display ground that is situated along the walking trails made along ridgetops or low mountain spurs and is maintained by regular walks during nightly excursions. A single male's "booming bowls" consist of a group of nearby hollowed out depressions in the soil (which might be considered as stages, using this book's terminology). These bowls are cleared of all debris and have bottoms consisting of a very fine soil tilth. They are organized in closely spaced groups only a few meters apart, connected by well-defined trails. Such a group of bowls may

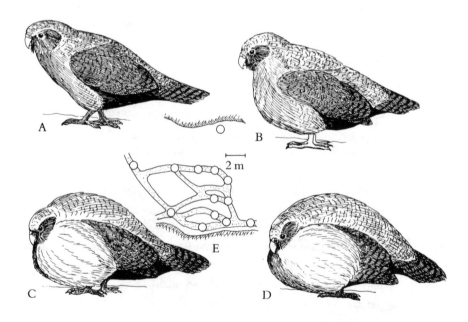

extend for 50 meters or more along a ridgecrest, or perhaps may occupy an area 10–20 meters in diameter on a hilltop (Figure 36E). Bowls are often excavated against banks, or below rock faces or other natural protruding objects that may serve as effective sound reflectors. The distance between such bowl clusters of neighboring males may range from 15 to several hundred meters, and at least historically groups of as many as 50 males' bowl systems might extend over several square kilometers. Aggressive interactions between males at bowl systems during the display period suggest that the sites are territorially defended. The male's home range may be separated by several kilometers from its bowl system. Bowl systems may be used in successive years, but booming does not necessarily occur every year, and indeed several years may pass between active booming periods.

While standing in a bowl, the displaying male utters a low-frequency, resonant call that under ideal conditions can be heard as far as 5 kilometers away. Starting its display in a standing position, it gradually lowers its head and progressively inflates its thoracic region until the anterior body has assumed a grotesque balloonlike shape (Figure 36A–D). At the same time it utters a series of about three grunting notes that then give way to deep booms. A maximum booming intensity is reached after about six to eight booms, which are uttered at a rather constant rate of about one every two seconds. These loud booming notes may be repeated up to 20 times, after which the volume gradually diminishes while another 10–20 booming notes are uttered. Thus

36. Display behavior of male kakapo, showing booming sequence (*A–D*), sketch map of a male's booming sites (*E*), and male wing-spreading ("dancing") behavior (*F, G*). *A–E* after sketches in Merton, Morris, and Atkinson (1984); *F* and *G* after photos in Morris and Smith (1988).

F

G

an entire calling sequence of 35 or so notes might last about 70 seconds. Such boom-ing sequences are repeated after about 20–30 seconds of silence. Booming can be in-duced by playing tape recordings of booming, and males booming in groups display longer each night and over a longer booming season than do single males. Normally the birds call for 6–8 hours per night, starting about an hour after dark, and leave their display sites about an hour before daylight.

Males have also been seen displaying posturally at bowls. These displays include side-to-side rocking movements, and walking backward while slowly raising and lowering the wings (Figure 36G). Other possible displays have been seen, such as bill-whetting, wing-flapping, and wing-stretching movements (Figure 36F), and the birds often have been observed carrying vegetation about at this time. A variety of shrieking and screaming calls have also been reported at display bowls. Attempted copulations have been observed when females have been introduced, but no detailed descriptions of this behavior appear to be available. Apparent females have also been seen visiting bowls and interacting vocally with the males, but such visits appear to be infrequent and brief.

Little else is known about this elusive species. Why it alone among parrots should be nocturnal, polygynous, and lek forming remains a mystery. Perhaps the species' flightless condition and limited mobility make it desirable that females ready for egg laying can depend on finding reproductively active males without undue searching. Additionally, the large size of the birds in an (at least originally) relatively predator-free environment may have facilitated the development of effective long-distance low-frequency acoustic signaling. The elaborate postural displays would seem to be unexpected in a nocturnally displaying bird, unless perhaps they have been "acciden-tally" carried over from a time when it was more diurnally active. The species' small clutch size of only one or two eggs, and its apparently very long egg-laying interval (and subsequent hatching intervals of the chicks), perhaps make unisexual care by the female more feasible than in most parrots. The possible role of male-male competi-tion, the potential effects of age on individual male signaling behavior and breeding success, the maximum numbers of females that might be attracted to or fertilized by a single male, are among the many basic questions about kakapos that are still wholly unresolved.

Lekking Hummingbirds: "Glittering Garments of the Rainbow"

It should not be surprising that lekking behavior has arisen in the hummingbirds, inasmuch as male emancipation from nesting and brood-rearing responsibilities is characteristic of the entire family, and male polygyny or promiscuity is seemingly rampant. It would thus seem but a short step from the highly territorial and distinctly spaced distributions of male hummingbirds to one that is more highly clumped. This

1. *(top)* Male mandarin performing burp display toward female. Photo by author.

2. *(bottom)* Male mallards performing down–up (foreground) and head–up–tail–up (background). Photo by author.

3. *(top)* Male red-breasted mergansers performing synchronous salute displays toward a female. Photo by author.

4. *(bottom)* Male common goldeneyes performing synchronous head throw displays toward a female. Photo by author.

5. *(top)* Male Temminck's tragopan at peak of advertisement display, after suddenly rising from behind a boulder. Photo by David S. Rimlinger.

6. *(bottom)* Male Bulwer's pheasant in strutting display. Note tips of rectrices scratching substrate. Photo by author.

7. *(top)* Male Malay peacock-pheasant in lateral display. Photo by author.

8. *(bottom)* Male Palawan peacock-pheasant in lateral display toward female. Photo by author.

9. *(top)* Male great argus in frontal display, as seen from the side. Photo by author.

10. *(bottom)* Male great argus in frontal display, as seen by the female. Note location of male's eye. Photo by author.

11. *(top)* Male ruffed grouse in drumming display. Photo by author.

12. *(middle)* Male black grouse in advertisement display posture. Photo by author.

13. *(bottom)* Two male black grouse in confrontational display at territorial boundaries. Photo by author.

14. *(top)* Male greater prairie-chicken displaying to female. Photo by author.

15. *(middle)* Two male sage grouse displaying to a female. Photo by author.

16. *(bottom)* Male wild turkeys in social display before females. Photo by author.

17. *(below)* Dark-morph male ruff in upright posture. Photo by author.

18. *(top)* Medium- and light-morph male ruffs in oblique and squatting postures. Photo by author.

19. *(middle)* Light-morph male ruff squatting. Photo by author.

20. *(bottom)* Dark-morph male ruff in squatting display toward female. Photo by author.

21. *(facing page, top)* Great bustard, adults in alert posture. Photo by author.

22. *(facing page, bottom)* Great bustard, adult male in high-intensity threat posture. Photo by author.

23. *(below)* Kori bustard, male performing balloon display. Photo by author.

24. *(top)* Male long-wattled umbrellabird with extended wattle. Photo by Ken Fink.

25. *(middle left)* Adult calfbird. Photo by Ken Fink.

26. *(bottom left)* Male bearded bellbird. Photo by author.

27. *(middle right)* Male Peruvian cock-of-the-rock. Photo by David S. Rimlinger.

28. *(bottom right)* Male Guianan cock-of-the-rock. Photo by Ken Fink.

29. *(top)* Male white-bearded manakin. Photo by Ken Fink.

30. *(middle)* Male round-tailed manakin. Photo by Ken Fink.

31. *(bottom)* Male long-tailed manakin. Photo by Ken Fink.

32. *(top)* Male great gray bowerbird constructing his bower. Photo by Ken Fink.

33. *(bottom left)* Male king bird-of-paradise. Photo by David S. Rimlinger.

34. *(bottom right)* Male red bird-of-paradise. Photo by author.

35. *(top)* Male raggiana bird-of-paradise (Salvadori's race) in wing-beating display. Photo by David S. Rimlinger.

36. *(bottom)* Male greater bird-of-paradise. Photo by author.

37. *(top)* Male Jackson's widowbird displaying in his arena toward a visiting female. Photo by Steffan Andersson.

38. *(bottom)* Male straw-tailed whydah. Photo by author.

might occur at sites where male visibility is unusually favorable, where females might perhaps regularly visit, or around particularly successful individual males. Any of these circumstances might be mechanisms favoring incipient arena behavior in this group, as might others.

The only theoretical contribution to the evolution of lek behavior in humming-birds is that of Stiles and Wolf (1979). These authors suggested that although the most common male hummingbird mating strategy is to control a rich food (flower) supply on the mating territory, lekking tends to evolve in those hummingbirds where this is impossible. Such is the case when defensible flowers are controlled by some other more dominant species, or where sufficiently rich and defensible nectar sources do not occur within the species' habitat. The hermit group of hummingbirds (*Phaethornis* and related genera) includes several species that regularly engage in lekking behavior; these birds specialize in exploiting nondefensible flowers, whether scattered in distri-bution or primarily insect-pollinated species that produce rather little nectar.

Most non-hermit hummingbirds fall into the opposite category of male mating strategy, in which males territorially defend local and rich nectar sources, to which females may be attracted and individually courted whenever they appear. However, changes in the distribution of flowers can result in local variations in mating systems, which can range within the same species from controlling such food-centered mating territories to essentially leklike systems (Stiles, 1973). It is therefore apparent that fac-ultative arena behavior is to be expected in many hummingbird species, and since most of the more than 300 species are still unstudied, many more examples of hum-mingbird lekking are likely to be reported in the future.

The hermits are unusual among hummingbirds in that males are relatively dull-colored, and they furthermore consume many insects together with the nectar that they obtain from flowers. It is generally believed that they are structurally and behav-iorally more like the ancestral hummingbirds than are the non-hermits. Hermits tend to be tropical forest-inhabiting species, where bright colors are of rather little value for advertisement but where vocalizations might be quite effective. On the other hand, open-country non-hermit hummingbirds are prone to have male plumages that are much richer in iridescent colors, while they are generally weaker in male vocalizations.

Thus, in contrast to most lekking birds, the hummingbirds that are most prone to arena behavior tend not to be the most brilliantly plumaged species. Additionally, male:female mass ratios in hummingbirds are related to their overall relative body size rather than to mating system variations. In the larger hummingbird species the males tend to have a somewhat greater adult mass than females, as would be expected in any polygynous/promiscuous mating system. However, among the quite small species this ratio is reversed. The females of these species apparently must maintain a rela-tively large mass in order to be able to cope with the energy demands associated with egg laying and unisexual care of the dependent young, whereas a very small body size might be simultaneously selected for among males of the same species for diverse reasons, such as ecological isolation (Johnsgard, 1983b).

Two species of hermit hummingbirds have been especially well studied in terms of their lek behavior. These are the green (Guy's) hermit (*Phaethornis guy*) and the long-tailed hermit (*P. superciliosa*). The little hermit (*P. longuemareus*) has also been observed to a lesser degree (Wiley, 1971; Snow, 1968). All of these species apparently form leks throughout their entire range. Some other hermits may form leks in parts of their range (e.g., the red hermit *P. ruber*), whereas others seemingly do not form leks at all. Whether of a lek-forming species or not, the primary advertising behavior of male hermits is singing, which ranges from simple and monotonous in some species to quite complex and melodious in others. The hermits also regularly perform "tail wagging" (vertical tail pumping), and all of the species so far studied perform hovering displays that seem to exhibit the bird's throat pattern and/or its gape, sometimes supplemented by tongue extension. At least in the best-studied species, there is no good evidence that the displays of resident males are qualitatively different toward other males than toward females, which may be a reflection of the great similarity between young males and females in the sexually dichromatic species and the lack of clear-cut adult sexual dichromatism in others (Stiles and Wolf, 1979).

The long-tailed hermit is especially well studied, mostly as a result of the extensive observations by Stiles and Wolf (1979). It is a fairly large hummingbird, adults weighing about 6 grams, with males averaging about 4–5 percent heavier than females. The sexes are identical in plumage, being mostly bronze-colored above with buffy head striping and with a long and pointed tail, the individual rectrices contrastingly tipped with white or buffy. Females undertake all the nesting duties, and males rarely even appear at the nest site. It is a relatively common species, occurring in forests from southern Mexico south to Brazil, and apparently is a lekking species throughout its range. Early observers have reported that some leks of this species may have a diameter of as much as about 200 meters, and may contain more than a hundred participating males. The activity period for a particular lek often lasts for most of the year (July to April in Guyana), and individual active lek locations may persist over at least as long as 20 successive years (G. Stiles, pers. comm.).

In a Costa Rican study area, four long-tailed hermit leks that averaged about a kilometer apart were studied by Stiles and Wolf. Three of these were located along streams, which provide gaps in the forest canopy as well as flight paths, abundant food plants, and dense thickets. Numbers of males on a single lek ranged from 5 to 25, with individual territory size ranging from about 20 to 500 square meters. The smallest territories and greatest number of resident males were located in very dense thickets, and the largest territories were in open forest. Only a slight tendency favoring centrality in territorial dispersion was evident, and residents were usually not in visual contact. Central territories nevertheless were the most strongly contested and the most stable over time. They also typically were located in the densest vegetation. Singing from dense thickets is perhaps an antipredator adaptation, countering the possible predator attraction effect produced by the collective singing of all the males, which can be heard for up to at least 100 meters.

As might be expected, peripheral territories (or empty central territories) tended to be occupied by younger and presumably more subordinate males. Surprisingly, however, peripheral males appeared to participate to some extent in mating, in contrast to the typical situation in most lekking birds. Birds returning in successive years either reoccupied their old territories or occupied vacant territories, often moving closer to the lek center. Nevertheless, Stiles and Wolf judged that the apparently very high population turnover rate (estimated at about 50 percent annually) and consequent short average lifespan of these hummingbirds make ascendance up the dominance hierarchy unlikely to be clearly related to age. Participating in lek behavior thus probably increases any male's lifetime reproductive success.

Each male's lek territory consisted of up to five song perches and the area around and between these perches. These song perches were the focal points of male activities. They were usually located 1.5–2 meters above ground, typically on bare twigs surrounded by small open spaces amidst thick vegetation, providing both for excellent vision and sufficient room for aerial displays. The territories seldom contained flowers that were used as nectar sources, and thus nearly all feeding was done off-territory. Flowers occurring on a male's territory were not defended interspecifically or intraspecifically. Other males were sometimes permitted to fly though the resident's territory, but could use a song perch or sing there only if the owner was absent.

Several vocalizations are uttered by males at the lek, of which the advertisement song is the most characteristic. This species' "song" is a simple and monotonously repeated single phrase, which is often combined with vertical tail pumping, especially when an intruder approaches (Figure 37E). Songs were uttered about 60–70 times a minute when singing alone, and 90–100 times a minute when intruders approached. The entire song sequence may be continued for up to as long as 30 minutes. The rate, volume, uniformity, and rhythm of the song provide an acoustic index to a male's territorial status, general aggressive state, and also his immediate stimulus input. A male might spend half of his daytime hours on the lek early in the lekking season, and most of that time may be spent in singing. Individual variations in song types were evident, with individuals occupying a particular lek subunit tending to sing essentially the same song type. Apparently each male learns its own song during the first year, and sings that same song type thereafter.

Several other calls or mechanical sounds are also uttered by males at the lek, of which the most significant is perhaps the "bill pop." This is a dry, snapping sound that is produced with a sudden snapping open of the bill during the male's visually impressive "gape display," and is normally produced by resident birds during aggressive encounters at song perches.

The visual displays of the male long-tailed hermit while at its song perch are numerous. One of these is the "float" display (Figure 37A). In this display the male flies parallel to a perched bird while his bill is tilted upward and opened, exposing his throat and gape to the perched individual. The float display is most often performed by the resident male toward an intruder. The "gape and bill pop" display was also most often performed in this context, but sometimes was performed by the intruding individual. It consists of flying toward the perched bird in a shallow upward arc (Figure 37B), with the bill opened at the top of the arc, flashing the orange gape and producing the bill-popping sound. When closest to the perched bird the bill is closed and the head is lowered. This display often preceded or followed a perch exchange. Such exchanges (Figure 37C) consist of a bird circling above and behind a perched bird and supplanting it from its position. This activity was sometimes repeated as many as ten times in quick succession, each time reversing the positions of the two birds. Occasionally mutual hovering (the hover-up display) occurred as the birds faced one another while hovering some distance apart, apparently representing a high-intensity

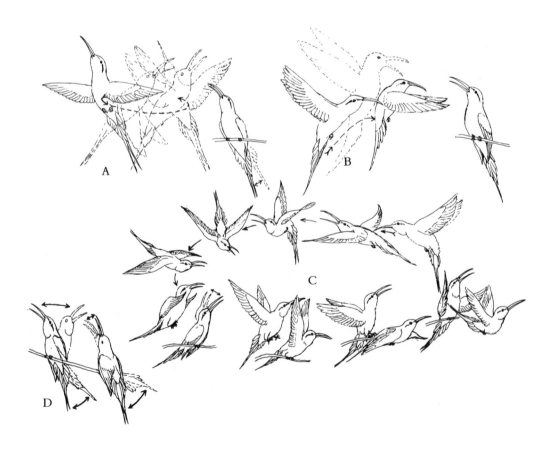

aggressive interaction. Direct chases by the resident bird until the intruder has left the territory were also common.

A side-by-side display (Figure 37D) occurs when a hovering bird alights near a perched one, usually facing in the opposite direction. Each of the two birds then moves the head through a horizontal arc, with the bill lifted and opened widely. The tails are also spread and vigorously pumped through a vertical arc of 60–90 degrees, and the wings are vibrated or fluttered. This display is used in both aggressive and sexual encounters. Copulation occurs when one bird mounts the other, typically immediately after performing a gape and bill pop, a float, or a side-by-side display. During treading both birds twist and vibrate their tails, and the superior bird also vibrates its wings, apparently to help maintain its balance (Figure 37F). Homosexual mountings between males sometimes occur, but these mating attempts are brief. The female's primary means of identifying her sex is thus perhaps simply to remain perched

37. Display behavior of male long-tailed hermit, including float display (*A*), aerial gape and bill pop display (*B*), perch exchange sequence (*C*), side-by-side display (*D*), singing with aggressive tail wagging (*E*), and copulation (*F*). After sketches in Stiles and Wolf (1979). Also shown are two male marvelous spatuletails in aerial display (*G*, after description by Taczanowski and Stolzmann, 1881).

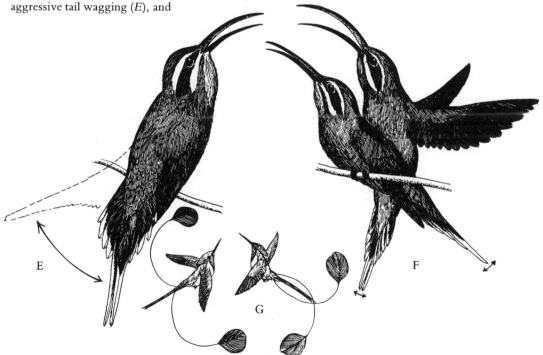

long enough to allow copulation to be successfully completed. No specific female precopulatory displays were evident, although females were never observed to perform the float display or the gape and bill pop sequence. Precopulatory behavior thus seems to differ little from that associated with aggressive male-male encounters. Matings appear to be fairly rare and limited to the lek itself; in the judgment of Stiles and Wolf perhaps an average of only a single female visits a 14-male lek per day for mating. False mountings, involving a male attempting to copulate with inanimate objects such as leaves, are apparently fairly common in this species as well as in other hermit hummingbirds.

The only other hermit that has been fairly intensively studied in terms of its lekking behavior is the green hermit (Snow, 1974; Harger and Lyon, 1980). This species ranges from Costa Rica to Peru, and occurs in montane forests at elevations slightly above those used by the long-tailed hermit. The sexes are slightly dichromatic as adults, with young males closely resembling females. It is slightly heavier than the long-tailed hermit (adults averaging about 7 grams), and male territories also average slightly larger, or about 250 square meters. Within each male's territory several perches are regularly occupied and strongly defended, and the intensity of male activity at the lek largely coincides with the chronology of the breeding season. Up to about 50 percent of the daytime hours may be spent by a male at the lek. Each year a new group of males, evidently consisting of yearling birds, was observed to join the lek, but about 70–80 percent of the resident birds seemed to be adults. Five of six adult birds observed on the lek by Snow in early 1959 were still there and still occupying the same territories several years later. This observation, plus the 20–30 percent apparent annual recruitment rate, suggests a much lower population turnover rate in green hermits than was estimated by Stiles and Wolf for long-tailed hummingbirds, and a lower, perhaps more typical rate of annual mortality (10–30 percent), inasmuch as it may require two years for the fully adult male plumage to be attained.

The displays and songs of the green hermit are similar to those of the long-tailed. The usual male song is a brief, monosyllabic *waatch* or disyllabic *waatchee* note, with the latter song type being louder and seemingly more typical of adult birds. Minor acoustic variations occurred within local lek groups, even among groups located only about 50 meters apart. The song is uttered from low perches on the territory at an average rate of about 50 songs per minute (when no special stimulus is present) to more than 70 (when a female is present). Other calls are also uttered by males, such as during aggressive chases or more generally while in flight.

The two major visual displays observed among males were the "tock display" and the "gape." The tock display is aerial, involving quick lateral dashes in front of a visitor of either sex, accompanied by gaping and uttering of a mechanical *tock* note. These notes are uttered in synchrony with gape flashing at each turning point during the lateral dashes, with the notes thus occurring in rapid-fire machinegun-like succession. This display evidently functions as a precopulatory signal as well as a territorial proclamation display toward other males; in the latter situation the perched bird often alternates with the airborne bird in the course of their territorial contests. Two types of gaping displays were observed, nonaggressive and aggressive. Nonaggressive gaping is performed when a perched male is approached from above, the direction from which females and young males always approach. In the aggressive gaping display the male's gape is directed toward another bird approaching at the same level, normally another male, or at the approach of some other animal species. Although females were often observed visiting the lek, only two attempted matings were seen, and both occurred at an adult male's perch. However, "false matings" with vegetation were very frequently observed.

In the tiny (averaging under 3 grams!) little hermit, the two sexes are indistinguishable in the field, and singing assemblies very similar to those of the two just-described species occur. On one such lek in Trinidad, 23 males were present, and each sang regularly from not only the same twig within its territory but also the same position on the twig. Each of the 23 males could be assigned to one of seven different song types, the song types almost always paralleling the spatial groupings of the males within the larger lek assemblage (Wiley, 1971).

In several non-hermit tropical hummingbird species lekking behavior has also been reported, although it would seem to be distinctly less frequent and is less well studied than in the hermits. For example, the crimson topaz (*Topaza pella*) of northern South America is a common lowland forest bird of primary tropical forests, in which there is very marked sexual dichromatism, with males being spectacularly patterned with iridescent coloration. Davis (1958) observed at least 20 males of this species, plus a female or immature male, at a lek in Guyana. This lek was situated along a river channel and extended for about 50 meters, with the birds perched near the tops of rather tall trees. At that distance, few details of postural display could be seen, but a mechanical, repeated song note was heard.

Apparently a leklike arena behavior occurs in at least one other brilliantly and remarkably plumaged species of South American hummingbird, the marvelous spatuletail (*Loddigesia mirabilis*). In adult males of this rarely observed Peruvian species the two outermost rectrices are racket-shaped, and additionally their shafts are curved in a broad semicircle so that they cross over doubly (thus their enlarged tips are situated on the same sides as their points of insertion). The flight feathers also have enlarged quills that evidently allow for clicking noises to be produced in flight, as in some manakins. In females and immature males the spatulate condition of the tail is only slightly evident, and the plumage is much less iridescent. Groups of both mature and immature males evidently gather to display before females. Males hovered in pairs or larger groups opposite each other, their bodies positioned vertically. With tails spread they then flew from one side to the other, simultaneously producing clicking sounds (Figure 37G). Sometimes one bird would "hang" below a branch as another performed, only to suddenly change places with the performer, again in a manner somewhat resembling manakins. During display the remarkable racket-shaped outer tail features were spread to varying degrees, and at times their spatulate tips were bent forward so that they were almost in line with the bird's iridescent violet crown (Taczanowski and Stolzmann, 1881).

Only a few reports of possible lekking behavior have been provided for North American hummingbirds, and these also involve species that show high levels of sexual dichromatism and normally exhibit well-dispersed territories. Barash (1972) briefly observed a leklike situation involving the broad-tailed hummingbird (*Selasphorus platycercus*), during which three males were territorially situated within about 7 meters of one another in a Colorado pine woodland. The males occupied identical locations over a period of at least four days, and aggressively interacted with one another as well as displaying sexually toward a female that appeared once.

Tamm, Armstrong, and Tooze (1989) described what they regarded as an exploded lek territorial pattern among calliope hummingbirds (*Stellula calliope*) in British Columbia. They believed that females visited the males specifically for mating purposes rather than to forage within their territories, since females were sometimes observed visiting male territories that contained no profitable flowers. However, these authors found no evidence that female mating preferences were influenced by possible individual differences in male display rates or by apparent male territorial quality, as judged by relative available nectar supplies within their territories. Dive displays were directed by territorial males toward all nearby females, but rival males typically were directly chased and rarely were dived at. This observation suggests that male diving behavior, with its attendant sounds and often species-specific visual components, may be an important part of courtship display. Nevertheless, male calliope hummingbirds have at times dived at me when I intruded into their territories, and diving behavior has also been observed as an apparent aggressive territorial display among several other North American hummingbirds (Johnsgard, 1983b).

The Australian Lyrebirds and Their Forest Courts

Two species of endemic Australian birds are among that island-continent's most fascinating creatures. These are the superb lyrebird (*Menura novaehollandiae*) and the Albert's lyrebird (*M. alberti*), which together comprise the distinctive "lower" (suboscine) passerine family Menuridae. Beyond the remarkable tails of adult males, both species are notable for their complex and spectacular advertising displays and their equally complex and spectacular vocalizations, which are famous for their mimicking characteristics. Avian vocal mimicry is not of direct concern in this book, since it is not yet known to bear directly on the question of individual male fitness with respect to mate choice by females. However, inasmuch as it might provide a mechanism for producing more effective individuality during male advertising, and to the extent that it may increase either the acoustic range, amplitude, or territorial "valence" of a particular male, it should not be wholly overlooked in the context of intersexual and intrasexual aspects of sexual selection.

The superb lyrebird is rather large for a passerine bird, with adult males sometimes exceeding a meter in total length, over half of which is comprised of the tail. Adult birds reportedly weigh about as much as a red junglefowl (*Gallus gallus*) (ca. 1 kilogram), but specific comparative mass data are seemingly still unavailable. The tail consists of 16 rectrices, of which only in adult males are the outer pair elongated and lyre-shaped. These lyrate feathers are mostly dark brown above, but exhibit golden brown and black markings on a conspicuous silvery white background on their undersides. There are 12 highly filamentous white feathers between the two lyrate ones. The central pair of feathers, or "medians," are very narrow, considerably longer than the filamentous ones, and are brownish black above and silvery below. Apart from the female's much less well developed tail, the sexes are similar in appearance, and juvenile males closely resemble females. At three years of age young males still exhibit rather femalelike rectrices, but in succeeding molts these feathers become increasingly like the adult male's. By the fourth or fifth year their inner rectrices are becoming partially filamentous, but a fully adult tail may not be attained in males until their seventh or eighth year. Some wild males are known to have survived more than 20 years, and the remarkably small clutch consists of only a single egg (Smith, 1968), suggesting that average mortality rates in lyrebirds must be very low.

The mating system of the superb lyrebird has often been debated. The work of Kenyon (1972) indicated that the males may maintain simultaneous polygynous pair bonds (harems). However, Lill (1979) believed Kenyon's evidence was inadequate for this claim, and judged that pair bonding either is lacking or at most involves a very limited time commitment on the part of the male. Certainly the male plays no known role in nest defense or chick feeding.

One male studied by Kenyon in Sherbrooke Forest, Victoria, had a territory that during a three-year study period enclosed the separate but overlapping "territories" (home ranges) of three marked females, plus that of at least one and perhaps two other unmarked females. Near the middle of the male's territory were his favorite display mounds, which also served as copulation sites. Within one male's territory more than 90 display mounds were plotted. Later, Robinson and Frith (1981) reported that two other males in this same general area had territories that respectively encompassed the ranges of four and six females. These same authors mapped the territories of nine male lyrebirds in another location near Canberra, and estimated them to have an average size of 2.4 hectares (range 0.9–3.7) and an average of 42 display mounds (range 20–76) present. Territories typically extended from a creek upward to a ridgecrest and often had naturally defined boundaries, such as tracks or trails. So far as males are concerned such territories must certainly be resource-inclusive, but female home ranges evidently do not coincide with the boundaries of male territories.

Within their territories males establish cleared and scratched areas of soil about 1–2 meters in width and up to about 1.5 decimeters high, on which they display and sing. These male-altered advertisement sites thus readily qualify as "ground courts," or perhaps might be better described as "display mounds" as earlier defined. The birds may also occasionally sing from tree perches, rocks, or fallen logs. Some display mounds are used for several years in succession, whereas others may be abandoned after only a few weeks. The total number of active mounds is greatest immediately prior to the egg-laying period, but some male mound-scratching activity extends nearly throughout the year. Mound construction tends to be concentrated along the more open ridges, where males can challenge rivals on opposite sides of the valley by loud singing. However, females tend to spend more of their time in lower and wetter areas, where food is more abundant and where more suitable nesting sites occur. Males seemingly choose sites for displaying that not only provide acoustic advantages but also are fairly cool and have a dense ground cover. Robinson and Frith surmised that females might be attracted to the scratched earth of the mound because they associate it with food, and copulations probably also normally occur at or near display mounds.

Males regularly advertise on their display mounds during a period of about four months. Throughout that period they often display and call several times a day, depending largely upon the weather. At the beginning of the display the male raises his tail from its normal position (Figure 38B), simultaneously spreading the outer lyrate feathers and bringing the filamentous and median feathers forward and progressively

38. Display behavior of male superb lyrebird, showing various stages of tail erection and tail spreading. After photos by L. H. Scott.

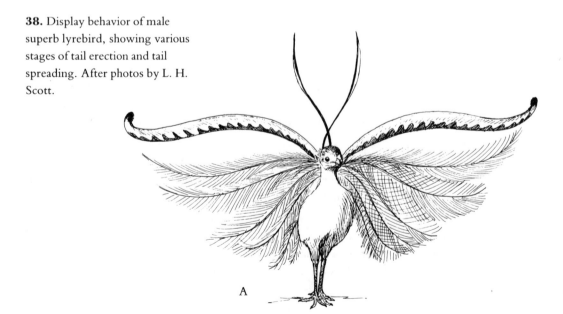

A

over his back (Figure 38A). Eventually his entire body and head are effectively hidden by the veil of overhanging filamentous plumes (Figure 38C). Standing in this posture, the male performs a loud, highly variable array of vocalizations. These include not only a basic lyrebird vocabulary but also "borrowed" notes and sounds of many other birds, and rarely even incorporate sounds from nonbiological sources. Males are especially prone to incorporate those notes that are relatively loud and clear, which probably serve particularly well for advertisement.

A male may sing for four hours or more per day during the peak of his display activity, which adaptively corresponds with the peak period of egg laying. A male's first songs are typically uttered at dawn, from a nocturnal tree roost high in his territory. He then gradually moves downhill through his territory, singing successively from a variety of display sites (Robinson and Frith, 1981). Apparently young lyrebirds gradually learn their songs from older birds, and thus local song characteristics often develop. As compared with typical passerine birds, lyrebirds have relatively simple syringeal structures and associated musculature (Figure 39E), so their diverse song vocabularies are evidently more dependent on individual learning and memory abilities than on complex syringeal anatomy.

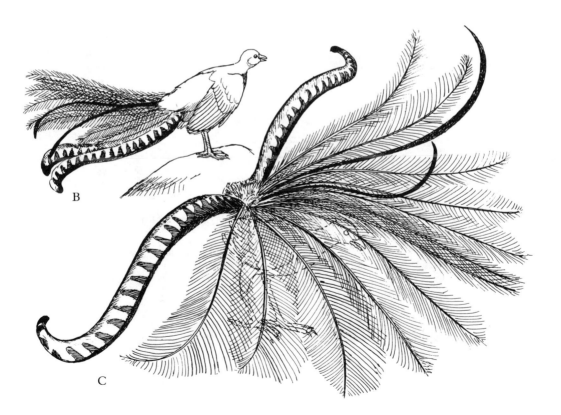

39. Display behavior of Albert's lyrebird, including side (*A*) and rear (*B*) views of male singing. After photos in Curtis (1972). Also shown are rear views of outermost rectrices of Albert's (*C*) and superb (*D*) lyrebirds, and syringeal anatomy of superb lyrebird (*E*). After Garrod (1876).

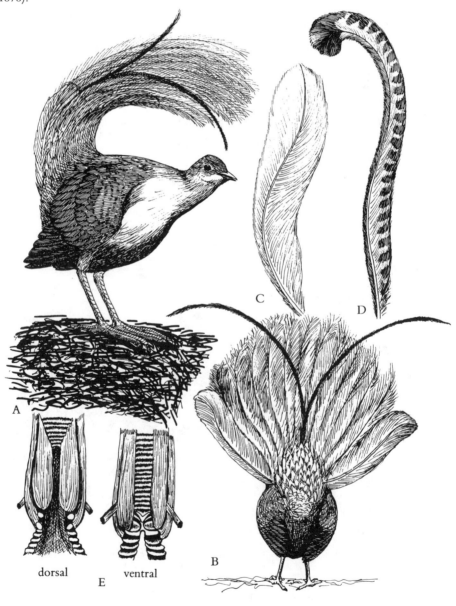

dorsal　　ventral

E

A

B

C

D

Nothing can yet be said with certainty about the possible differential attraction of females to individual males according to their age, vocal complexity, or territorial attributes. However, it is very probable that some kind of individual fitness gradient is present among males in this species, given the species' apparent average longevity, the high degree of individual variation in male songs, and the equally great variations in territorial size and number of display mounds known to be present among males in wild populations.

The other species of lyrebird, the Albert's lyrebird, is far less well studied than is the superb. Although similar in general size and appearance, the tail of the adult male Albert's lyrebird is much less specialized. This is especially true with regard to the length and patterning of the outermost rectrices, which in the Albert's are shorter than in the superb, are scarcely modified from the adjacent ones in shape, and are unpatterned (Figure 39C, D). Similarly, as compared with the superb, the 12 filamentous plumes in the Albert's are also much shorter, are fully webbed for a greater proportion of their total length, and are distinctly curved upward throughout. The paired narrow median feathers are quite similar in shape and structure in the two species, but also are somewhat shorter in the Albert's (Curtis, 1972).

Unlike the superb lyrebird, the Albert's does not produce scratched-out display mounds. Rather, the male typically uses a preexisting display site usually consisting of several ground-level vines, or a mixture of vines and fallen branches that are seemingly lying at random on or near the ground and that cross over each other so as to form a crude platform. The vines often lie slightly above the ground, but usually can be depressed by the bird's weight to touch it. The soil around and between the vines may be variably scratched and thus somewhat modified. At times the male may also occasionally call from rocks, logs, fallen branches, or bare ground, as does the superb lyrebird. Also like the superb, male Albert's lyrebirds may sing for as long as four hours a day at the peak of the display season, with single sessions sometimes lasting for more than an hour. The male's vocalizations are loud and prolonged and, like those of the superb, often include the mimicking of other species' vocalizations. The singing season extends from about May to August, centering on the June-July nesting season.

The posture assumed by a male Albert's lyrebird during his song display is much like that of the superb lyrebird. His tail is initially lifted to the vertical and the outermost rectrices spread about 120 degrees, so that the more filamentous feathers sweep forward almost horizontally above his head (Figure 39A). When seen from behind, the male's bright rufous under tail coverts are quite conspicuous. The silvery undersides of his filamentous feathers are also quite evident, and his narrow median feathers extend upward in graceful curves (Figure 39B). His tail is then more fully spread to cover about 180 degrees and is drooped sufficiently to cover the top and front of his body, the filamentous feathers forming a feathered dome that is tipped with a lacelike curtain. In this posture he sings and also vibrates his tail occasionally, producing a shimmering visual effect. During the last stage of his display the male regularly performs prancing movements that cause the surrounding vegetation to oscillate or vibrate. Curtis (1972) speculated that the use of a vine platform, largely surrounded by screening vegetation, may announce the bird's general location through the shaking of nearby vegetation. However, a stalking predator may also be less likely to locate the precise position of the displaying male because of his relative invisibility. Given the effective vegetational screening and reduced tail feather specializations, the primary advertisement in this relatively inconspicuous species is probably achieved through acoustic rather than visual signaling. As in the superb lyrebird, copulation probably occurs at or near the display platforms.

In summary, the hummingbirds, kakapo, and lyrebirds represent three very different but nevertheless ecologically appropriate approaches to male advertisement efficiency. The hummingbirds are primarily diurnal and use mostly visual, often highly colorful or iridescent signals in well-illuminated habitats. The kakapo is primarily nocturnal and is visually inconspicuous, the males using mainly low-frequency and far-carrying acoustic signals for their advertising. The lyrebirds are predominantly diurnal and forest-dwelling, with males using both acoustic (efficient for distant signal transmission) and visual (for close-up signaling) displays. In contrast to the arena-forming birds so far discussed, in all three of these groups the young are altricial rather than precocial at hatching. However, male emancipation from parental duties is probably facilitated in all of them by the evolution of reduced clutch sizes of one egg (lyrebirds) or a maximum of two eggs (kakapo and hummingbirds). There are also unusually long incubation periods (2–3 weeks in hummingbirds; 6 in lyrebird) and nestling periods (about 3 weeks in hummingbirds; 6 in lyrebird). Long incubation periods (slow embryonic growth rates) may allow females more off-nest foraging time, and by having no more than two nestlings to feed as well as slow nestling growth rates (reduced daily nestling food demands), females alone can handle all parental responsibilities. This allows males to spend nearly all their nonforaging time establishing, defending, and advertising their mating station territories.

In all the groups of arena birds still to be considered, the young are also altricial. However, parental foraging demands are typically ameliorated in them by the evolution of diets involving plentiful foods (fruit or seeds) that can be obtained by females without time-consuming and perhaps unproductive hunting. The groups still to be considered also are entirely tropical or subtropical in distribution, with associated long breeding seasons and warm nights. Such conditions reduce the ecological demands on females for continuous brooding of newly hatched young, and also allow for leisurely fledging periods. Except for the relatively prolific finches, the clutch sizes in these groups are generally of only 1–2 eggs, and both incubation and fledging periods often tend to be unusually long. Again excepting the finches, delayed sexual maturity in males as well as relatively low adult mortality rates are probably also fairly common if not yet documented.

8

Cotingids: Forest Bells and Feathered Umbrellas

Among the groups of birds that engage in arena behavior, few are more diverse in appearance than the cotingids. This colorful group of tropical New World birds lacks a common name and includes such remarkable and exotic-looking forms as the bellbirds, the umbrellabirds, and two cock-of-the-rock species. Indeed, this family of "lower" passerine birds is so variable anatomically that its taxonomic limits are still obscure. Some species usually considered as cotingids perhaps better belong with the manakins (Pipridae) or the tyrant flycatchers (Tyrannidae). All three of these groups are taxonomically regarded as "suboscines," which are characterized in part by a less complex syringeal musculature than occurs in the oscines, or typical passerine songbirds. The syringes of cotingids generally lack intrinsic musculature, and, in contrast to the closely related manakins, exhibit a rather low level of anatomical variation (Prum, 1989).

Snow (1982) has suggested that the highly diverse suboscine birds now occurring in South America originated from primitive stock that was already present when South America split from Africa more than 100 million years ago. According to him, one subgroup of these primitive passerine birds, the tyrannoids, evolved into a group of fruit- and insect-eating birds, mostly occupying the canopy and middle layers of the tropical forest. From this stock, the insect-eating specialists eventually became the modern-day tyrannid flycatchers. Among the fruit eaters, the birds that adapted to eating the smaller fruits of shrubs and lower trees differentiated into the present-day manakins, and the birds that adapted to the larger forest fruits evolved into the modern cotingids. This last-named group expanded geographically and ecologically into the tropical and temperate forests of the Andes during Cenozoic times.

Of these three groups, the insect-catching tyrannids are almost entirely monogamous, with males sometimes participating in incubation as well as regularly assisting in caring for the young. Nonmonogamous matings in this family are evidently limited to only a few genera (*Myobius, Terenotriccus, Oncotoma, Rhynchocyclus, Colopteryx,* and *Pipromorpha*), whose females construct roofed nests in which the young can remain dry while their mothers forage. In manakins and cotingids, however, the fruit-eating syndrome and the capacity of females to regurgitate fruit at the nest has

165

apparently more fully emancipated males of at least some species from parental responsibilities, with resultant increased potential for nonmonogamous matings and
arena-forming behavior (Snow and Snow, 1979).

According to David Snow, mating systems in the cotingids range all the way from
monogamy to polygamy and promiscuity. In the nonmonogamous species, males
may display either solitarily or in aggregated or leklike groups. Of the latter category,
the cocks-of-the-rock (*Rupicola*) are by far the most famous as exhibiting "classical"
lek behavior. Other variably studied species, including the two species of red-
cotingas (*Phoenicircus*), the calfbird (*Perissocephalus tricolor*), and the red-ruffed fruit-
crow (*Pyroderus scutatus*), probably also should be regarded as typical lekking species,
but the screaming piha (*Lipaugus vociferans*), bellbirds (*Procnias* spp.), and umbrellabirds (*Cephalopterus* spp.) may be better described as having a more dispersed or
exploded lek male spacing pattern. The rufous piha (*L. unirufus*) might also be regarded as having an exploded lek pattern (Skutch, 1989). The aberrant sharpbill (*Oxyruncus cristatus*), sometimes taxonomically included in the Cotingidae but often placed
in a monotypic family, reportedly exhibits exploded lek behavior (Stiles and Skutch,
1989). The tyrannid-like ochre-bellied flycatcher (*Pipromorpha oleaginea*) likewise
shows exploded lek behavior (Skutch, 1960; Snow and Snow, 1979).

Degrees of sexual dimorphism and sexual dichromatism vary greatly in the cotingids, and this variety is perhaps at least in part related to the diversity of mating systems present in it. Weight data for the species that appear to qualify as either classic lek
or exploded lek species are still quite limited. However, wing length measurements
suggest that the greatest degree of sexual dimorphism occurs in the umbrellabirds
(which have almost no sexual dichromatism but show marked sexual differences in
adult plumage structure and in tracheal anatomy), and in the four species of bellbirds
(which, contrariwise, show high degrees of adult sexual dichromatism but reduced
plumage structure differences).

As to the cocks-of-the-rock, extremely limited weight data suggest that the male:
female mass ratio may be about 1.2:1. In these classic lek-forming and carotinoid-rich
cotingids there is also a high degree of sexual dichromatism, as well as marked sexual
differences in adult plumage morphology. In these ways the cocks-of-the-rock represent an extreme behavioral and structural type among the cotingids, and at times they
have been placed in a separate family, based on some additional internal anatomical
peculiarities.

In the similarly carotinoid-rich red-cotingas, considerable sexual dichromatism
likewise occurs. However, sexual dimorphism in the red-cotinga genus is reversed,
the adult females having significantly larger measurements than males and perhaps
weighing about 20 percent more. David Snow speculated that this aberrant genus
may behaviorally as well as structurally be allied with the manakins rather than with
typical cotingids, in that the males probably perform aerial maneuvers during display
that may result in selection for small body size. Males also have specialized primaries

that probably generate species-specific wing noises in flight, as similarly occurs in the cocks-of-the-rock.

In the calfbird, one of the largest of the cotingids, there is a reduced degree of adult sexual dimorphism, and furthermore virtually no sexual dichromatism exists in these rather dull-colored, predominantly coffee-brown birds that are named for their calf-like calls. In the red-ruffed fruit-crow and the screaming piha there appears to be little or no adult sexual dimorphism (based on very limited weight data). There is also very little sexual dichromatism in these birds; in the case of the fruit-crows both sexes are rather brightly patterned, while in the screaming piha both are mostly an inconspicuous gray.

In part because of rather limited information, it is impossible to track the progressive changes in cotingid behavior and spacing patterns from monogamous, inclusive-territory species at one extreme to promiscuous species with mating station territories at the other. Nevertheless, it is possible to find some representative types that seem to lead toward the latter extreme.

The Exploded Lek Species of Cotingids

The bellbirds are a group of four medium-sized (100–250 grams) cotingids that are highly specialized for fruit eating. They are notable for their extremely wide gapes, which may not only be very useful for swallowing large fruit, but certainly also play a visual signaling role during social display. Adults are sexually dichromatic, and are also extremely diethic with regard to sound production. Although the females are perhaps silent, the males utter clear, bell-like or hammer-and-anvil-like calls that are sometimes audible for up to a kilometer or more, and that Snow (1982) believed to be the loudest of all avian vocalizations. In all four bellbirds the adult males have one or more curious extendable wattlelike structures (present in three species) or bare throat skin (in one species) (Figures 40, 41). Adult males of all the species are variably white in plumage (one is entirely white), whereas the inconspicuous females are mostly forest-green above and streaked with yellowish and green below. In the white bellbird (*P. alba*) and the three-wattled bellbird (*P. tricarunculata*) curious elongated caruncles emerge from the base of the bill (Figure 41C, E). However, in the bare-throated bellbird (*P. nudicollis*) the male's throat is bare and colored a brilliant blue (Figure 41D), and the bearded bellbird (*P. averano*) has only rudimentary feathers among an array of short and stringlike blackish throat wattles (Figure 40). In males of both the latter two species there is a strange patch of bare pink skin on the lateral side of the tibia. At least in the bearded bellbird it is likely that three years are required for males to attain their full plumage.

The displays of the bearded, white, and three-wattled bellbirds are now fairly well known, mainly owing to the work of Barbara Snow (1961, 1970, 1977b), and she has also observed (1978) to a limited degree the social behavior of the bare-throated bell-

40. Behavior of male bearded bellbird, including display jump postures (*A, B*), calling posture (*C*), submissive posture (*D*), wing–up during display-preening and thigh exposure (*E*), and precopulatory posture (*F*). After sketches in Snow (1970) except for *C,* which is from a published photo.

41. Behavior of bellbirds, including changing place and calling by three-wattled bellbird (*A–C*), calling by male bare-throated bellbird (*D*), and various stages of wattle elongation by male white bellbird (*E*). After various sources, including sketches in Snow (1982) and (*D*) photo by author.

bird in captivity. David Snow effectively summarized (1982) the available information on the biology of the bellbird species.

Male bellbirds of all species occupy territorial sites high in trees above the canopy, usually perching on the topmost dead branches. They spend most of their daytime hours there at least during the breeding season, leaving their perches only briefly to feed. From these sites they utter loud territorial calls. In the white bellbird this call is a bell-like *ding-ding,* but in the other three species it is a less pure and perhaps more resonant *bock* note, similar to that produced by a hammer striking an anvil. The bill is opened widely during calling, perhaps thereby maximally exposing the opened gape as well as drawing attention to the outer surface of the throat.

Male bearded bellbirds may call from as close as about 45 meters from other males (Snow, 1970), but probably in most cases the birds are much farther apart. Snow concluded that typically a group of males may compete for a particularly favored territory, which invariably contains a traditional mating site, namely a specific branch of a sapling or "visiting perch," to which females regularly come for copulation. One dominant male, probably an experienced and older bird, is able to maintain this territory and its favored mating site only by his constant advertisement calling and successful aggressive interactions with other males. Other males, usually subadults and subordinate adults, gather close to this highly favored location and constantly try to take it over whenever the circumstances allow. Temporary takeovers may occur when the dominant male is in full molt, but these interlopers are quickly expelled at the end of the dominant male's molt. Thereby an aggregation of males apparently develops, and an exploded lek breeding system is attained.

Much of the male bearded bellbird's time is thus spent in calling from a conspicuous site high in the tree (Figure 40C). When another bellbird lands on the dominant male's territory, the resident male leads the visitor down toward the visiting perch of the display sapling. Like the other bellbirds, the dominant male bearded bellbird then performs a short flight display or display jump, which consists of a silent flight or jump from the visiting perch of the display sapling to an adjacent perch slightly over a meter away. The bird lands in a crouched position, with his tail fanned (Figure 40A). He remains there for about eight seconds, and then quickly turns and leaps back. While thus perched, he closely watches the visitor, often turning his head to do so (Figure 40B). If the visiting bird has reached the display sapling, the resident male moves to the mating branch and begins to display-preen his underwing surface, making brief preening movements of the under wing coverts or axillary area (Figure 40E). During this display the wing facing the visiting bird is raised and preened, and the corresponding leg (with the bare skin patch) is exposed to view. He also lowers his head between preening movements, seemingly to display his brown crown. Finally, the male stops preening and stares intently at the visitor. If the visitor is a female she may now move to the inner end of the mating branch, but if it is another male he remains in the upper branches of the display sapling or in another tree. He typically then soon flies off, perhaps after assuming an upright and thin-bodied submissive stance (Figure 40D). Once a female reaches the mating branch she is watched intently by the crouching male (Figure 40F). He may then jump toward her with a loud, explosive *bock* note, landing either on her back or, if she has moved, on the place she had just previously occupied. If mounting is allowed, the female normally crouches to receive the male, but if not she may move to another branch or another tree. Probably females visit all the calling males in any single area before selecting a mating partner.

Actual mating by bellbirds has been described only for the bearded bellbird, but at least the three-wattled and white bellbirds have similar premating displays that are directed toward visiting females at their perches. All four species are known to perform display jumps or display flights. In the three-wattled bellbird this display is a fairly long flight, inasmuch as the male's perches are located 2–5 meters apart. In this species the flight is preceded by a loud multiple call that lasts up to about five seconds. Additionally, the three-wattled bellbird has a "changing place" display (Figure 41A, B), which is performed either at his high song post or at the lower visiting perch. The male lands on the perch in a crouched position, and with his tail spread (Figure 41A). If the visiting bird occupies a position closer to the base of the branch than where the owner has landed, the latter quickly flies over and changes place with the visitor (Figure 41B). The owner then moves close to the visitor and calls in its ear with an extremely loud *bock* call with his black gape widely exposed (Figure 41C). If the visitor is a male, it typically then either flies off or even falls off the end of the perch, although it may soon return and repeat the entire sequence. During such calls the male's wattles are greatly extended, and are made conspicuous by occasional head-shaking movements (Snow, 1982). According to Stiles and Skutch (1989) the male's loudest calls may be heard for at least half a kilometer, and vary somewhat with locality. After singing, the bird may fly out a short distance, turn, and return to his perch, where he briefly spreads his tail and retracts his neck.

In the white bellbird the male's visiting perch is a horizontal vine or bare branch. When the male displays before a visiting bird here, he crouches and shakes his wattle silently (Figure 41E). This wattle increases considerably in length during display activity, and at least in Snow's observations apparently always hangs down the right-hand side of the beak, whereas the visiting female cautiously approaches from the left. When the female is fairly close, the male calls with a loud, double and bell-like *kong-kay* call, and simultaneously swings his body sharply toward the female. This evidently is a premounting movement, which typically causes the female to retire. Apparently keeping the wattle on the right prior to the body-swinging movement prevents it from getting in the way and perhaps entering the opened mouth. Snow (1976) suggested that perhaps the double vocalization, the two parts of which are acoustically quite different, may be the result of the two halves of the syrinx operating independently.

The umbrellabirds, another genus of unique cotingids, are usually placed in three

42. Male umbrellabird traits,
including trachea (after Sick,
1954) and elongated wattle of
long-wattled umbrellabird (*A*),
plus resting (*B*) and display (*C*)
postures of bare-necked
umbrellabird. After Crandall
(1945b). Also shown are pouter-
pigeon (*D*), moo-call (*E*),
fluffed-up (*F*), and motionless
(*G, H*) postures of male calfbird.
After sketches in Snow (1972).

species that geographically replace one another. They are similar in body size, the adult weights of about 350–450 grams making them perhaps the largest of the cotingids. In all species both sexes are a glossy bluish black, with the males being distinguished from females by their umbrella-like crest and a pendant throat pouch or "wattle" (Figure 42A–C). Unlike the wattles of the bellbirds, this structure (presumably consisting of the anterior esophagus) is clearly inflatable, and was described as an "air sac" by Crandall (1945b). In two species, the long-wattled (*C. penduliger*) and the Amazonian (*C. ornatus*) umbrellabirds, the "wattle" is entirely covered with black feathers. However, in the bare-necked umbrellabird (*C. glabricollis*) the throat area is nearly entirely bare of feathers and bright scarlet in adult males (females have a smaller orangish skin patch), with a tiny tuft of feathers at its lowermost tip. In at least one (*ornatus*) and probably all species the trachea of the adult male is doubly inflated, and the anterior syrinx has several large external tympaniform membranes (Figure 42A). These specializations of the throat and trachea probably contribute to the low-pitched booming call produced by males by providing air chambers for resonating low-frequency vocalizations. Calling is done from low perches in tall trees; the far-carrying, booming notes of the male are said to resemble those of a bull.

Males of at least the bare-necked umbrellabird appear to be organized in small, exploded leks, which in one case contained four adult males (Cordier, 1943). Each male controlled up to three branches, located only about 5–10 meters above ground, where it spent all its daylight time except when foraging. Before daylight the resident males begin uttering their booming calls, which are preceded (Figure 42B) by a filling of the air sac until it is as large as a large tomato, and the fleshy wattle at its base is extended from its normal length of about 2 centimeters to about 7–10 centimeters. When this air sac is fully distended (Figure 42C) the bird utters a booming call, while leaning forward with the bill closed. The accompanying initial sound is a loud *hoom* note, sounding like a heavy mallet striking a drum. He then throws his head back and opens his bill, uttering a dry, hacking *hik-ratch,* and then quickly brings the head forward again. Lastly he produces a second, softer *hoom* as the air sac is deflated, followed by a final, dry *k* sound, apparently without specific head movements (Stiles and Skutch, 1989). During the first phase of display the head is rapidly moved from side to side, causing the feather-tipped wattle to gyrate wildly, and the forward head movement during the second phase of display causes the umbrella-like crest to fall forward and cover the bill (Crandall, 1945b).

Other aspects of display and mating are still unreported for umbrellabirds, and perhaps there are also some interspecific differences in male spacing and advertisement behavior. Thus, in the Amazonian umbrellabird the male's head movements may be different from those just described, and perhaps only a single booming sound is uttered (Sick, 1954). The male long-wattled umbrellabird also utters a booming call, but details of this are unavailable. However, the general umbrellabird spacing and male advertisement pattern would seem to be rather similar to those of bellbirds, with a greater reliance on low-frequency sounds, as is acoustically appropriate for calls uttered from below rather than from above the forest canopy.

Probably at least one additional cotingid qualifies as an exploded lek species, namely the screaming piha. In this inconspicuously gray and greenish species, which only questionably belongs in the Cotingidae, leks may consist of up to 30 males. Such large leks may be scattered over several hectares, but most leks are smaller at about 4–10 birds. Males are spaced about 40–60 meters apart in rather thick forest, and thus are out of sight of each other but are in auditory contact. They call from perches about 6–16 meters high, well below the forest canopy. Their loud, ringing calls can be heard for some 300–400 meters in the forest, and males occupying adjacent territories may call alternately during the same general singing period, so that one male will often begin a call sequence just as its neighbor is ending its. Up to 75 percent of the daylight hours might be spent in calling, and the rather short silent periods are probably devoted to foraging (B. Snow, 1961).

The Typical Leks of Calfbirds and Red-cotingas

Of the species of cotingids believed to form fairly typical aggregated-male leks, those of the calfbird and red-cotingas (*Phoenicircus* spp.) are among the best studied, although even for these the available information is distinctly limited. Both display in similar forested habitats; indeed Trail and Donahue (1991) reported that a group of eight calfbirds and one of six to ten Guianan red-cotingas (*P. carnifex*) both displayed in separate leks at the same forest site in Suriname. There the calfbirds occupied display sites at about 15–25 meters above ground, and the red-cotingas displayed and foraged at somewhat lower heights of about 8–12 meters. The wholly frugivorous red-cotingas are considerably smaller in mass (ca. 75–85 grams) than the calfbirds, which as adults weigh about 250–400 grams, and have a male:female mass ratio of 1:1. Calfbirds forage both on quite large fruits and also on understory insects, obtaining most of the latter at a height of about 6–10 meters but sometimes coming down to forage in the shrub layer as well. In the red-cotingas the males are predominantly a brilliant red, whereas both sexes of the calfbird are dull-colored. Trail (1990) commented that a sexually monochromatic condition occurs in about 25 percent of all known avian lek-breeding birds, and suggested that in the calfbird this condition has resulted from social competition affecting both sexes, whereby each engages in sexual mimicry of the opposite sex. The calfbird has been studied by Barbara Snow (1972), David Snow (1976, 1982), and Trail (1990).

In one calfbird lek studied by the Snows, the number of males usually present included four probable adults and four immatures. This lek measured about 15 by 25 meters in diameter. Dominance on the lek was shared about equally by two adult birds, each of which controlled about half of the lek's total area. The four adults were present for most of the daylight hours, but the immatures regularly visited the lek only at dawn and dusk. The adult males established perching sites for calling and displaying that were situated about 0.5–15 meters apart and in direct view of one another. These perching sites were in a group of six subcanopy trees that were all of the same species. The perches were about 10–12 meters above ground, near the tops of these small trees, where there were long and horizontal branches that were only sparsely surrounded by vegetation.

In a lek studied by Trail, one alpha male controlled the "core area" of the lek, which consisted of a circular area about 20 meters in diameter around the alpha perch, which is where all copulations occurred and where nearly all female visits took place. As many as five other males competed for this area during female visits, but at other times these birds retreated to more distant perching sites from 26 to 119 meters away and in various directions; the latter did not represent defended territories as such. All the males typically called and postured most intensely around daybreak, followed by sporadic activity until late afternoon when a second but less intense period of display often occurred.

Aggressive attacks and chases among the male calfbirds are frequent on the lek, the alpha male typically driving other males from his perch but unable completely to prevent male intrusion into the core area. Coordinated male displays by males other than the alpha male also occur. These consist of side-by-side displays and antiphonal calls, with one bird calling precisely when the other stops and the two birds also posturing in precise unison, according to Trail. The Snows also reported nonoverlapping calling by the resident males and judged it to be a cooperative activity, although the birds were seemingly in a constant state of tension.

The primary advertisement call and display of male calfbirds is the "moo-call" (actually a three-part *grrr-aaa-oooo*), a vocalization that is begun from a normal perching posture. The bird first leans forward and inhales air while uttering a preliminary growling note. Then he moves back into an upright stance, straightening his flexed legs and uttering a louder *aaaa* note. In this phase he fluffs out his anterior body feathers into a large hood or cowl, and exposes two orange-colored tufts of under tail coverts on either side of the tail (Figure 42E). Then he sinks back down, utters the loudest *oooo* note, and depresses his tail so that the orange under tail coverts move to the upper surface of the tail and contrast with its black surface. Finally he leans back slightly beyond the vertical, simultaneously raising all his anterior feathers but lowering all posterior feathers except the under tail coverts.

Moo-calling is only performed when two or more males are present on the lek, or when another is at least within hearing distance; lone males may utter a "half-moo" that lacks the final lowing note. Adult birds mooing close together orient their backs to one another. At low levels of aggression between males, a partially fluffed "pouter-pigeon posture" (Figure 42D) is assumed. At higher aggressive levels display preening may be performed, or a "fluffed-up" posture (Figure 42F) is assumed in combination with display preening, with all the body feathers raised, and the wings often partly spread or even fully opened.

The most important postural display performed by assertive males is the stationary "motionless" posture (Figure 42G, H), which may be held for up to ten minutes as one male intently watches a rival. In this horizontally oriented body posture the neck is outstretched, the body feathers are sleeked, the wings are slightly drooped, and the tail is slightly raised. From both in front and behind, the orange under tail coverts are highly apparent. Two rival males may spend several minutes within a meter of one another in this posture without actually attacking (Snow, 1972).

Although females closely resemble males, they are relatively silent at the lek and also appear to be quite nervous. When they visit they typically fly directly to the alpha male's display perch, often with several males in close pursuit. They may be chased from the perch by non-alpha males, and additionally females may chase away other females. Copulation occurs on the display perch of the alpha male, either without specific preceding displays or following intense moo-calling by the male, pecking at the male's feathered "cowl" by the female, and occasional self-preening of the wing's ventral wrist feathers (Trail, 1990).

43. Behavior of male black-necked red-cotinga, including relaxed perching (*A, B*), alert postures (*C, D*), bowing while calling (*E–G*), and head bobbing between calls (*H*). After sketches in Trail and Donahue (1991). Also shown (*I*) are the male's specialized primaries. After Snow (1982).

Trail (1990) believed that non-alpha males are able to approach the dominant male on his perch by assuming pseudofemale stances, and additionally females sometimes adopted fluffed-out male postures when displacing other females from the alpha perch. Trail suggested that this species' monochromatism, rather than being a result of "phyletic inertia" (resulting from presumably inadequate time to evolve marked sexual dichromatism) or "behavioral transference" (the shifting of male-limited traits to visual and acoustic displays rather than to bright male plumages), is instead produced by social selection. Trail suggested that some of the selective factors possibly favoring sexual monomorphism in calfbirds include intersexual mimicry, intense interfemale aggression, the lack of a traditional lek site, apparently cooperative display partnerships among males, and unusually brief or "cryptic" copulations. These interacting factors tend to produce intense intrasexual competition in both sexes, and may favor monomorphism and monochromatism rather than dimorphism and dichromatism.

A few other cotingids probably display under typical lek conditions. One, the red-ruffed fruit-crow, is still only very poorly described. However, about 7–10 males apparently gather at tightly packed leks, the males sometimes only a few meters apart,

and utter loud booming or lowing calls. These low-pitched calls are apparently produced while inhaling and inflating a neck or throat pouch, which presumably provides a resonating chamber. Several features of this species' displays and social organization suggest strong similarities with those of calfbirds (Snow, 1982).

The extremely poorly known but spectacularly plumaged crimson fruit-crow (*Haematoderus militaris*) was recently closely observed for the first time by Bierregaard et al. (1984). They reported seeing an apparent male display flight but found no evidence of lekking behavior in the species. They did hear a low-pitched hooting call, suggestive of a long-distance acoustic advertisement display by males.

The other known lek-breeding cotingids are the red-cotingas, which consist of two very similar and mostly or entirely allopatric species. They are still rather little studied, but recent observations by Trail and Donahue (1991) confirm that they are fairly typical lek species. In an area of Peruvian forest measuring about 150 by 300 meters a group of 6–10 males of the black-necked red-cotinga (*P. nigricollis*) regularly displayed, the displays occurring about 8–15 meters above ground in understory trees. In another lek in Suriname two or three male Guianan red-cotingas (*P. carnifex*) displayed in a lek of about 50 by 100 meters. In this case the two resident males occupied slightly overlapping areas of activity, usually remaining at least 20 meters apart, whereas the third male was an irregular visitor. In the larger lek of the black-necked species the display areas seemed to be of about the same size, or somewhat larger.

The displays of these two species are quite similar, but their male vocalizations are quite distinct. Display activity mostly occurred around dawn but sometimes also in late afternoon. Between displays, males variously relaxed or assumed alert postures (Figure 43A–D). All the postural displays seemed associated with vocalizations or mechanical sound production. During their advertising call, the males would bow forward and ruffle their rump feathers conspicuously (Figure 43E–G). The males called from any of 3–5 (in *nigricollis*) or as many as 12 or more (in *carnifex*) favored calling perches, which consisted of horizontal branches or vines situated some 8–10 meters above ground. Between calls, bowing or head pumping also occurred in *nigricollis* (Figure 43H). In both species the most obvious advertisement display was a horizontal flight of 6–20 meters between calling perches. Associated with this flight was a mechanical whistlelike sound having a frequency of about 5 kHz. This sound resembled a cricket stridulation call, and probably was caused by vibrations of the males' highly modified seventh primary feathers (Figure 43I). Calling during flights also occurred in both species. Males of *carnifex* also produced a second wing-buzzing sound that was different in sound characteristics from the species' wing whistle, and that wasn't detected in *nigricollis*. Females were sometimes observed passing through the lek areas, and males then exhibited a higher rate of calling and flight displays, but no direct courtship interactions were observed.

The Cocks-of-the-rock and
Their Private Courts within Leks

There are two allopatric species of cocks-of-the-rock. These include the more lowland-inhabiting (at 150–1,500 meters elevation) Guianan cock-of-the-rock (*R. rupicola*), and the more montane-adapted (at 1,500–2,500 meters elevation) Peruvian cock-of-the-rock (*R. peruviana*). The former species is somewhat smaller, with a considerably shorter tail, but in both species the adult males are bright orange to red, whereas females and first-year males are mostly brown to reddish brown. Weight data are limited, but the two species probably collectively range from about 150 to 250 grams, with males somewhat larger than females. Both species are wholly dependent on rocky habitats for nesting, the nests being constructed of mud and vegetation and placed on diverse rocky sites that include caves, sides of boulders, and bare rock faces. Both species also are highly frugivorous, mainly plucking their foods while in flight. Brooding females carry such fruit, and occasionally also small vertebrates, back to the nest to regurgitate and feed to their young.

Males of the two species are quite different in many plumage traits. The crest of the Peruvian is much bushier and less laterally compressed, and the wing's inner secondaries are square-tipped, gray, and contrast strongly with the male's otherwise blackish wings and long black tail. In the Guianan species the corresponding five inner secondaries are barred brown and buff, like the short and inconspicuous tail, but have extremely broad vanes, with long and silky ornamental barb fringes (Figure 44H). The rump feathers and upper tail coverts are also unusually long, very broad, and, like the inner secondaries, are elaborately fringed, nearly hiding the tail below. In both species, and especially the Peruvian, the outermost primaries of males are highly attenuated at the tip (Figure 44H, I), which produces a distinctive whinnying whistle when the birds are in flight. Only in the Guianan species are the black wings contrastingly patterned with a white speculum near the base of the primaries. In both species a femalelike first-year male plumage is present; these young birds may visit leks but do not hold territories there.

The social and sexual behavior of the Guianan cock-of-the-rock is by far the better documented. However, the social behavior of these two species appears to differ considerably, as will be noted below.

The first detailed behavioral observations on the Guianan species were those of Gilliard (1962), who watched a lek in Guyana that contained three regularly attending males, plus two additional adult males that were present periodically. Several non-adult males and about seven females were also seen. Each of the three resident males maintained a defended territory that included a "privately owned display stage" or ground court, and a similarly "privately owned" low perch. All three resident males confined their activities within a long-oval area measuring about 12 by 22 meters. The

44. Social behavior of Guianan cock-of-the-rock, including territorial perching (*A, C*), ground-crouching (*B*), and precopulatory posturing (*D*). After photos in Trail (1985b). Also shown are male Peruvian cocks-of-the-rock in confrontational perching (*E*) and wing spreading (*F*), plus the outer primaries of male Peruvian (*G*) and Guianan (*H*) cocks-of-the-rock. After sketches in Snow (1982).

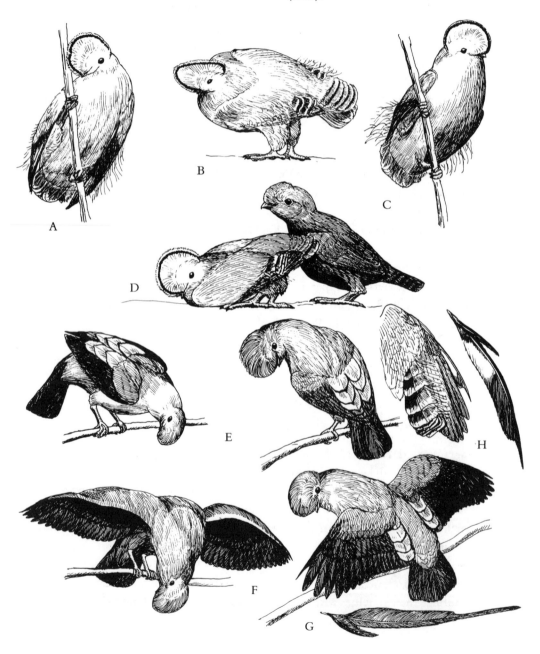

most centrally located of the three resident males had the largest display stage, and this was bounded on either side by those of the other two males, both of whose primary ground display sites were located about 2.5 meters away. Secondary and tertiary display stages and perches were sometimes also used by the males.

The leks observed by Snow (1971, 1976, 1982) were of varied and rather indefinite sizes; the individual male ground courts varied from about 0.5 to 2 meters in diameter. Individual courts ranged from being nearly contiguous to more than 10 meters apart. Courts in active use were kept free of loose vegetational debris, not by the males carrying it away directly, but rather indirectly by their powerful wing-beating action during display at the court.

More recently, Trail (1983, 1984, 1985a, 1985b, 1985c) described his studies in Suriname, where for five breeding seasons he observed a lek having an average of 55 territorial males. Resident territorial males not only dominated their individual ground courts but also controlled access to their display perches. These were often vertically oriented sapling stems or lianas located immediately above the court, within a collective air space roughly 2.5 meters wide and about 3–4 meters high. Trail observed over 400 matings, which occurred either on the ground court or, less frequently, on perches immediately adjacent to it. In one year, 61 territorial resident males were present. Of these only 43 percent successfully obtained matings, and the alpha male alone accounted for 30 percent of the total of 138 matings observed that season.

The specific basis for individual mate choice by females of this species is still uncertain, but Trail found that individually marked females typically appeared at a lek repeatedly over a period of about six days. Such females visited an average of three separate males during this period, but usually (in 76 percent of the cases) only mated with one of them, suggesting that significant female discrimination probably exists. However, interference by other males, including competing courtship efforts by yearlings and especially by territorial neighbors, often (in 28 percent of observed cases) disrupted attempted copulatory sequences. The preferred or alpha male typically occupied a central position in the lek, apparently not because of any topographic or physical characteristics of the lek itself but as a result of male-sampling "rules" seemingly used by females and of male responses to these female tactics. However, it was not clear whether this male clustering effect was caused by subordinate males simply gathering around the seemingly most attractive male (the "hotshot hypothesis") or, as was judged to be more probable, because such male clustering is most suitable for the associated pool comparison tactics used by females (Trail and Adams, 1989).

Vocalizations of resident males include loud, buglelike calls uttered upon the birds' first arrival at the lek. These calls are audible for considerable distances, and were termed the "assembly call" by Gilliard. The similarly loud combined chorusing by males whenever a female appears may be heard for up to 200 meters. Males also produce a variety of other vocalizations, primarily during aggressive encounters. Two mechanical sounds are also produced by males at the lek. One is a bill clicking produced during head bobbing, and the other is the whistling sound generated by the wings during flight or strong wing beating.

Postural displays by males at the lek are numerous, including several that are clearly limited to aggressive male-male interactions. Apart from these, a lateral fanning of the ornamental upper tail coverts, performed either on the court or on display perches, is highly conspicuous and is clearly female-oriented. When these coverts are fully spread, the entire upper surface of the bird resembles a fanlike feathered ornament (Gilliard, 1962). Typically this display is performed on the ground while the male is in a stationary and silent posture, with his upper body surface held almost parallel to the ground, his tail depressed, and his legs flexed. His head may be held upright but commonly is tilted sideways, so that one eye is directed upward (Figure 44B). The purpose of such head tilting is evidently to maximize the female's lateral profile view of the male's crest. When a male is on his arboreal display perch a more upright posture is typical (Figure 44A, C). His wing feathers are then often sufficiently spread to expose their white patterning, or he may beat his wings, especially during advertisement calling. As soon as a female arrives at the lek, the resident males fly down to their terrestrial courts, each landing with a loud call and with strong wing beating. Each male then crouches and remains immobile in the flattened stance just described, moving only to adjust his position in order to optimize the female's view, or perhaps to shift slightly in order to be seen by the female in full sunlight (Snow, 1982).

When a female lands on the court beside the male he becomes even more prostrate and maximally expands his dorsal plumage. He silently maintains a rigid posture facing away from her, with the bill directed downward in a posture rather similar to the ruff's squatting display (see Figure 44D). The female may then peck at the immobile male's fringed feathers or touch his rump area with her breast prior to copulation. However, copulation may occur without these preliminaries. Thus, if the female remains on the court for about 20 seconds or more the male begins a series of short hops that may allow him to circle around behind her and mount. At this time other males often attempt to interfere by their wing beating and calling, sometimes thereby frightening the female away. Copulation lasts about 10–15 seconds, and is accompanied by vigorous wing beating by the mounted male and a chorus of calls from rival males (Trail, 1985b).

The lekking behavior of the Peruvian cock-of-the-rock is evidently quite different, judging from a detailed description by Benalcazar and Benalcazar (1984), which had been summarized earlier by Snow (1982). On a lek in southern Colombia, six males were regularly present at dawn and again during late afternoon. These six males consisted of three rival "pairs," with each pair of males controlling small individual arboreal display sites ("courts") that were closely adjacent to each other. The distance between the courts of adjacent pairs was 6–9 meters, and the entire display area or lek was about 20–25 meters in diameter. One male in the lek was dominant to all of the other five, had the largest court, and also spent more time on the display area than any of the others. His arboreal court was in a tree about 7 meters high, and consisted of horizontal branches that were several meters above ground and that had been plucked nearly devoid of leaves. Only this male was observed to copulate with females, and females visited his court almost exclusively.

Pairs of rival males spent much of their time performing mutual confrontational displays, during which the two birds would face one another and call as they slowly flapped their wings, fanned their tails, and gradually lowered their heads until each bird's dorsal surface formed a regular curve (Figure 44E, F). Members of each rival pair performed such displays almost simultaneously, especially at their common court boundaries. When females arrived on the lek the level of male activities (calling and confrontational displays) was greatly increased. If a female appeared on a resident male's court, the male would chase her from perch to perch within his court area, performing his confrontational display toward her. He then would call and face her from about a meter away. If she remained still he would stop calling, fly to her, and mount her. Copulation evidently thus occurs on branches rather than on the ground. No disturbances from the other nearest rival male were observed at this time; instead this bird would remain silent or display alone in his own court. The difference between this behavior and that of the Guianan cock-of-the-rock is surprisingly great and rather inexplicable for such seemingly closely related species.

9

Manakins: Spectacular Soloists and Dazzling Duets

Other than the incomparable hummingbirds, probably no group of New World birds deserves the title of "feathered jewels" more than do the manakins. This family of mostly South American birds contains at least 40 species, many of which are scarcely known beyond what can be gleaned from museum skins and skeletons. Traditional taxonomies have accepted over 50 species in at least 17 genera, but Prum (1989) believed that of these only 40 species in 11 genera actually constitute the true manakins. The remainder are perhaps closer to the tyrannid flycatchers and the cotingas, which are also "lower" passerine groups that are endemic to the New World. Male manakins distinctly resemble some cotingids in that they typically sparkle both with structural colors, such as the sky-blue backs of the blue manakins (*Chiroxiphia*), and with feather pigment colors ranging from brilliant yellow and red carotinoids to intensely black melanins. Added to these diverse colors are a variety of tail shapes, wing and tail specializations for sound production that include some of the most remarkably modified feather shapes of all birds, and curious vocalizations. Finally, there are a bewildering array of display postures and movements, including somersaults, cartwheel flights, backward jumps, stationary pivots, and forward or backward "dances" along twigs. Certainly in manakin displays there are all the makings for drama, if not fiction.

Nearly all the manakins are sexually dichromatic in adult plumage, with males variably brighter than females. However, in the striped manakin (*Machaeropterus regulus*), whose males reportedly display in pairs, females closely resemble the rather dull-colored males. In two other genera (*Neopelma* and *Tyranneutes*), which very doubtfully belong in the Pipridae (Prum, 1989), the males display in exploded leks, and both sexes are also quite dull-colored. However, these are clearly exceptions to the usual sexually divergent manakin plumage condition. In spite of their usually brighter colors, males are generally no heavier than females. They may even be somewhat smaller in average adult mass, probably as a result of sexual selection for their aerial display abilities.

Manakins are largely frugivorous birds, foraging for fruit near the ground and in the lower forest canopy. They sometimes even pluck it in flight as do many cotingids,

184

45. Behavior of male manakins, including "tense" posture (*A*), normal appearance (*B, top*), grunt-jump (*B, left*), fanning (*B, middle*), and slide-down-the-pole (*B, right*) displays of white-bearded manakin (after Darnton, 1958, and Snow, 1962); upright posture (*C*), backward dance (*D*), joint display (*E*), and landing display (*F*) of *rubrocapilla* form of golden-headed manakin (after sketches in Sick, 1967); duet dance (*G*) and premating display (*H*) by blue-backed manakin (after Snow, 1976); plus duet dancing by long-tailed manakin (*I*) (after photo by G. G. Dimijian, in Foster, 1984).

46. Social behavior of male wire-tailed manakin, showing twisting display with female (*A, B*), side jumping (*C*), multistep side jumping (*D*), and tail-up-freeze display, followed by mutual approach (*E*). Also shown is the male tail structure (*F*). Mostly after sketches in Schwartz and Snow (1978).

but they also consume some insects. So far as is known the males never participate in nest building or brood care in any way (Sick, 1967). In all manakin species so far observed, polygynous matings and arena displays are typical, with male courtship displays having been described to varying degrees for about 27 species (Prum, 1989). Prum concluded that polygynous mating systems have been independently evolved in the manakins, cotingas, and a few tyrant flycatchers, and that few display similarities among them exist.

Several major and comprehensive reviews of male manakin courtship behavior have been written. One of these (Sick, 1959, 1967) was largely organized taxonomically and discussed the sexual displays of about 20 species. Another, by David Snow (1963c), emphasized the evolution of leks and communal displays, as well as display movements. Snow has included synopses of the male displays for 21 species. Most recently a phylogenetic (cladistic) analysis of male display behavior for seven genera and 25 species of manakins has been provided by Prum (1989).

Sick (1967) mentioned many of the basic movements and postures of male manakin displays that were later more fully analyzed and quantified by Prum (1989). These include aggressive or sexual chasing, crouching (Figure 45A), upright postures (Figure 45C), vertical and horizontal sliding (Figures 45E, 46E), body pivoting in place (Figures 46A–B, 47B), wing raising (Figures 45D, 47C), wing snapping (Figure 48C), jumping (Figure 46C), jumping with wing fluttering (Figure 45G), longer flight displays (Figure 49A), about-face displays and similar body reorientations after landing on a perch (Figure 45F), tail feather brushing movements (Figure 46A–B), and other less conspicuous postures or movements. Many of these displays are performed with great rapidity, making their interpretation and details very difficult to analyze without photographic or acoustic equipment. Prum (1989) recognized a total of 44 derived behavioral characters that he judged to be present in courting male manakins but are apparently absent in other tyrannoid birds, and that he used for establishing a proposed cladistic phylogeny. He thus identified several clades of allopatric or parapatric manakin species that were also similarly differentiated by his separate analysis of variations in syringeal anatomy.

In the suboscine manakins the male's syringeal anatomy is not highly complex in its associated musculature (Figure 47D). Vocal sound production by manakins is apparently limited to vibration of the internal tympaniform membranes, which evidently operate interdependently. However, there is a high level of intraspecific and interspecific diversity in manakin syringeal structure, including major variations in relative muscle development (Prum, 1989). These differences must have large acoustic effects. Sick (1967) judged that the male's advertising call is the most important of manakin auditory signals. Among the "dancing" species of manakins the male's advertising call typically ceases when the dancing display begins, replaced then by postural displays and a variety of mechanical sounds.

47. Displays of male red-capped manakin, including starting posture (*A*), pivoting phase (*B*), and backward dance sequence (*C*). After photos in Crandall (1945a). Also shown are the trachea (*D,* after Garrod, 1876) and male wing (*E,* after Chapman, 1935) of the golden-collared manakin, plus the outer primaries of yellow-thighed (*F*), gray-headed (*G*), lance-tailed (*H*), white-collared (*I*), and white-ruffed (*J*) manakins (after drawings in Ridgway, 1907).

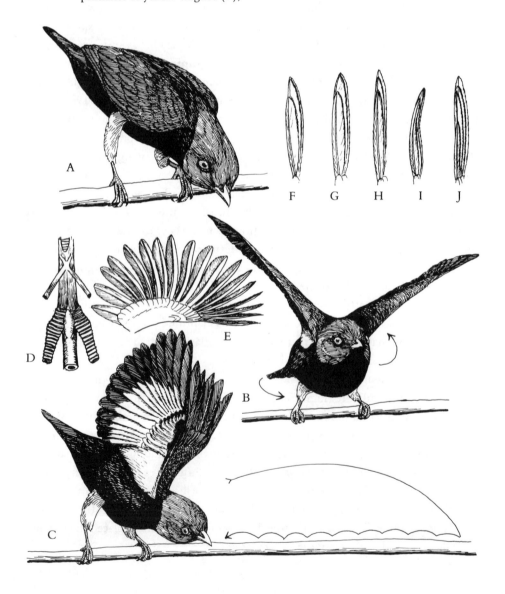

48. Displays of male manakins, including pivoting display by fiery-capped manakin (*A*), looping jump-flight by pin-tailed manakin (*B*), and perched display by club-winged manakin (*C*). Also shown are the male club-winged's fifth through seventh secondaries, and the female's fifth secondary (*D*). *A* and *B* after sketches in Sick (1967), *C* after a description by Willis (1966), and *D* after drawings in Newton (1896).

Among the mechanical sounds generated by male manakins are wing-snapping noises of the secondaries, wing-whirring sounds of the primaries while in flight, tail noises caused by vibration of the rectrices, and stamping sounds produced by the feet. In many manakins the outer primaries are variously attenuated for producing vibratory noises while in flight (Figure 47E–I), whereas in others the secondaries may be modified for generating mechanical sounds through vibration while the birds remain on the display perch (Figure 48D). In the band-tailed manakin (*Pipra fasciicauda*) the vegetation immediately around the displaying bird may also be set into motion; as with the Albert's lyrebird, this movement may provide a visual or acoustic accessory to the display. Snow (1963c) believed that manakin display flights are almost certainly derived from the male's approach flight toward the female prior to copulation. He suggested that the horizontal and vertical sliding movements on perches may be ritualized versions of ordinary locomotory movements while perched, and that upward jumping as well as darting to and fro may also be derived from unritualized locomotory movements, such as those associated with foraging.

Display perches of male manakins are usually horizontal twigs or small branches (Figure 48B) that are free of side limbs and leaves and have clear aerial approaches. In some species the birds may also display while clinging to vertical vines or branches (Figure 48A), or while standing on horizontal buttress roots or fallen logs (Figure 49A). In some manakins (especially species of the genus *Manacus*) a ground court is established and maintained by debris removal, and in others (such as *Chiroxiphia pareola*) the male's display perch may be transformed into an arboreal "court" by similar vegetational modification.

In some manakin species, such as several species of *Pipra* as well as in *Corapipo, Machaeropterus,* and the golden-winged manakin (*Masius chrysopterus*), the males typically display in exploded or dispersed arenas out of visual contact but within hearing of each other (Prum and Johnson, 1987; Prum, 1989). Exploded or dispersed arenas also are apparently typical of the aberrant genera *Neopelma* and *Tyranneutes*. Prum (1989) excluded these from the true manakin group, and he believed their male displays to be independently derived. As in other avian groups, it is impossible to make hard and fast distinctions between the male spacing patterns of exploded arenas and those of typical leks on the one hand or dispersed individual male territories on the other. These spacing patterns may be subject to local ecological or population variables even within a single species.

Manakins frequently assemble in typical leks. There the males may be within sight of each other but operate independently and competitively (*Manacus*), or they may perform as synchronized groups of two or even more closely interacting males. In some species of the latter type, such as in the golden-headed manakin (*Pipra erythrocephala*) (Figure 45E) and the fiery-capped manakin (*Machaeropterus pyrocephalus*) (Figure 48A), the two males also clearly interact as sexual rivals. The same general situation seems to apply to the band-tailed manakin (*Pipra fasciicauda*), where

49. Displays of male manakins, including sonogram (*A*, above) and postures (*A*, below) of white-throated manakin during flight display and log display sequence (after Prum, 1986).

Also shown (*B–F*) is the wheel-flight display by male swallow-tailed manakins (after sketches in Sick, 1967). Numbers identify individual male participants.

the alpha and beta males may display simultaneously but usually independently on the same court (Robbins, 1983, 1985). However, in the "blue" manakins (*Chiroxiphia* spp.) the males often appear to play equal, even seemingly cooperative roles during preliminary joint display (Figures 45G–I, 49B–F).

The Joint or Communal Displays of Manakins

Snow (1963c) called attention to the remarkable situations in which two or more manakin males take an apparently equal part in a joint display performance, which he termed "communal displays." He suggested that two males displaying jointly are twice as likely to attract a female than solo performers, but that they must nevertheless represent sexual rivals. He judged that the display sites used by such interacting males are seemingly "communal property." However, a dominant bird may actually control the site, so that neighboring males are allowed to display with him only until a female is attracted, after which he is able to exclude his neighbors and court the female individually. Snow also suggested that some of these neighboring males might "own" display perches elsewhere, or perhaps are younger birds that are subordinate to the dominant male, at least for the time.

The potential benefits of communal male displays in the *Chiroxiphia* manakins have also received the attention of Foster (1977, 1981, 1984) and McDonald (1989a). Foster argued (1977) that the subordinate male's apparent helping behavior is probably not the result of kin selection or pure altruism, but instead results from individual selection pressures. Not only might his courting technique be improved by participating in a more dominant male's displays as a "helper," but also his mating opportunities include both stealing occasional copulations from the dominant male while still a subordinate and perhaps sometimes outliving him and taking over his display site. If females exhibit "lek faithfulness" to a particular display site, a young male's chances of later attracting females may be enhanced by participating long before he has any real prospect of individual reproductive payoff.

McDonald (1989a) confirmed the role of the age-related effect on relative male dominance in the long-tailed manakin (*C. linearis*). He found that of 50–60 active males present in leks containing 3–15 males each, only 6–8 males were of alpha rank. Males may be at least eight years old before obtaining beta status, and alpha tenure for dominant males might last for 2–4 years or more. Lower-ranking males may affiliate with other males on as many as six different display sites or perch "zones," each of which serves as a display activity hub. As a male increases in age and status, he reduces the number of perch zones in which he participates but progressively increases his probability of copulatory success. Four of 85 males accounted for more than 90 percent of the 117 copulations observed by McDonald over a four-year period, and a total of only eight of the 85 males obtained any copulations. Foster (1977) reported that, in all pairs of displaying males of this species that she observed, it was always the alpha male that obtained the copulations. Only in the absence of the alpha male did the beta male copulate. In one case she observed that the alpha male copulated with both of two females that were present, and in another case three females were mated by a dominant male while the subordinate individual remained in the vicinity and watched. In the swallow-tailed manakin (*C. caudata*) a similar situation applies (Foster, 1981), with two or rarely three males displaying simultaneously, but with all copulations (of 22 completed or partial copulation sequences observed) being performed by the dominant male. Similarly, in the blue-backed manakin (*C. pareola*) copulations are probably restricted to the dominant male of each displaying group (Snow, 1963a). The situation in the last of these four species, the lance-tailed (*C. lanceolata*), is still essentially unstudied, but the limited available information for it suggests that it is much like the other three species in its general social and sexual behavior patterns (D. Snow, 1977b; Foster, 1978, 1981).

Snow (1977b) judged that female choice is the primary selective agent that has shaped the secondary sexual characteristics in this genus, rather than overt intermale aggression, and furthermore noted that interspecies behavioral divergence has been greatest in the early advertising phases of male display, which serve to attract females to the display site. The four species have diverged only slightly if at all in the later stages of courtship. If female choice, rather than male aggression, is the primary selective factor, females must make their choice based on the traits of both the dominant and subordinate males, since both play equal roles during early stages of courtship. Ultimately, the dominant male "dismisses" his subordinates, thus eliminating possible female choice as to her final mating partner. Foster (1977, 1983) believed that the joint coordinated male displays in this genus represent ritualized supplanting behavior by one male toward another while both are competing for space on a display branch, which probably serves to control direct intermale aggression and thus decreases the probability of precopulatory disruption.

The Solo Displays and Mobile Leks of *Corapipo* Manakins

As summarized by David Snow (1963c), solo displays are characteristic of a few manakin species, including the Central American form of the white-ruffed manakin (*Corapipo [leucorrhoa] altera*). According to Skutch (1967), in this species a male displays alone, performing slow flights from one mossy fallen log perch to another. When approaching the log, the male may make a rapid descent and land with a call, or he may perform a slow, undulating and mothlike flight to the log. He erects his white ruff both while in flight and when on the log, and also occasionally flicks his wings while on the log, although he lacks special wing markings. The outer primaries of this species are distinctly attenuated (the Central American forms less so than the South American) and are somewhat curved (Figure 47I, J), which suggests that loud wing noises are probably produced in flight. Skutch reported that individual display logs may be used by as many as four males, but these are apparently rivals that either use the log at differing times or perhaps compete simultaneously for its control.

More recently, Prum (1985) has described the behavior of a related species, the white-throated manakin (*C. gutturalis*). He reported that some males display habitually in a few sites, in a typical solo territorial manner, but that at times up to six males may attempt to use the same log simultaneously. Not only is *Corapipo* different from most other manakins in using a fallen log as a display site, but also perhaps in that several males may compete for control of a single display site rather than displaying in nearby contiguous territories. Male competition for display sites at several sites has resulted in a kind of "detached" or "mobile" lek that somewhat resembles the mobile courting groups of ducks described earlier.

The white-throated manakin is a small manakin with marked sexual dichromatism. The male's advertisement call consists of a series of up to ten *seeeu-seee-ee-ee-ee-ee* notes that become shorter and more rapidly uttered toward the end of the sequence, which lasts about 1.5 seconds (Figure 49A, left segment of sonogram). These calls typically precede a display bout. They are also uttered while the bird is approaching a display log during a log approach display, and while flashing his colorfully patterned primaries (Figure 49A, left). When about half a meter above the log the male suddenly stalls and drops to it, producing a muffled, mechanical "pop" sound during the stall and while flashing the white wing markings (vertical middle segment of sonogram). He then immediately rebounds back into the air while uttering a sharp, squeaky *tickee-yeah* (right segment of sonogram) and lands again a short distance away, this time facing the other direction (Figure 49A, center and right). Perched displays on the log consist of the same upward bill-pointing posture, often with quick about-face movements involving a short hop. The male may also make vey short flights from one perch to another and back again, seemingly maintaining the bill-up posture even in flight.

Exploded Leks and the *Pipra* Manakins

In the small, sexually dichromatic white-crowned manakin (*Pipra pipra*) of Central America, adult males are entirely black except for their white crowns. They display in an exploded lek distribution, the birds being situated about 60 meters apart, or just within hearing range of each other. Like the *Corapipo* manakins, the males perform slow butterfly-like flapping display flights, during which they evidently do not vocalize but a soft, probably wing-produced sound is evident. They also have a poorly developed display flight to their various display perches, which are located a few meters apart and up to 9 meters above ground. About-face displays on the perch and quick back and forth darting movements are also performed. Little else is known of this species' displays (D. Snow, 1961).

In the white-fronted manakin (*P. serena*) the sexes are also dichromatic, with the generally black males having white on the forecrown, yellow on the underparts (at least in the nominate race), and light blue on the rump. In the fairly closely related blue-crowned manakin (*P. coronata*) males are variably green to nearly all black, except for a blue crown. Males of both these species defend territories about 30–40 meters in diameter. They advertise them by calling incessantly from, and flying back and forth between, horizontal and vertical perches near the ground that are scattered throughout these territories. In one case observed by Prum (1985), two males of the white-fronted manakin thus held immediately adjacent territories. Two other rather nearby groups held three and four males respectively, the situation seemingly representing an exploded lek. Males would call from their perches while erecting their yellow chest patches, and would often fly from perch to perch in rapid wing-whirring flights, following a vertical or horizontal sigmoid path. On low meter-wide "courts" composed of 5–10 vertical saplings near the forest floor, the birds would display by flicking open their wings, exposing their blue rumps. They would also utter loud descending notes in countersinging contests with intruding males, and would fly in an upright hummingbird-like manner in buzzy arc-shaped flights from perch to perch within their courts, frequently as a coordinated display with an intruding male. One male was observed to perform 20 court displays at five different courts within his territory. However, the males were also observed at times to abandon their own display courts and "display loosely" throughout the entire territory, such as by leap-frogging from branch to branch while calling excitedly.

This coordinated display activity appeared to Prum to be more probably competitive than cooperative in nature. He reported that some type of coordinated display activity has similarly been reported for five species of *Pipra* and two *Machaeropterus* species, the striped manakin (*M. regulus*) and the fiery-capped manakin (*M. pyrocephalus*). He judged that such coordinated displays could not only serve as a competitive interaction between males, but also might provide an excitatory or solicitation stimulus to visiting females.

In the South American wire-tailed manakin (*P. filicauda*) the males have uniquely elongated outer tail feathers whose shafts are extended into long wirelike filaments (Figure 46F). The tail, wing, and rump areas are blackish, contrasting with bright yellow underparts, a crimson crown, and a whitish iris. Males defend display areas that range in size from 10 by 20 to 25 by 35 meters, and within such areas there are many display perches, including a primary display site. These perches are horizontal to slightly inclined branches about 1–2 meters above ground. They are always free of twigs and foliage for some distance, and have clear flight pathways in front and for a short distance behind. Each display site is controlled by a single male, who may perform alone or in coordinated display with another male partner. This partner may be either an immature or another adult bird (Schwartz and Snow, 1978).

Male wire-tailed manakins utter a variety of call notes and produce various mechanical noises apparently using their wings. Most of the male display elements are similar to those of other *Pipra* species. They include a slow, butterfly-like flight away from the perch, a "dipping" or sigmoid display flight to the perch, an audibly or visually conspicuous landing, single-step (Figure 46C) or multistep (Figure 46D) side-to-side jumping or sliding on the perch, and a backward sliding movement toward another bird on the perch. Males also perform a "stationary display" with the head lowered, the tail raised, the back and rump feathers erected, and the wings slightly drooped and vibrating. However, the "twist" display is perhaps unique to this species. It requires two cooperating birds, one of them active and one passive. The active bird, facing away from its partner, lowers its head, raises its back and rump feathers, and edges toward the passive bird. It approaches the passive bird with jerky backward movements, simultaneously lifting and twisting its tail (Figure 46E). When close enough (Figure 46A, B), the wirelike feathers rhythmically brush against the partner, typically at its throat. Experienced partners may approach the active bird in an appropriate receptive stance (Figure 46E, left). Males may also perform a static "tail-up-freeze" posture (Figure 46E, right) prior to approaching the female in the twist display, with the male's yellow underparts framed in black and his tail strongly lifted.

In one case copulation followed after a male had performed the twisting display for about ten seconds. He took off for a swoop-in flight, during which the female assumed a receptive posture and the male looped back, landing, and immediately hopped up on her back. After copulation the male performed a stationary display, followed by a tail-up-freeze. At this point the female departed, but the male continued to display.

The Arena Displays of the Red-headed *Pipra* Manakins

Displays of the closely related red-capped manakin (*P. mentalis*) and golden-headed manakin (*P. erythrocephala*), respectively of Central America and tropical South America, are evidently quite similar to one another. Males of both species have golden to orange or red heads and yellow to red thighs, as well as pale bills and eyes that contrast with otherwise mostly black bodies. However, the red-capped manakin has white underwing markings, and the golden-headed has a black underwing area. In the eastern South American form, the "red-headed manakin" (*rubrocapilla*), males have a red rather than golden head, red thighs, and a distinctive white underwing patch, probably representing a separate species. The male displays of the red-capped manakin (Figure 47A–C) have been described by Crandall (1945a) from captive birds, and by Skutch (1969) under natural conditions. The displays of the golden-headed manakin (Figure 45C–E) have been extensively studied in nature by David Snow (1963b) and Lill (1976). Both species typically display in what appear to be typical leks (or "courtship assemblies") in which males occupy perches several meters apart, although pairs of males may sit together midway between their individual display perches. Horizontal display perches situated up to about 15 meters above ground are used in both species. Display flights to the perch, backward slides along the perch, stationary pivoting about-faces, back and forth darting between the display perch and an adjacent one, and wing extension displays are present in both species, as are mechanical wing noises (Snow, 1963b).

In the red-capped (or "yellow-thighed") manakin, males produce several distinct vocalizations, but these are perhaps less interesting than their mechanical sounds. The rectrices' shafts are thickened in males, and those of the secondaries are both thickened and curved. The snapping sounds made by the males are evidently produced by the feather shafts striking one another. Wing-whirring and wing-rustling noises are also made during flight. A male performs on horizontal branches, starting with his legs well apart, his yellow thighs exposed, and his body tilted forward (Figure 47A). He then about-faces as rapidly as possible, as one foot is held on the perch and the other is moved quickly from one side to the other, and his wings are opened with a resonant flap (Figure 47B). He then may jump ahead on the perch and slide backward with the wings lifted, exposing their white undersides (Figure 47C), or may simply raise his tail and move quickly backward with very short and mincing steps, sometimes approaching a female in this manner. He also often flies rapidly between his main display perch and another nearby one, or makes a circling flight around the display perch, ending the flight with a shrill call and a loud mechanical wing noise as he suddenly breaks his flight. When two males join at some midway point between their primary display perches, one will often "dance" or glide toward the second, with no clear dominance evident on the part of either (Skutch, 1969).

As compared with the red-capped manakin, the very closely related golden-headed and red-headed manakins are much better known. Sick (1959, 1967) described some aspects of the behavior of the red-headed form *rubrocapilla* in Brazil. David Snow (1963b) and Lill (1976) both studied the nominate golden-headed form *erythrocephala* on Trinidad, where it is highly abundant. However, it is rather difficult to observe, since it displays on perches about 7–14 meters above ground, in subcanopy trees. The display sites consist of horizontal branches that have been pulled and pecked at until they are free of leaves and tendrils, and thus would seem to qualify as arboreal courts. Different males may occupy display perches only about a meter apart in the same tree, or up to several meters apart in neighboring trees, and 6–12 males are usually present in a single courtship assembly. These display grounds persist from year to year, and the very same display perches may also be used in successive years.

Snow (1963b) described this species' advertising and other male calls, and identified seven postural displays. These included darting back and forth between nearby perches and its main perch, an upright posture (Figure 45C), a rapid about-face pivoting, a backward slide (Figure 45D, E), vertical wing flicking, a frenzied flutter while jumping and uttering an excited *zeek* note, and a display flight. The usual display flight is of about 25 meters, from the major display branch to a higher perch, followed by a quick return and a series of in-flight calls. The male approaches his perch with an upward and then downward swoop, and after landing may slide backward (Snow, 1963b). In the case of the Brazilian "red-headed" manakin, the male may try to land on the female after such a flight, but if she doesn't accept him he will land at her side with raised wings (Figure 45F). He may then repeat the flight. Differences between these two forms of uncertain taxonomic rank are not very great, and some apparent differences may only reflect differences in observers' personal observations and descriptions (Sick, 1959, 1967; Snow, 1963b).

David Snow (1963b) observed copulation in the golden-headed manakin only twice, and was unable to relate individual copulation effectiveness to differences in male ages or dominance status. However, Lill's (1976) multiyear study provides important information on age-related factors influencing male mating success, as well as a detailed description of sexual behavior patterns. In addition to the male displays described by Snow, Lill also reported behavior patterns associated with nearly 200 observed mating sequences. Copulation follows a period of mutual display, in which the male performs (in diminishing approximate observed frequency) darting back and forth, full display flights, upright posture, about-face, short, silent display flights, and backward slides. The female usually pecks lightly on the male's head at this precopulatory stage, and both sexes may "freeze" rigidly for as long as 15 seconds. In nearly all cases the male landed beside the female immediately prior to mating and then jumped onto her back, and in all but one case copulation occurred on a male's display perch.

Females were attracted to and preferentially mated with clustered males rather than isolated males. Lill concluded that a close spacing of males was an important factor in attracting receptive females, as well as in promoting a high level of territorial defense. Clustering thus conferred an apparent selective advantage on clustered males through their enhanced overall mating success. Two large leks exhibited higher overall rates of total matings, mating bouts, and mating female visits than did a neighboring small lek. However, individual male mating success distributions were not significantly different between residents of large versus small leks. Indeed, on a per-bird basis, males at the small lek actually received more female visits on average than did those at the large leks. Therefore, Lill concluded that enhanced sexual stimulation of females by male clustering has probably not been a major factor promoting lek displays in this species.

The Manic Displays of *Manacus* Manakins

One of the first detailed descriptions of manakin displays was the classic account by Chapman (1935) of the courtship of the golden-collared ("Gould's") manakin (*M. vitellinus*) on Barro Colorado Island, Panama. Adult males are mostly black, with a yellow throat, breast, and nuchal collar, yellowish underparts, and dull scarlet feet. Groups of males use the same leks year after year, with four to seven individual display sites present per lek. Males are separated by distances of from only a few to about 60 meters, averaging about 9. Leks are separated by distances of several hundred meters. Territorial boundaries of individual males extend for a well-defined distance beyond their display courts, the floors of which are kept clean of leaves and other debris. Courts are located in areas having an undergrowth of saplings, and average about half a square meter in ground dimensions.

While waiting for females, the male perches near his ground court, often within a few inches of a neighboring male, and calls occasionally. Two principal vocalizations, the *pee-you* and *chee-poooh,* are uttered. The syrinx of the species is relatively simple in structure (Figure 47D), but the intrinsic muscles are greatly hypertrophied (Lowe, 1942). Several additional mechanical sounds are produced by males. These include single snaps or cracks, a snipping sound, plus snapping and reedy whirrs. All of these sounds are apparently made by the wing feathers, of which the outer primaries are strongly narrowed and the secondaries are laterally curved, stiffened, and have thickened shafts (Figure 47D). They are inserted into tendons that lie above the radius in a unique manner that allows for muscular vibration (Lowe, 1942). The male's elongated and yellow feathers are erected into a beardlike shape during display, and his other yellow-feathered areas are also expanded. Up to ten crackling snaps of the wings may be delivered, at a rate of two per second, as the male stands on his court, or he may point his bill up and assume a rigid pose for a minute or more. He also darts back and forth across his court, turning in the air at the last instant so as to alight facing his point of departure. One male was seen sliding down a slender sapling while holding his body horizontal, with his bill touching the sapling in a dirigible-like manner (Chapman, 1935).

The displays of the white-bearded (or "black-and-white") manakin (*Manacus manacus*) are apparently quite similar to those of the golden-collared. They were first described by Darnton (1958), and later studied more completely by David Snow (1962). The white-bearded manakin is sexually dichromatic, with males having black crowns and upperparts and white underparts, including a well-developed white "beard" that can be strongly erected. As in the golden-collared manakin, the male's outer four primaries are very narrow and stiffened, and the shafts of the secondaries are thickened and inserted into tendons of the wing in a highly flexible manner. The stiffened outer vanes of the secondaries can thus rub over one another and perhaps also over the shafts of adjacent secondaries in a way that produces a dry, snapping noise that can be a single and short pulse of sound or a more prolonged or "rolled" snap.

The white-bearded manakin is a fruit-eating species that spends most of its time around the forest floor, or at least no more than 10 meters above it. Each male constructs a cleared ground court similar to that of the golden-collared, and groups of such courts similarly constitute leks that are used perennially. Each male has a roughly circular court, ranging in diameter from about 0.25 to 1 meter. Two or more upright saplings are present, always including a primary display site sapling. Most display occurs on these saplings, and thus they usually become worn smooth by the action of the bird's feet. All loose vegetation at the foot of such saplings is quickly removed by the owner. Courts are usually about 1–5 meters apart. On one large display ground there were 70 separate courts, many of them almost in physical contact, in an overall area about 9–18 meters in diameter. On a study area of about 180 hectares, some 205 courts were present, representing a manakin population estimated by David Snow (1962) to be about 250 adult males and a comparable number of females. Lill (1974b) found that resident males had a restricted home range around their arena. Some males survived for as long as 14 years, and were still breeding when at least 9–11 years old. Lill concluded that more aggressive males can establish courts at arena sites that are favored by females, and that at least during the second half of the mating season the females selected a specific mating site rather than merely a particular habitat for mating.

Male white-bearded manakins have a syringeal anatomy much like that of the golden-collared manakin (Lowe, 1942), and utter a variety of generally simple *peer, chwee, chee-poo,* and other notes. These notes generally seem to be subordinate to the mechanical noises made by the birds, especially during courtship display. Sexual activity occurs almost throughout the year but is reduced during the molting period. Although males display throughout the day, most activity occurs shortly after dawn, with a second early afternoon peak.

Displaying males exhibit a "tense" posture (Figure 45A), with the crown feathers flattened and the beard extending forward beyond the tip of the bill, as the bird extends his body out horizontally in preparation for further display activity. When not engaged in display, males exhibit little if any "beard" (Figure 45B, above). Snow (1963b) recognized six major postures of male display. The most frequently seen was the "snap-jump," which is a darting to and fro from one sapling to another on his court, accompanied by a loud snapping sound. The "rolled snap" is a series of snapping noises produced by the perched bird with his wings raised and vibrated over his back. The "grunt-jump" (Figure 45B, left) is a jump from the primary perch to the ground, where he lands momentarily and then leaps back to a higher point on the perch, producing a mechanical grunting sound as he leaves the ground. The "slide-down-the-pole" (Figure 45B, right) is a quick movement down the sapling with wings beating and the head held downward until he reaches the bottom of the sapling. The grunt-jump followed by the slide-down-the-pole seems to be the final phase of courtship that immediately precedes copulation in this species. In this situation the male slides down directly onto the female's back, but keeps one foot on the perch and one on the female's back during treading. At this time the female is perched crosswise on the sapling. During "fanning" (Figure 45B, middle) the male leans forward as in the tense posture, but sways his body from side to side while retracting his neck and tilting his bill upward slightly, while simultaneously beating his wings slowly and raising and lowering his tail. The feathers of the hindneck are also moved forward and upward so as to resemble two flickering white powder puffs on each side of the bird's head. This display is most frequently directed toward a female when she is approaching a male's court. When displaying toward rival males, the bill is tilted upward and the body is also held in an upright orientation.

There is a considerable diversity in the reproductive success of different males; David Snow (1963b) reported that one old male who displayed vigorously and was constantly at his court was observed copulating with two banded females and 15 times with unmarked females. Similarly, Lill (1974a) reported differential male mating success, with female preference for older resident males.

The Cooperative Displays of the *Chiroxiphia* Manakins

Of all the manakin groups, none are of greater biological interest than the blue manakins of the genus *Chiroxiphia*. This group of four rather closely related species, believed to be fairly closely related to *Pipra,* extends allopatrically from Central America south to southeastern Brazil and adjacent Argentina and Paraguay. In all, the males have sky-blue backs, red crown patches, and otherwise are largely to entirely black. In the most widespread species, the blue-backed manakin (*C. pareola*), the tail is usually quite short and square-tipped. In the other three species the central tail feathers are variously elongated. These include the swallow-tailed manakin (*C. caudata*), the lance-tailed manakin (*C. lanceolata*), and the long-tailed manakin (*C. linearis*). Of these three species the southernmost, the swallow-tailed, is distinctive in that the males have almost entirely blue bodies rather than just blue backs. In all these species the males live in groups that display jointly on various perches within their collective territory. Males also produce highly synchronized duets, and pairs (or sometimes larger numbers) of males display in perfect coordination during the early stages of courtship (D. Snow, 1977b).

One of the first persons to describe the displays of the *Chiroxiphia* manakins was Gilliard (1959), who studied the blue-backed manakin on Tobago Island. David Snow (1963a) subsequently studied this same species on Tobago, and later observed the swallow-tailed (or "blue") manakin in Brazil (1976). Although the lance-tailed manakin is still relatively little-studied, the long-tailed has received a great deal of research attention in recent years from McDonald (1989a, 1989b) and from Foster (1977, 1984). Foster has additionally observed and described the behavior of the swallow-tailed manakin (1978, 1981, 1987). Males of this species occur in groups of 4–6, with each group occupying a rather large area (about 35 hectares) within which up to five active display courts may be present. In each group there is a dominant or alpha male, and a linear hierarchy below that is probably age-related and is determined by the outcome of aggressive interactions.

According to Snow (1977b), male displays in this genus consist of three major phases. The first occurs when two males come together and utter a repeated series of calls above a display perch in a duetting fashion, with the two birds singing in variably overlapping phrases or with intervals between phrases. This duetting is the most variable element of display among the four species, especially with regard to the acoustic characteristics of each species. In at least two of the species (blue-backed and swallow-tailed), duets are initiated by one male (probably normally the dominant one), who is soon joined by another, and the same perhaps applies to the other species. However, it is rare that more than two birds participate in such coordinated calling.

The second display phase begins when both males fly down to the perch and sit side by side, alternately jumping up and fluttering their wings in a "joint dance" and each bird calling during his jump (Figure 45G). At times one might jump over the other bird, or it may turn while in the air and land facing in the opposite direction. If a female should arrive at the perch the males immediately face her and begin a "Catherine wheel" dance. The male nearest to her jumps up, calling once with a twanging note while hovering, and then flies back to land beyond the other, as the other simultaneously moves forward to take his place near the female and perform the same display (Figure 45I).

At least in the swallow-tailed manakin as many as three or even four males may sometimes participate in this remarkably coordinated behavior (Figure 49B–F), although two appears to be the usual number in most species. The nonhovering birds stand closely side by side, facing the same direction while perched crosswise on the branch, and gradually shuffling up to take their turn. As the active bird hovers, he utters a buzzing call while facing the female (Figure 49C, F), then flies to the end of the line. A single display sequence may include up to 150 jumps, with the birds becoming more excited and their movements progressively faster. The males thus execute smaller hovering circles while in flight, and the female hops up and down, watching the displaying male closely. This second phase of the display ends in the swallow-tailed manakin when the dominant male jumps into the air, and rather than facing the female, turns toward the other male(s) and utters a high-pitched "final call" that causes them to assume a bent-over submissive posture. The subordinate individual(s) then fly off to another nearby perch and play no further direct role (Foster, 1978). A similar termination of the second phase apparently also occurs in the other three species (D. Snow, 1977a).

The third phase of the display consists of the dominant male performing a largely silent display around the female, flying back and forth in a floating flight that displays his blue plumage. During brief landings between such flights he displays his brilliant red crown to the female, bowing toward her with stiff and jerky movements (Figure 45H). This behavior serves as a precopulatory display, and if the female is receptive copulation soon occurs on the display branch (Foster, 1978; D. Snow, 1977a).

Apart from the obvious male plumage differences, there are some marked acoustic differences in the duetting calls of the four species, and perhaps also some local "dialect" variations within species. In species where the two calling birds normally sit close together, as in the blue-backed, lance-tailed, and long-tailed manakins, the calls tend to be short, more rapidly repeated, and better synchronized. However, in the swallow-tailed manakin the birds may be up to several meters apart and there are longer intervals between the two birds' calls, suggesting that visual cues may be important in call synchronization. Wing-clicking sounds are seemingly less important in this genus than in most manakins, but a mechanical wing click has been observed in the precopulatory display sequence of the blue-backed manakin (D. Snow, 1977b).

As Snow (1977b), Foster (1977), and others have suggested, the remarkable social system of *Chiroxiphia* has probably been derived from a rather typical lek system in which each male occupied and defended his own display perch. By a gradual reduction of intermale aggression, and a corresponding increase in the capability of a single male to gain and maintain dominance over the other males in the lek, it has become increasingly beneficial for nondominant males to avoid aggressive encounters with the male and instead accept a subordinate but nearby position. In this situation the subordinates might enhance the mate-attracting power of the dominant male and perhaps either be able to "steal" copulations occasionally when the dominant male is temporarily distracted, or eventually outlive and replace him, thereby ultimately gaining his position and rank. In this way the seemingly altruistic and "cooperative" behavior of the subordinate males might be explained on the basis of simple individual selection rather than through kin selection or some other more elaborate explanation.

10

Bowerbirds: Courts, Maypoles, and Avenues

The late ornithologist E. Thomas Gilliard (1969) once remarked that all birds should perhaps be split into two
broad categories, namely bowerbirds and all other birds. This view gives some sense
of the remarkable attributes of the Australian and New Guinean bowerbirds. They
often are not particularly colorful, but usually construct intricate and elaborately decorated bowers that strain the credulity of people when they are told that a bird has
built them. Indeed, the first European naturalist to find a bower, Odoardo Beccari,
believed it to be of human rather than avian origin.

Thomas Gilliard's great interest in bowerbirds and birds-of-paradise, which he believed to be closely related groups, led him to speculate on the evolution of bower
building, which is confined to males in this distinctive group of passerine birds. He
believed it to be derived from generalized avian arena behavior (which he defined as
meaning mating station behavior, not specifically also implying clustered male lekking behavior). He functionally related the constructions of male bowerbirds to the
court-constructing activities of various arena species. These included such birds as the
Guianan cock-of-the-rock, males of which use cleared courts under arboreal perches
as display sites, and various birds-of-paradise such as the magnificent (*Diphyllodes
magnificus*) and six-wired birds-of-paradise (*Parotia* spp.), which similarly clear forested courts on which the males regularly perform advertisement displays.

Gilliard judged that the construction of bowers may have gradually evolved
through the male's incorporation of picking-up and depositing movements associated with his handling of vegetation and other objects during court clearing. Or perhaps it represents a retention of now-antiquated nest-building impulses on the part of
the male. Gilliard argued in favor of the second of these explanations, which had been
originally suggested by A. J. Marshall (1954, 1955). Gilliard thus redefined (1969:58)
avian arena behavior to include a polygynous species' "nondiscardable nesting tendencies." He further suggested that, through their court-clearing behavior, some species of birds other than bowerbirds may eventually also become bower builders, and
judged that the generally rather dull-colored species of bowerbirds that typically
build the most complex bowers are at the "leading edge of avian evolution."

Gilliard concluded that nearly all bowerbirds engage in typical female-attracting arena behavior, rather than pair-bonding behavior. He suggested that, in directing the bowerbirds' capacities for constructing ornamental bowers, sexual selection has transferred the functions of female mate choice from selecting colorful or complex male plumages to choosing among variably colorful and complex male-gathered or male-constructed objects. He called this hypothetical evolutionary process the transferral effect. After such a transfer has occurred, the male may again evolve more protectively colored and femalelike plumages, such as losing its bright mating colors or crests. Gilliard suggested that a specific example of this process is the exhibition to females of their unmarked nape feathers by males of some bowerbirds (*Chlamydera*) in a manner directly comparable to the nape exhibition display used by other closely related species that actually do possess brightly colored nape patches but build somewhat simpler or less highly decorated bowers. Additionally, in the three species of the genus *Amblyornis* there is an inverse relationship between the relative complexity of the bower that is constructed by the species and the male's relative breeding plumage brightness.

Another view of the possible evolutionary origin and function of bower building was provided by Borgia, Pruett-Jones, and Pruett-Jones (1985). They suggested that, rather than reflecting outmoded male nest-building tendencies or irrelevant motor activities, bowers represent an example of directed sexual selection that directly serves a "marker" function, namely by showing individual male ownership of a defended and variously decorated display site. The bower then becomes a functional proximate symbol of relative individual male status and quality. Females can thus potentially compare and evaluate every available male's relative status according to the quality of his display site, as well as his actual display characteristics and his ability to display at the site without disruption from other nearby males. These authors suggested that cleared courts probably served as an intermediate stage in the evolutionary history of bower construction, just as Gilliard hypothesized, but that by clearing such a display site a dominant male might prevent the clearing of nearby courts by other males, as well as limiting access to his own court by other males. The additional advantage of actually constructing a bower might have resulted from some accrued protection to females visiting the court. If hidden from the sides or above, the female might be less visible to predators but might still be able to escape from the front or rear. In support of this idea they pointed out that bowerbird species that build relatively open bowers tend to occur in closed forests, whereas those that build more enclosed bowers are mostly associated with open forests, where predator visibility and thus the average danger to visiting females might generally be greater. One might alternately suggest that the bower simply represents a sexual lure to the female, which when entered additionally becomes (especially in the case of avenue bowers) a kind of enclosed trap that makes it physically more difficult to escape when the male attempts to mount her (see Figure 51C).

50. Representative male
bowerbirds and their bowers,
including tooth-billed (*A*),
Archbold's (*B*), MacGregor's
(*C*), golden (*D*), streaked (*E*),
satin (*F*), spotted (*G*), and

yellow-breasted (*H*) bowerbirds.
Typical bower shape as seen from
above is shown in *C–H*. After
varied sources including Gilliard
(1962).

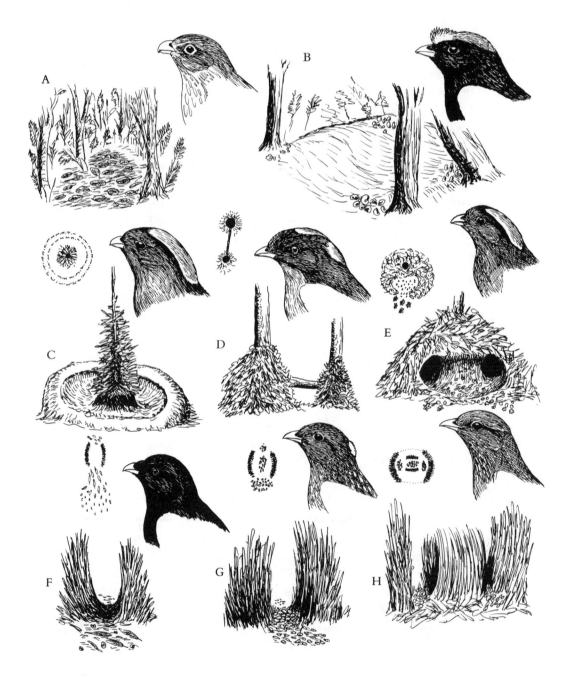

It is of course impossible to reconstruct the evolutionary history of bower building with any degree of assurance. Nevertheless, sufficient interspecies variability in present-day bower construction and bowerbird morphology exists to generate an "ethocline" of contemporary bower variability, and allow for some educated guesses about the ecological and taxonomic interrelationships of bowers and their makers.

A General Comparative Survey of Bowerbirds and Their Bowers

A very useful classification of bowerbird bowers according to their mode of construction was suggested by Marshall (1954, 1955). One group of arboreal bowerbirds, the catbirds (*Ailuroedus* spp.), construct no bowers at all. Rather, they show virtually no sexual dichromatism and are evidently monogamous, with the males of at least two of the three species known to assist in brood rearing. They are probably territorially dispersed, and the green and inconspicuously colored males utter very loud calls and songs from strongly defended singing perches low in the forest. Gilliard (1969) pointed out that the catbirds use large leaves that are laid flat in a special manner to provide a nest foundation, and that this activity may represent the behavioral basis for the earliest stages in court establishment and bower making among bowerbirds.

The simplest court known to be constructed by bowerbirds is that of the tooth-billed bowerbird or "stagemaker" (*Scenopoeetes dentirostris*) of tropical Australia. Males of this monomorphic species (sometimes included in the catbird genus *Ailuroedus*) closely resemble the typical catbirds, and like them have very loud vocalizations that are uttered close to their courts. Courts are made by clearing an area of forest floor up to nearly 3 meters in diameter. The male then cuts through the petioles of tree leaves of certain favored species using its unique tooth-edged bill, and arranges these leaves on the ground with their paler sides uppermost (Figure 50A). The leaves, averaging about 40 but ranging up to about 100 in number, are replaced as they wither (or are removed) with freshly cut ones, on an almost daily basis. Snail shells have sometimes also been found on the court surface or beside it, but these perhaps represent accidental inclusions that simply have not been removed by the resident male.

A somewhat similar display court is used by the Archbold's bowerbird (*Archboldia papuensis,* including the "Sanford's bowerbird" *A. p. sanfordi*). In these two allopatrically distributed forms (regarded as separate species by some authorities) the sexes are somewhat dichromatic, and males are spaced well apart in the forest. Their display courts are inconspicuous mats of dead fern stems that have been cleared of most other debris. At the edges of the mat are a variable number of decorative objects arranged in small groups or piles, such as snail shells, insect remains, resin chips, and bits of vegetation (Figure 50B). The mat itself is usually kept nearly free of such decorations, but the saplings and other woody vegetation around the edges are typically draped with orchid stems, or with fronds of ferns or bamboos, sometimes producing

51. Bower displays of Archbold's bowerbird, including male-female interactions (*A*, after sketches in Gilliard, 1959) and male "groveling" display (*D*, after photos in Coates, 1990); also copulation in satin bowerbird (*B, C*; after photos by N. Chaffer).

a curtainlike effect. The "bower" of the male was described by Gilliard as resembling the flattened and disheveled bed of some mammal, with little or no apparent "design." A secondary court was largely covered with snail shells, arranged in a disorderly manner, as well as some insect parts, blue berries, and centipede skeletons. Five different courts averaged about 2–3 meters long by about 1.5 meters wide, and all were located on gently sloping montane forest floors. All had snail shells, beetle parts, and resin chips present on the floor. Bits of charcoal and green berries were also present. Three active sites were within an area of about 5 kilometers. In other areas active display sites have been found as close as a few hundred meters apart, or separated by as much as about 2 kilometers (Coates, 1990).

All of the remaining species of bowerbirds are evidently also polygynous, and construct more complex bowers that fall into one or the other of two broad categories, namely "avenue-type" and "maypole-type" bowers. Avenue-type bowers were the first to be discovered by European naturalists, and were initially believed to be toys constructed by aboriginal mothers to amuse their children. The bowers generally consist of two walls of twigs, separated by enough space to barely allow an adult bowerbird to pass through, and variably decorated at one or both ends with items such as snail shells, colorful feathers, dried bones, pebbles, and the like (Figure 50F–H). Avenue-type bowers are typical of all the five species placed in two variably dichromatic and predominantly Australian bowerbird genera (*Ptilorhynchus* and *Chlamydera*), and in the three species of a highly dichromatic genus (*Sericulus*) that occurs mainly in New Guinea. It has also been reported among at least seven of the eight species of these typical avenue builders (all but the little-studied *Sericulus bakeri*) that the twigs of the bower's inner wall may be stained or "painted" by the male, using plant juices, charcoal, or pulverized mixtures of dried grasses and saliva, in varied colors that include black, gray, green, yellow, and reddish brown.

Maypole-type bowers, the other general bower category, are constructed by relatively little-studied species of two mostly dichromatic bowerbird genera (*Amblyornis*, *Prionodura*). In all these species' bowers, a central axis is established by a vertical sapling, which provides them with physical support. Around this central tower a cone of vegetational fabric is packed at the base. Additionally an assortment of decorations such as twigs, mosses, and lichens may be associated with and attached to the central sapling in a manner resembling a decorated Christmas tree (Figure 50C, D), and the tower may support a hutlike roof and walls (Figure 50E). A secondary vertical tower is present in the golden bowerbird (*Prionodura newtoniana*), with the two spires invariably connected by a roughly horizontal vine or stick (Figure 50D). In some of the maypole-building species the area in front of or around the bower may additionally be decorated with freshly plucked flowers, small fruits, colorful insect parts, or snail shells (Figure 50E). In the golden bowerbird the maypole itself may be decorated with living orchids. This latter characteristic of using live plants for decorations has given some of these maypole-building species of *Amblyornis* the general name of "gardener bowerbirds." However, in contrast to nearly universal painting behavior of the avenue builders, only one population of a single species (*Amblyornis inornatus*) of maypole builders is so far known to paint the sides of its bowers (Diamond, 1987).

Court Behavior of the Tooth-billed and Archbold's Bowerbirds

Relatively little information is yet available on the court-related behavior of the tooth-billed or "stagemaker" bowerbird. Gilliard (1969) believed that such courts are widely spaced but nevertheless clustered as "clan arenas" or exploded leks in various parts of the jungle. He also believed that the distribution of these courts may be related to the breeding hierarchy of their owners, which in retrospect seems remarkably insightful. In some areas five or six males can at times be heard calling simultaneously, suggesting an exploded lek situation. The species is known to have an exceptionally well developed voice, and has also been described as a vocal mimic. The birds sing in their private court perches, usually on horizontal limbs up to about 6 meters above ground, probably thereby attracting females to their courts. In one case a male was seen advancing forward over his court in a crouching posture, his beak gaping and his upper breast feathers ruffled, and as he moved forward he flicked his wings out repeatedly and jerked his tail upward. Later the same display was seen, and the bird also snatched up a dead leaf and stood for several minutes holding it, uttering a soft continuous whistling note in synchrony with head-bobbing movements (Gilliard, 1969). Although no other bird was visible, it seems highly probable that one was hiding nearby.

From their forested courts male Archbold's bowerbirds likewise utter a variety of vocalizations, including powerful whistles, deep growls, and crowlike notes. The behavior of the male also includes the manipulation of objects such as snail shells on the court, and the carrying in of vegetation to decorate the court. When a female arrived on a court observed by Gilliard she landed on a lower perch at the court, frequently flying from one perch to another at the edges of the court and often flying within a fraction of a meter of the male, which lay churring on the fern mat below. The female was never more than about 1.5 meters from the male, and while perched her head was always directed toward him. The male performed what appeared to be highly submissive, begging postures, crawling on the ground with his wings partly opened, his body and tail pressed to the ground, and his crown depressed, but the forehead feathers elevated. His beak was often widely opened, exposing his gape, but for part of the time a strand of vegetation was held crossways in his mouth (Figure 51A). As the male approached the female, she would fly over him, often hovering briefly just above him, and then continue to another low perch at the opposite side of the court. At times the male would begin to hop toward the perched female, but would then immediately stop and again assume the "whipped dog" attitude. Churring vocalizations were almost continuously uttered, except during hopping. At one point the male made a complete circuit of the court in about five minutes, following the female as she moved around the court edges. Eventually both birds flew off silently. Excellent photographs of a displaying male Archbold's bowerbird and his bower have recently been published (Coates, 1990); the wings of the male in his groveling or "whipped dog" posture (Figure 51D) are somewhat less extended than as shown by Gilliard, but otherwise the postures appear to be essentially identical.

Male Advertisement and Sexual Behavior
of Maypole-Building Bowerbirds

Of the several species that construct maypole-type bowers, perhaps the one with the simplest bower type is the MacGregor's bowerbird (*Amblyornis macgregoriae*) (Figure 50C), a thrush-sized species that has a wide distribution throughout much of New Guinea. Except for its stunning orange crest, which when erected seems much too large for its body (Figure 52A), the adult males are generally olive-brown and inconspicuously colored. Their bowers are placed on ridgetops (Figure 52B) and typically consist of a circular, saucer-shaped pedestal about a meter in diameter and surrounding a sapling, bounded by a mossy rim up to about 30 centimeters high and with moss also banked up on the central sapling column (Figure 52C). This sapling, about 1–2 meters high, is decorated for most of its length with as many as 800 or more slender, short twigs and rootlets, all organized horizontally, mostly under 20 centimeters long. Additionally, fern shafts, bamboo leaves, white lichens, as well as occasional bits of spider webbing ("animal silk") and small orchids may be present. Small black and white objects, such as charcoal, animal dung, seeds, dried berries, and similar items, may be present in small quantities on either the central sapling or the mossy parapet. At times smaller, apparently subsidiary bowers are nearby, which may be the remains of earlier, mostly demolished bowers or may represent new, seemingly abortive and temporary constructions (Gilliard, 1969).

Recent studies by Pruett-Jones and Pruett-Jones (1982) have shown that this species' bowers are regularly spaced within the available habitat, especially along ridgecrests. Males in their studies usually maintained single bowers, but 10 percent maintained two bowers simultaneously. The males spent the majority (54 percent) of the daylight hours at or within 20 meters of their bowers, defending them from conspecific males, which rather frequently tried to disrupt mating attempts (39 percent of 18 observed attempts). The species appears to be entirely promiscuous, or perhaps more accurately shows male dominance polygyny. Some of its dispersion features (such as the regularly spaced bowers) suggest typical territoriality, whereas other features (the defense by males of a small and resource-poor display area, a possible if not probable female choice mating system, and the close proximity of sexually competing males) resemble a leklike organization. Two nests were located, both of which

52. Male MacGregor's bowerbird, showing male crest raising from the side and behind (*A*) and male standing by his bower (*C*). After sketches in Cooper and Forshaw (1979) and photos in Coates (1990). Also shown is a distribution map (*B*, after Pruett-Jones and Pruett-Jones, 1982) of active bowers along ridgecrests (dotted lines).

were situated along ridgecrests and about midway between adjacent bowers. No males were seen near these nests at any time.

Intruding males often attempt to destroy bowers, especially when the resident male is absent, and on average a male suffered the marauding effects of other males once every ten hours. Besides working to build and maintain their bowers, which sometimes required up to five or more hours to repair following damage by rivals, males also vocalized frequently. Such vocalizations averaged 1.4 times per hour in the just-mentioned study, and the males not only mimicked other species, but also imitated various other environmental auditory stimuli. They usually sang from traditional song perches at or near the bower, and occasionally from as far as 60 meters away. Countersinging between males associated with adjacent bowers was sometimes also heard. Together with bower quality, the ability of a male to protect his bower from marauders and its optimum ridgetop location may be important indicators of his relative fitness and genetic quality (Pruett-Jones and Pruett-Jones, 1983).

The bower of the golden bowerbird (*Prionodura newtoniana*) of Queensland is rather different from that of the other maypole builders, especially in that it is always constructed between two trees, or between a tree and a bush. Small twigs and sticks are piled up around, or "glued" (by fungal growth) to, these supports, reaching a height of about 2 or even as much as 3 meters around one support and usually a lower height around the other, often about a half meter lower. Between these two pyramidal piles a "bridge" is invariably present, consisting of a more or less horizontal vine or stick that was already in place and that can serve as a runway or display site (Figure 50D). This horizontal perch may be up to more than a meter above ground, or only a short distance above it. The bower's decorations usually include grayish green lichens, flowers, and less frequently mosses, ferns, berries, and other vegetation. Unfortunately there are still no detailed descriptions of male display behavior at these bowers (Gilliard, 1969; Cooper and Forshaw, 1979); but evidently the horizontal bridge plays a significant role in this.

Schodde (1976) has suggested that the bower of *Prionodura* may actually be the least specialized of all bower types, and may represent a basic design from which both avenue and more complex maypole bowers could have been derived. Bowers of the avenue type might have evolved simply by extending the two end pyramids, discarding the supporting saplings, and substituting a central avenue for the connecting display stick. The *Amblyornis* type of maypole bower might have been derived by constructing only a single pyramid and making a "display runway" around it. This argument would seem to be less persuasive than Gilliard's suggestion that the earliest bowers were little more than variously decorated forest-floor courts, which then gradually developed around a central focal point (the maypole) or a focal pathway (the avenue).

One of the largest and most elaborate of all bowers is that of the streaked (also called the striped gardener) bowerbird (*Amblyornis subalaris*) of southeastern New Guinea, which occurs in the same areas as but at lower altitudes than the closely related Mac-Gregor's bowerbird, and on slopes rather than ridgecrests. Both species construct stick towers around a central upright sapling, but in the streaked bowerbird there is a teepeelike roof of sticks less than a meter high, with an opening that faces a front "yard" (Figure 50E). This yard in turn may be enclosed by a perimeter of twigs, and is largely covered by a "flowerbed" of blackish plant fibers taken from the stems of tree ferns, into which the bird inserts a variety of freshly plucked flowers, berries, brightly colored leaves, and dead beetles. From this "gardening" activity the common descriptive name "gardener" derives. Yellow-, red-, and blue-colored objects are apparently favored by the males as bower decorations, and adult males have yellowish underparts and a bright orange crest. Spacing between adjacent bowers is said to be several hundred meters (Coates, 1990). Little is known of the behavior of the male at the bower, but Gilliard (1969) judged that the species' complex bower may be an example of an evolved species-specific reproductive isolating mechanism, serving to help separate it from the closely related MacGregor's bowerbird. He furthermore suggested that the streaked bowerbird may have been ecologically displaced downward to lower altitudes by the widely ranging MacGregor's.

A very similar hutlike bower is built by the Vogelkop bowerbird (*A. inornatus*) of extreme northwestern New Guinea. This relatively isolated species' bower is quite large, reaching heights of at least 1.5 meters and as wide as about 2 meters in diameter. Although it is not in contact with any of its congeners it produces perhaps the most elaborately decorated of all bowers. The front "garden" of mosses or ferns is decorated with colorful plant or animal materials (often of color-segregated brown, green, blue, yellow, or red items) on a usually neatly tended bed. Diamond (1987) reported that in two different mountain ranges less than 200 kilometers apart this species' bowers were quite different. In one area the bowers had a much smaller central tower that was woven rather than glued together, and decorations were mostly colorful and grouped by color, whereas in the other area the decorations were black, brown, or gray, placed on a shiny black mat that had perhaps been painted with excrement. Additionally there were individual and age-related variations in bower structure.

Males of the maypole-building bowerbirds utter an extraordinary variety of calls, and evidently display to females from the entrance of the bower while squatting or crouching among the bower ornaments. Of these three species, males of the Vogelkop bowerbird build the most elaborate bowers but have the least ornate plumage; the streaked bowerbird is intermediate in both these features; and males of the MacGregor's bowerbird build the simplest bower but have the longest, most elaborate crests, thus supporting Gilliard's transferral hypothesis (1969).

Male Advertisement and Sexual Behavior
of Avenue-Building Bowerbirds

Of the avenue type of bowers, those of the golden regent (or flame) bowerbird (*Sericulus aureus*) of New Guinea and the Australian regent bowerbird (*S. chrysocephalus*) appear to be among the simplest in structure. Males of these two species are quite colorful, with much yellow and black especially on the nape, upper back, and wings. Males of the third congeneric and similar species, the extremely rare fire-maned bowerbird (*S. bakeri*), have only recently been found to construct a simple avenue-type bower (Coates, 1990).

For the golden regent bowerbird, a small avenue-type bower about 25 centimeters high and nearly 20 centimeters long has been observed. This had cleared areas at both ends of the avenue, and small blue berries and a black mushroom placed within the avenue (Gilliard, 1969). In one or both of two other bowers the decorations consisted of such items as blue and brown berries, purple and white flowers, and a snail shell. One of these was about 20 centimeters in length and with walls up to 35 centimeters high, nearly meeting at the top. Coates (1990) described and provided photographs of additional bowers, and noted that apparent bower painting by the male has been observed. The bowers of the Australian regent bowerbird are similar, averaging about 25 centimeters wide, 20 centimeters long, and about 20–30 centimeters high. They often have a rudimentary platform or floor of twigs between the walls, on which a few decorations such as leaves, snail shells, or berries may be placed. They are inconspicuous and are often hidden under shrubs or other undergrowth. In this species the male is known to paint the interior walls of his bower with a mixture of saliva and masticated fruit pulp or vegetable matter (Cooper and Forshaw, 1979).

The fire-maned bowerbird's bower was discovered in 1986 by B. J. Coates and R. D. Mackey. It was only about 20 centimeters long and 20 wide. It was a crudely constructed and slightly arched-over avenue of thin twigs, simply decorated at one end and in the interior with blue berries and fruit. Display behavior of the male has not yet been observed (Mackey, 1989).

The satin bowerbird (*Ptilonorhynchus violaceus*) of eastern Australia is a well-studied and attractive species; the males are generally a uniform purplish black, with a contrasting iris that is lilac-colored to dark blue but becomes rose-red in display. The bower (Figures 50F, 51B, C) is typically larger and more elaborate than those of the regent and fire-maned bowerbirds, with a stick foundation and two parallel walls of sticks about 20 centimeters in length and 30 centimeters high. These are separated by the central avenue, about 10 centimeters wide at the base but arching over somewhat toward the top and sometimes coming into contact there. The structure is placed under a shrub or tree branch, in a clearing about a meter long and nearly as wide that is extensively decorated. These decorations often include blue feathers (frequently parrot feathers), flowers and berries (especially blue, violet, or purple ones), human artifacts (especially blue or bluish items, such as glass, paper, or plastic, plus aluminum foil), and various other items such as grayish fungi, brownish snail shells, greenish flowers or fruit, and the like. Blue (the plumage and iris color of the adult male) is clearly the male's favorite color, followed by yellowish green (the color of immature males and females) (Gilliard, 1969). The inside walls are usually coated with macerated charcoal, wood pulp, or other vegetable matter, which is either applied directly by the male with his bill or using a saturated wad of bark as a paintbrush. The axis of most bowers is oriented north-south, perhaps so that the male can watch for females without staring into the rising sun (Marshall, 1954; Cooper and Forshaw, 1979).

Male satin bowerbirds attain their full adult plumage only when they are about seven years old, although birds in immature plumage are known to court females. Some individual bowers have been maintained for more than 30 years, and some males are known to have reached ages of more than 20 years. Permanent bower sites of dominant males may be used by more than one male over time, as old males are replaced or displaced by others. Dominant or "senior" males may stimulate the building of nearby bowers by subsidiary males, producing a clustering of bowers and an exploded lek type of male distribution (Vellenga, 1980). Male satin bowerbirds exhibit a highly skewed mating success pattern, with females showing a consistent preference for particular males as mating partners. Females especially favor those males having well-constructed and highly decorated bowers, suggesting that, at least in this species, bower quality alone may provide a direct index to individual male quality (Borgia, 1985b, 1986; Borgia, Pruett-Jones, and Pruett-Jones, 1985). Males also frequently steal feathers from other nearby bowers, and the rate of feather stealing is directly correlated with the number of feathers already present in that male's bower. Feather stealing may thus be related to both the degree of bower decoration and male mating success (Borgia and Gore, 1986). Females may additionally be able to identify and choose high-quality males based on their level of ectoparasite infection, as reflected in the relative brightness of their plumage (Borgia and Collis, 1989, 1990).

Most male display by satin bowerbirds occurs during morning hours, starting in late winter (or about July in New South Wales) with more intense activity later in spring (September, October). Contrary to some earlier opinions, males are entirely promiscuous, having been observed to mate with as many as five different females and to court several others within a single breeding season (Vellenga, 1970, 1980). Males sing from and posture at their bowers. When a female has been seen by the displaying male, he begins a series of constant vocalizations accompanied by occasional poses with his tail elevated, bill pointed toward the ground, and body held on stiff legs. Or he may hold a feather or other bower object in his bill, with wings outstretched and tail raised before the watching female, and may prance and flap his wings to the beat of his calling (Figure 51B). During this time the female typically stands within the bower's avenue, nipping at some of its sticks and watching the male attentively but not directly participating otherwise. Suddenly, however, the male may rush into the bower, mount the female (Figure 51C), and copulate. Afterward she may remain in the bower for a time, or even return to it repeatedly after the male has forcibly driven her from it with severe pecking and clawing movements (Chaffer, 1959; Cooper and Forshaw, 1979). Some copulation attempts may be interrupted by other males, who may also hide in the surrounding vegetation and try to court visiting females during the owner's absence. Typically a female copulates only once before starting her egg laying (Borgia, 1986).

Among the four species of the genus *Chlamydera* the bowers constructed are always of the avenue type. In the spotted (*C. maculata*), great gray (*C. nuchalis*), and fawn-breasted (*C. cerviniventris*) bowerbirds these are of fairly simple construction, consisting only of two parallel walls with decorations that are usually placed at either end and also in the bower's central avenue (see Figures 50G, 53, 54). In the yellow-breasted or Lauterbach's bowerbird (*C. lauterbachi*), the bower is additionally enclosed at both

53. Display behavior of male great gray bowerbird, including twig and grass insertion during bower construction (*A, B*), singing (*C*), central arena displays (*D–F*), wing-drooping display (*G*), and male's approach route relative to female and bower (*H*). After sketches in Veselovsky (1978).

ends, and the associated male display behavior is also somewhat more complex (Figures 50H, 54C). Painting of the inside bower walls has been reported for all four of these species (Cooper and Forshaw, 1979; Coates, 1990).

In the avenue bower of the great gray bowerbird (Figure 54A), green plant materials are the primary decorations, but sometimes gray, glassy white, and silver objects are also used, such as stones, bones, and shells. More than 1,200 stones were counted at one bower. Bowers are often placed in the shade of trees or under bushes, and frequently are located close to water. The central avenue is usually about 15 centimeters wide, and its walls are up to about 50 centimeters high. Most bowers are oriented on a north-south axis. The male great gray bowerbird occasionally paints the inner walls with saliva and probably fruit pulp, sometimes staining the walls slightly darker but at other times having no apparent visual effect (Warham, 1957; Gilliard, 1969; Cooper and Forshaw, 1979). However, Veselovsky (1978) observed no painting or traces of bower-painting behavior in his observations. The majority of the bowers he observed were oriented northwest-southeast; each bower had 700–920 twigs per wall, the male carrying an estimated total of 4,000–5,000 twigs to the bower site. The total weight of

54. Nape exhibition displays of male great gray (*A*), fawn-breasted (*B*), and yellow-breasted (*C*) bowerbirds at their bowers. After photos in Gilliard (1959) and Coates (1990).

the decorations (snail shells, bones, glass, stones) ranged from 6.2 to 12.1 kilograms per bower, and from 5,000 to 12,000 total items were present. White or gray objects were evidently preferred; experimentally introduced blue, red, black, yellow, and green objects were thrown out of the bower by the male. In a sample of 26 bowers, the average wall length was 55 centimeters, the average maximum height 46 centimeters, the maximum wall-to-wall width about 60 centimeters, and the total cleared bower area about 2 square meters.

Males of the spotted bowerbird resemble those of the great gray in their plumage and bower configuration (Figure 50G), and in usually placing their bower under a bush near water. They also have a similar lilac-pink nape tuft and paint the interior of their bower a reddish brown color. In one bower the avenue was about 50 centimeters long and its walls about equally high, with grass stems along the inner walls arching toward the center and almost forming a roof. This species too seemingly favors white, gray, pale green, and a few other colors for its decorative bower objects, but evidently rejects red, yellow, and blue. There is no documented indication of a directional bower orientation in this species (Marshall, 1954; Gilliard, 1969; Cooper and Forshaw, 1979; Borgia and Mueller, 1992).

The fawn-breasted bowerbird's bower also consists of a mat of sticks bounded by two parallel walls (Figure 54B). The mat is often quite thick, perhaps reflecting an adaptation for coping with periodic flooding. In contrast to the last two species, the fawn-breasted seems to select only fresh green plant materials for decoration, which

C

are regularly replaced as they wither. Clusters of berries may also be hung from the tops of the avenue walls, or placed on the bower mat (Gilliard, 1969). Males paint the inner walls of the bower with a mixture of saliva and green plant materials, which dries to a dull gray and reddish brown color. In some areas of New Guinea there is a definite approximately north-south directional orientation of this species' bowers, but curiously in one area all of the bowers were found to have an east-west orientation (Cooper and Forshaw, 1979).

The yellow-breasted bowerbird has a more complex bower structure, with a somewhat four-sided arrangement and with the walls of the main avenue angled outward rather than oriented vertically or arched over (Figure 50H). The inner walls are usually the longest and are lined with fine grass stems. Bowers are decorated with such items as gray or bluish gray stones and red, blue-green, or blue berries, which are typically placed at the end passages and in the center of the main passage. Green or blue fruits or berries are favorite decorations for both ends of the avenue, while stones may often be placed in the main passage. Collectively the bower proper may weigh 3–7.5 kilograms, with up to nearly 5 kilograms more of stone decorations (Gilliard, 1969). There is no apparent directional orientation to this species' bower. Although painting of the bower by this species has at times been questioned, such behavior has recently been noted by several observers (Cooper and Forshaw, 1979; Coates, 1990).

Males of all the *Chlamydera* species have now been observed displaying at their bowers, as extensively summarized by Gilliard (1969), Cooper and Forshaw (1979), and Coates (1990). Among the most complete observations on a single species are those of Warham (1957) and Veselovsky (1978), both on the great gray bowerbird. Veselovsky provided considerable information on bower-building behavior. He found it required about three weeks for a male to construct a completely new bower, although bowers that had been destroyed were often repaired in less than three days. Items were brought to the bower by the male at an average rate of 90 twigs and 28 decorations (glass fragments) per hour. The twigs and grass were individually inserted by the male, using pushing and prodding movements of his head and bill (Figure 53A, B).

Displays of the great gray bowerbird were described by Warham (1957) as consisting of two general types. These were "central" displays performed toward females while the male stands among ornaments in and around the bower, and "peripheral" displays performed while circling the outer boundary of the bower platform. During peripheral display the male droops his wings, raises his colorful nape feathers, and cocks his tail while running or walking in a strutting manner around the bower. The female meanwhile moves evasively, in such a way as to remain on the opposite side of the bower from the male (Figure 53G). The male also utters loud ticking or hissing vocalizations while holding a bower ornament in his bill, or while standing fairly upright and holding all his feathers closely compressed against his body (Figure 53C).

Singing may also occur on various branches about 6–8 meters high, at varied and sometimes considerable distances from the bower itself.

During initial stages of central displays, the male assumes a variety of postures, including the actual or ritualized picking up of bower objects. Often he will assume an upright posture, with bill either gaping or holding a bower ornament, and then call excitedly as he attempts to approach the female. When very near her, he stretches his neck forward and turns his head to the side, so that his expanded lilac-colored nape tuft is fully visible to her (Figure 53E, F). By bounding or leaping movements the male may circle the entire display area, thereby apparently trying to keep the female within the confines of his bower (Figure 53H). As display progresses the female tends to move to the end of the avenue opposite to that where the male is displaying, or she may watch him from a low branch directly above (Figure 54A). In either case, the male orients his head so that his nape tuft is directly exposed to her view, at times holding a bower ornament in his bill and frequently jerking his head rhythmically up and down. Mating frequently occurs within the walls of the bower, but has also been observed just outside the bower entrance (Warham, 1957; Veselovsky, 1978).

Gilliard (1959) has pointed out that although the male fawn-breasted bowerbird lacks a colorful nape tuft comparable to that of the great gray, it performs essentially the same kind of nape exhibition display when a female is visiting the bower (Figure 54B). This display is often performed while the male is holding and shaking a large green berry or spray of berries in his bill. Gilliard suggested that a once-present colorful crest has probably been secondarily lost in this species, although the associated postural orientation persists, and that the use of a colorful object in the bill represents a case of transferral effect in sexual signaling. A similar nape exhibition display occurs in the yellow-breasted bowerbird, which also lacks a nape tuft. When a female enters and stands in the bower's central avenue, the male may call while hidden from her view behind one of the avenue walls. He may also stand in a cross passage and directly exhibit his nape to the female's view while calling and often holding objects in his bill (Figure 54C), may partly fan the tail and turn it toward the head, may flick up one wing, or may face the female while hissing, gaping, and flicking up both wings simultaneously (Coates, 1990).

Borgia and Mueller (1992) have determined that mating success in males of the spotted bowerbird is correlated with the quality of the male's bower and with several measures of bower decoration, such as significant accumulations of reddish pink, reddish, or purple glass (colors that approximate the male's reddish pink nuchal crest color). These results suggest that bower quality influences mate choice by females. Well-built bowers are typical of older males, so by choosing such bowers the females are also choosing age indicators as a mating criterion.

In a review of bower building, Diamond (1986, 1987) noted that bowerbirds vary interspecifically as to their general bower structure (maypole versus avenue), and exhibit both interspecific and intraspecific variations within these broad structural types. Bowerbirds also vary in the kinds of decorations that are gathered (snail shells, leaves, berries, acorns, fungi, flowers, stones, sticks, bones, human artifacts, etc.), and in the way the decorations are distributed within and around the bower. Bowerbirds also vary as to their apparently preferred or accepted ornament colors. Schodde (1976) suggested that maypole-building species seem to prefer reds and yellows, while avenue builders tend to select green to violet objects. However, Diamond concluded that there are no clearly evident correlations between preferred colors and those that are present in, or contrast with, the adult plumage or softpart colors of either males or females, nor colors that are specifically rarity- or habitat-related. Bowerbirds vary as to whether they paint their bowers, and perhaps in painting methods or materials. Lastly, they vary in the topographic locations they select for bowers (ridgecrests, hillsides, river bottoms, etc.), and sometimes also in the bower's directional positioning (such as oriented relative to cardinal directions or to uphill-downhill gradients). Diamond suggested that individual learning must play some role in bower building, as is indicated by the relatively rudimentary bowers built by young and inexperienced birds and by their lack of decorations or inappropriate ones. Changes in bower complexity with increasing age may result either from individual trial-and-error learning or perhaps from observing and imitating bower building by older males. The latter method is perhaps at least partly responsible, and as such the cultural transmission of bower-building "styles" becomes increasingly plausible.

11

Birds-of-paradise: Surreal Visions of Paradise

Few if any groups of birds set a naturalist's imagination on fire as do the birds-of-paradise. Their very names suggest our sense of their ethereal and unearthly mystery, while their remote homes in New Guinea place them outside of the personal experience of all but a few intrepid ornithologists. Charles Darwin was fascinated by birds-of-paradise as seeming classic and remarkable examples of sexual selection, although he never was able to study them in nature. More recently, they have been the subject of several monographs (Gilliard, 1969; Cooper and Forshaw, 1979). They have offered ecologists and ethologists some remarkable examples of adaptive radiation in foraging behavior and associated variations in bill morphology. They have also provided wonderful examples of ecological influences, such as interspecific variations in foraging niches, on such interrelated behavioral adaptations as social systems and dispersion patterns. Diamond (1986) has recently reviewed the biology of the birds-of-paradise from these standpoints, and also reviewed the corresponding biology of bowerbirds, which are now believed to be only a rather distantly related assemblage. His organization provides a convenient framework around which to build a discussion of the social and sexual behavior of these remarkable birds. His classification of birds-of-paradise social systems is as follows, together with his representative examples of each type:

I. Monogamous; frugivorous species whose males do not maintain exclusive territories and do participate in parental care. Typical of two species of manucodes (*Manucodia*) and MacGregor's bird-of-paradise (*Macgregoria pulcra*).

II. Polygynous (or promiscuous); males emancipated from parental duties.
   A. Males dispersed and territorial. Typical of three species of sicklebills (*Epimachus*) and the superb bird-of-paradise (*Lophorina superba*). (Males of all these species are probably in vocal but not visual contact.)
   B. Males dispersed but nonterritorial. Typical of magnificent riflebird (*Ptiloris magnificus*), magnificent bird-of-paradise (*Cicinnurus magnificus*), and twelve-wired bird-of-paradise

(*Seleucidis melanoleuca*). (Males of the first two species are probably in vocal contact.)

C. Males often in dispersed pairs, but sometimes solitary or in larger groups. Typical of king bird-of-paradise (*Cicinnurus regius*).

D. Male dispersion variable, from leklike congregations to dispersed; displaying on terrestrial courts. Typical of parotias (*Parotia*).

E. Males congregated in typical leks, displaying on arboreal perches. Typical of six of seven species of *Paradisaea*, the standardwing (*Semioptera wallacei*), and Stephanie's bird-of-paradise (*Astrapia stephaniae*).

In a paper published a few years earlier, Beehler and Pruett-Jones (1983) had similarly examined all the nonmonogamous bird-of-paradise species with regard to major variations in their male advertisement display dispersions and their diets. They recognized four major behavioral categories of male dispersion and relative territoriality, namely (IA) males dispersed and territorial, (IB) males dispersed but nonterritorial, (II) males organized in exploded leks, and (III) males organized in true leks. They listed 25 species as falling into these categories, with varying levels of confidence based on the then–available information (level 1 = confident placement, level 2 = moderately firm placement, and level 3 = available information not definitely reliable), as follows:

I. Dispersed males species
  A. Males territorial
    Level 1. Superb bird-of-paradise; black-billed, brown, and pale-billed sicklebills (*Epimachus albertis, E. meyeri,* and *E. bruijnii*) (Beehler and Beehler, 1986; Diamond, 1986)
  B. Males nonterritorial
    Level 1. Magnificent bird-of-paradise
    Level 2. Loria's bird-of-paradise (*Cnemophilis loriae*), magnificent riflebird, twelve-wired bird-of-paradise, splendid astrapia (*Astrapia splendidissima*)
    Level 3. Paradise riflebird (*Ptiloris paradiseus*), Victoria's riflebird (*P. victoria*)
  C. Male territoriality still uncertain
    Level 1. (No species included in this category)
    Level 2. Black sicklebill (*Epimachus fastuosus*), blue bird-of-paradise (*Paradisaea rudolphi*)
    Level 3. Wilson's bird-of-paradise (*Cicinnurus respublica*)
II. Exploded lek species
    Level 1. Carola's parotia (*Parotia carolae*), Lawes' parotia (*P.*

lawesii), king bird-of-paradise, Goldie's bird-of-
paradise (*Paradisaea decora*)
    Level 2. King-of-Saxony bird-of-paradise (*Pteridophora alberti*)
  III. True lek species
    Level 1. Standardwing; greater (*Paradisaea apoda*), lesser (*P.
minor*), raggiana (*P. raggiana*), red (*P. rubra*), and
emperor (*P. guilielmi*) birds-of-paradise
    Level 2. Stephanie's astrapia (*Astrapia stephaniae*)

Besides the questions of relative territoriality and male spacing tendencies in birds-of-paradise, another point of interest concerns the male's possible modification of the physical display site into one (a "court") that is especially suitable for performance of his self-advertisement displays. Beehler (1987a) used the term "court" to mean any traditionally used display site for polygynous birds, but as mentioned earlier I have restricted its use in this book to those sites that have been in some way adaptively modified by the resident males themselves. In at least some of the *Paradisaea* species, males may pluck at twigs, bark, leaves, mosses, etc., while "waiting" idly at their territorial sites. These seemingly random actions can produce significant alterations of their display sites, which perhaps in most cases are only reflections of a male's general internal state. However, it is known that males of all four of the commonly accepted species of *Parotia* (all but *helenae*, which is often regarded as only a subspecies of *lawesii*) display on what would seem to be purposefully cleared ground courts a few meters in diameter, or on display perches just above such cleared courts. In at least one of these species, the Lawes' parotia, a bowerbird-like behavior occurs in which display objects are brought in and placed on the court (Pruett-Jones and Pruett-Jones, 1988a; Coates, 1990).

    Court-clearing behavior evidently also occurs in the Wilson's bird-of-paradise, but is not well documented. In the congeneric magnificent bird-of-paradise these courts similarly consist of more or less circular areas of ground that are about 5 meters across, within forest interiors. Several mostly dead saplings (which eventually are killed by the males' plucking behavior) are probably always present on the courts, but leaves, twigs, and small plants are regularly removed from the ground and from these associated saplings. Leaves are also plucked from nearby saplings and lower trees in the court's immediate vicinity, and the bark of such saplings and trees is also plucked at. Such behavior gradually kills the vegetation and allows light from the sky to penetrate to the court below. Gilliard (1969) described the court of the magnificent bird-of-paradise as representing a "bower," seemingly comparable in his view to the simplest bowers of bowerbirds. He believed that such "bower building" has probably evolved convergently in several lines of birds-of-paradise, as a result of the incorporation of court-clearing activities into courtship ceremonies through evolutionary ritualization.

Beehler (1987a) has attempted to compare the known behavior of birds-of-paradise with predictions based on contemporary theories of vertebrate mating systems as reviewed by Emlen and Oring (1977), Wittenberger (1979), Oring (1982), and others. He reviewed the three stages of lek evolution (male emancipation, development of an arena display system, and finally the spatial clustering of males). He noted that clutch reduction is present in birds-of-paradise (which average two eggs per clutch), and that many but not all species are fruit eating, both of which facilitate male emancipation. He also reported that little information exists on female choice and male dominance as mating factors for birds-of-paradise, although males have a long delay in plumage and sexual maturation, resulting in a high proportion of femalelike "floater" males in the population. In two lekking species (raggiana and lesser birds-of-paradise) one or two dominant males were observed to mate repeatedly, whereas others did not mate at all. Males of raggiana may remain in full femalelike plumage for up to six years, and several more years may be needed to attain full male plumage. For such birds it is in the interest of a young male to become a subordinate to an older and more successful male, with the possibility of eventually replacing him over a long lifetime. Beehler suggested that lek clustering in arena-displaying birds tends to occur among species in which females show high levels of range overlap, so that the pool of available females at any one site would counterweigh the obvious disadvantages to males in clustering. Thus, the females can benefit from their widely ranging food-gathering behavior in such situations, but participating males must share food resources in the vicinity of the display site with their mating rivals.

The manifold displays, plumages, and relationships of the birds-of-paradise still pose a host of unresolved questions for evolutionists, ethologists, and ecologists. However, the classification of the display dispersion traits of nonmonogamous birds-of-paradise that was provided by Beehler and Pruett-Jones (1983) can serve as a convenient basis for functional organization and further discussion of individual species in the remainder of this chapter.

Nonmonogamous Birds-of-paradise Having Dispersed, Territorial Males

Only a few species of birds-of-paradise are known to maintain mutually exclusive male territories. Most of these species are primarily insectivorous, which is an extremely unusual foraging niche for a polygynous/promiscuous tropical passerine bird. In one or two of these species (superb and perhaps brown sicklebill) the female nests within the limits of male territories (Beehler and Pruett-Jones, 1983), but in another (the black-billed sicklebill) females may have exclusive territories (Beehler, 1987b).

One remarkable species that is fairly well studied and that falls into this behavioral category is the superb bird-of-paradise, which is widespread throughout much of the

55. Displays of male superb
bird–of–paradise, including
normal posture (*A*), gorget
spreading and partial crest raising
(*B*), frontal display toward
female (*C*), and full display as
seen from the front (*D*). After
sketches in Frith and Frith (1988).

midmontane zone of New Guinea. It is a small (about 25 centimeters long, averaging about 90 grams), mostly insectivorous bird, adult males of which are almost entirely black except for an iridescent blue breast shield and glossy black, highly erectile capelike head plumes. Smaller erectile blackish feathers occur at the base of the lower mandible and above the nostrils at the base of the upper mandible. Like many birds-of-paradise, females are inconspicuously patterned, with blackish barring below and brownish to gray above, and with rufous-tinted wings.

Male superb birds-of-paradise occupy defended territories as small as about 1.5 hectares, and display solitarily from arboreal perches (Beehler and Pruett-Jones, 1983). Advertising males frequently call from rather high and concealed perches in treetops, with rasping or hoarse and fairly loud notes. Although some observations of display in captive birds have been published, the only good descriptions of displays in wild birds are those by Frith and Frith (1988). One display site was on a dead tree trunk about a half-meter in diameter and 10 meters long, below a break in the forest canopy on a rather steep slope. The male's initial display posture (Figure 55A) consisted of crouching with the breast shield compressed, the bill directed upward toward a female, and the nasal tufts projecting upward. Interspersed with this posture is a forward flicking of the cape over the head, together with similar breast shield flashing and a downward movement of the head, exhibiting the iridescent crown to the female's view (Figure 55C). In this stage the cape is not fully expanded. On two occasions when the female approached the male he backed to the edge of the log, sleeked his body feathers, but held the breast shield fully erect and gaped to expose his yellow mouth lining. Wing-clicking noises accompanied this display.

In high-intensity display, which the Friths saw four times, the breast shield is fully expanded and thrust forward, the nasal tufts are erected, and the cape is not only flicked forward but is also spread laterally, forming a complete semicircle above the bird's head and extending down behind the breast shield, producing together with the dark wings and underparts a complete black circle that is broken by the iridescent blue breast shield. At the base of the cape (just above the bird's actual eyes) a pair of iridescent blue-green "eye spots" become visible, although the bird's actual eyes are quite inconspicuous. In this posture, with the mouth remaining closed, the male proceeded to dance around the female in short, sharp steps (Figure 55D). In separate observations of a captive male a similar display was observed, but with the bird crouching on his horizontal perch, with the breast shield near the perch and the cape not raised so far. The nasal and lower mandibular tufts were erected, the mouth was initially opened (Figure 55B), and loud wing clicking occurred occasionally during the display. Swaying from side to side and slight hissing sounds may also occur during high-intensity display.

The identity of the "eye spots" that become visible during the high-intensity display, which were also described by Manson-Bahr (1935) and Morrison-Scott (1936), is still somewhat uncertain. They probably consist of specialized crown feathers that are not evident upon handling preserved study specimens of these birds.

As further examples of the dispersed-and-territorial-males category, the several species of *Epimachus* offer additional information. One of these is the previously little-studied pale-billed sicklebill of the lowland forests of western New Guinea, which was the subject of recent research by Beehler and Beehler (1986). It is a medium-sized bird (about 35 centimeters long), both sexes with decurved bills and rufous tails. The adult males differ from females in being blackish brown above and mostly smoky gray below, with erectile, iridescent pectoral feathers around the base of the greenish neck, short erectile plumes behind the nostrils, and a pale yellowish gape. Both sexes have a partially bare bluish face and pale yellowish bills; females are slightly smaller in mass (male:female ratio 1.2:1) (Beehler, 1987b).

The birds primarily occur in older forests having some large canopy trees, but sometimes also use selectively logged forests or forest edges. Males defend small foraging territories (but home ranges may be at least 15 hectares), with territorial boundaries defended by countersinging or chases. They forage on a mixture of insects and fruits, mostly the former. Displaying males are spaced out at distances of several hundred meters. They advertise from trees in early morning and again in late afternoon, calling at maximum rates of about once per minute. The song is a series of hoarse or hollow whistles, with an average content of eight notes and length of about five seconds. This vocalization is fairly high-pitched and carries considerable distances. Displays are preceded by advertisement calls, but the display itself is silent. It consists of perching upright on a branch, with the upper pectoral plumes (edged with coppery and purple tones) erected to form an Elizabethan ruff around the base of the head, and the longer, lower breast feathers fluffed out into a wide skirt, with the tail spread (Figure 56D). The high point of the display is marked by extended bill rattling, and the posture may be held for up to 10 seconds (Beehler and Beehler, 1986).

The brown sicklebill of the central New Guinea cordillera is rather different in appearance, and is somewhat larger than the pale-billed. Both sexes have black bills and greatly elongated tails that in adult males may increase the bird's total length to about a meter (Figure 57). Males also have a pale blue iris contrasting with a mostly blackish head, a yellow gape, a sooty brown breast, and longer iridescent purple- and violet-tinted pectoral plumes that are longer than those of the buff-billed sicklebill. Males sing loudly from dispersed arboreal perches, the bursts of notes sounding like the sounds of a pneumatic hammer or a machine gun. These males are certainly in auditory contact with other males in the forest, given their very loud calls. The male's posturing is so far described only from captive individuals (Crandall, 1932, 1946). Crandall stated that from a resting position (Figure 57A) the bird initially expands his pectoral plumes, which then extend outward and upward, eventually forming a complete fringe around the sides of the body as seen from the front. He then utters his

56. Displays of male birds-of-paradise, including ventral and lateral views of twelve-wired bird-of-paradise (*A,* after sketches in Crandall, 1932), Victoria's riflebird (*B,* after photo in Beehler, 1983c), magnificent riflebird (*C,* after photo in Gilliard, 1969; rear view inset sketch after Coates, 1990), and pale-billed sicklebill (*D,* after painting in Beehler and Beehler, 1986).

57. Displays of male brown sicklebill, including resting posture (*A*), pectoral raising with gaping as seen from the side (*B*), and pectoral raising with tail spreading as seen from the front (*C*). Adapted from photos in Crandall (1946).

rattling call and turns his breast upward, spreading the breast feathers so that his body appears to be flattened in front. The upper pectoral feathers closely encircle the throat, making the iridescent black of that area conspicuous. The bill is closed, the wings are folded and the tail is spread slightly. In another, perhaps more intense form of the display the body is suddenly drawn erect, the breast feathers are widely spread, and the breast shield is thrown up so as to completely enclose the head and frame the pale bluish eyes. The beak is widely opened, exposing the yellow gape, and the tail is somewhat spread (Figure 57B, C). This stationary position may be held for about five seconds, or the bird may begin to rotate his body in a series of short jerks, until he is at right angles to his perch, and then jerks back in the other direction until he is again at right angles to the perch. This rotational movement may continue for up to five minutes.

Birds-of-paradise with Solitary but Nonterritorial Males

There are several species of birds-of-paradise in which the males have established, dispersed display sites but do not appear to establish exclusive-use foraging territories, at least in the case of the magnificent bird-of-paradise. Some of these species (such as the magnificent riflebird and twelve-wired bird-of-paradise) are mostly insectivorous, and others (such as the magnificent bird-of-paradise) appear to be mostly frugivorous (Diamond, 1986).

Adults of the magnificent bird-of-paradise weigh about 90 grams, or nearly the same as the superb bird-of-paradise, but the species ranges from hills near sea level to about 1,750 meters elevation. Males are small (about 18 centimeters long), with very short tails except for the two wiry and iridescent green central feathers that gracefully spread outward from each other in opposite curves. The male has a golden yellow dorsal collar, mostly orange wings, and a large iridescent green ventral breast shield that hides most of the underparts. The lower neck and edges of the ventral breast shield are marked with iridescent blue barring. The bill is pale blue, there is also a patch of pale blue skin around and behind the eyes, and the mouth interior is pale green. Females are slightly smaller (male:female mass ratio 1.2:1), and are mostly dark brown above and barred with brown below. They also have bluish bills and a small bluish patch of skin around and behind the eyes.

Males maintain environmentally modified display sites (courts) that they attend for about seven months of the year, and they maintain ownership of these courts between seasons. Home ranges are not defended, and males may visit and even occasionally display at a neighbor's site when the latter is absent. Females interact with males only at display sites. The sites of nearest-neighbor resident males averaged about 200 meters apart in one study, and one male maintained two nearby display sites simultaneously. The estimated home range was 15–25 hectares (Beehler and Pruett-Jones, 1983; Beehler, 1987b). The courts of each male consist of well-tended areas of forest floor about 5 meters in diameter from which all loose vegetation has been removed, on sloping and eroding ground covered by low forest regrowth, and on which several vertical or nearly vertical stripped saplings are present. These sites face various directions, but always have a clear cone of light penetrating from the canopy vegetation to the ground, owing to leaf removal by the bird (Rand, 1940).

Males are present at their sites each morning, but may leave them near noon to return again in the afternoon. Much of a male's time there is spent preening, in clearing the display ground, or in plucking leaves or bark fragments from living vegetation. He frequently calls from a perch near the edge of his court. These notes are loud, strident, and far-carrying, and often are given in a trilled and rolling series. The male's wings make a loud rustling sound during flight, and sometimes produce clacking mechanical noises similar to those made by striking two pebbles together.

The male's displays have been described for a captive individual by Seth-Smith (1923), and for three wild birds by Rand (1940). Rand's description is most complete and probably most representative of the species. He described three distinct male displays. The first of these, pulsing of the male's breast shield, was quite variable. The body was held in a normal perching position and the shield variously expanded, from only slight expansion (Figure 58A) to an extreme in which the upper part of the shield extended up to each side of the head (Figure 58B). Undulation of the shield sends glossy shimmers across it, and an iridescent spot just in front of each eye becomes highly conspicuous. This display was only seen when the bird was near the ground on a perch, and when displayed toward another bird the breast was usually oriented toward it.

The second display was a very stereotyped horizontal display (Figure 58D). It was always performed when the bird was clinging to a vertical sapling a short distance above ground. He would suddenly extend his body horizontally, with the breast shield maximally expanded and flattened in a heartlike shape. The yellow cape was raised and brought in line with the body axis. In Seth-Smith's account, the pale green mouth lining is at times exposed by gaping, and the head is thrown back and the cocked tail directed so far forward that the two long tail plumes reach the head and sway from side to side. In one case, Rand observed that copulation followed the horizontal display after a female had paused above a displaying male. The male performed the horizontal display, with the crest straight out, his tail held in line with the body and its long feathers vibrating. This posture was held for about 30 seconds as the female gradually hopped closer. The male then rather deliberately hopped up and mounted her.

The third display, called pecking by Rand, consisted of the tail being cocked to a vertical position and the mouth opened widely to expose the greenish lining (Figure 58E). This posture was seen only once and immediately followed copulation. The male vigorously and repeatedly pecked at the nape of the female, then drew back after each peck, exposing his gape, soon after which the female flew off.

58. Displays of male magnificent bird-of-paradise, including breast shield display (*A*), later phase of breast display (*B*), and "final" (cape) display (*C*). After sketches in Seth-Smith (1923). Also shown is a male in "horizontal" (cape) display (*D*) and in postcopulatory gaping and tail-cocking display (*E*). After sketches in Rand (1940).

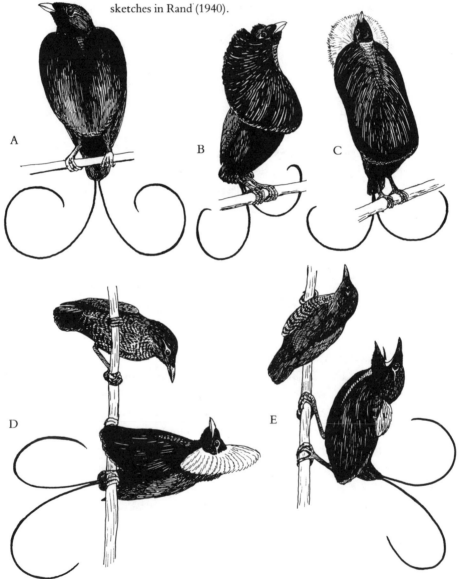

Coates (1990), who has described and photographed these remarkable displays, recognized a total of five separate display elements. These include (in usual sequence) back display, breast display, intense breast display (= horizontal display of Rand), cape display, and dancing display. Back display is a silent display performed initially, while in an inverted posture. This is followed by a pulsing of the iridescent breast in a horizontal posture, with low calling, and directed toward any female that may be present. When a female is present, an intense form of this display is performed with maximum breast spreading and urgent calling. This serves to attract the female; when she is quite close the cape display is performed, with a fully erected and expanded cape, drawing her even closer. When the female is very close, the male performs a remarkable dancing display, with his tail cocked and quivering, his bill opened, his neck jerked from side to side, and a clicking, buzzing song uttered. This last display may last up to 12 seconds, and probably precedes copulation.

Another group of birds-of-paradise whose dispersed displaying males are not known to exhibit foraging territoriality are the riflebirds and the twelve-wired bird-of-paradise. The twelve-wired bird-of-paradise is a fairly large and rather long-billed species (males weighing about 200 grams) that occurs along the New Guinea lowlands and on Salawati Island. The sexes are similar in size but very different in appearance. Males have a glossy purplish head, neck, breast, and upperparts, and lemon-yellow underparts. The breast shield is edged with glossy green, and on each side six of the flank feathers have greatly extended shafts that protrude and curve back toward the breast. The birds consume a mixture of arthropods (mostly insects and their larvae) and plant materials, the latter including fruits, berries, and nectar. These birds display solitarily, males apparently calling daily from the same area of forest but also ranging widely from such display sites while foraging; Beehler and Pruett-Jones (1983) reported a mean nearest-neighbor distance of 730 meters.

The displays of the twelve-wired bird-of-paradise were first described by Crandall (1937a, 1937b) from a captive bird that survived more than 22 years in the Bronx Zoo, and which had required seven years to attain its fully adult plumage. When perching normally, the "wires" for which the species is named were not organized in any special way, but during display posturing they were brought into alignment in the horizontal plane (Figure 56A, right). The bird first stood horizontally relative to its perch, with its green-edged breast shield widely extended but the yellow underparts compressed and the wings tightly closed (Figure 56A, left). It then jumped sideways to the vertical "trunk" of the perch and turned around it slowly, repeatedly uttering sharp, metallic notes and gaping widely during each note, thus exposing its green lining. The bird then leaped back to the perch from which it started, and again stood with the body in a horizontal axis. It then rapidly opened and closed its wings 10–12 times. Once the bird spiraled slowly down the trunk of the perch, head downward, instead of going directly around it.

In a rare observation of wild birds, Bergman (1957b) observed similar displays on and near a withered and leafless tree branch about 20 meters above ground at the top of a live tree on Salawati Island. In this case a female was present, and the two birds "twirled" around the tree's top branch, the female above the male. She then climbed to the top of the branch and mating occurred. Later, two females appeared and display continued, but this time it did not lead to mating. Coates (1990) also observed one copulation, which was preceded by the male facing the female above him, expanding his breast feathers, and pecking at her body. Evidently most male display occurs on one or more display posts (the tops of bare and broken-off branches high in a tree) that are dispersed through the male's range, with one such post being favored.

The riflebirds (so named because of the explosive, riflelike calls of the males) occur in both New Guinea (one species) and eastern Australia (three species, the magnificent also occurring in New Guinea). The magnificent riflebird is about the same size as the twelve-wired bird-of-paradise (males about 200 grams), and is somewhat similar in appearance. It has a long bill and lives on a mixed diet of insects and plant materials, mostly fruits and berries, with fruit comprising about 70 percent of the diet (Beehler, and Pruett-Jones, 1983). Adult males are mostly glossy bluish black, with a large iridescent blue-green breast shield that extends forward to the chin and is edged below with bronzy-green and reddish tints. The male's crown is also iridescent greenish blue. The flank plumes are extended as filamentous hairlike feathers, but none of these recurve forward as in the twelve-wired.

The magnificent riflebird occupies primary rain forest habitats up to about 1,600 meters of elevation; Beehler and Pruett-Jones (1983) found males to be dispersed at average nearest-neighbor distances of 175 meters. Males called from fixed areas of forest, suggesting that like the twelve-wired they are widely ranging but perhaps nonterritorial. The very loud whistling notes of males carry great distances. Crandall and Leister (1937) described the displays of a captive male that displayed to a stuffed female. Perched on a sloping branch, he would perform two types of display, one long and one short. The short display began when he jerked his head and neck a few times, then threw his wings open to their fullest extent. He simultaneously extended his neck and moved his head to just behind the bend on one wing. He then moved his head back and forth (Figure 56C) with increasing speed until this was brought to an abrupt stop after about a dozen such swaying movements. The wings were then folded. In the longer version the head movements were punctuated with rustling-plopping sounds made by the wings as they were repeatedly snapped back from a somewhat relaxed and elevated position to one of maximal extension. After up to about 35 such swings and wing snaps the display ended.

Similar displays have been seen in two wild males, who performed mutually to one another while perched on a nearly horizontal vine about 3 meters above ground, and in another case by a lone male displaying on a sunlit woody vine about 4 meters above ground. In both of these cases blowing or croaking sounds were produced, in contrast to the observations of Crandall. It would seem that this display may serve both to attract females and as a male–male agonistic display (Cooper and Forshaw, 1979). Displays have been observed in immature as well as adult males (Coates, 1990). True vocalizations by displaying males may be lacking, but a hissing or swishing sound was produced during the head-swinging phase, and a deep blowing sound was associated with the preliminary neck-stretching phase.

The displays of the Victoria's riflebird of northeastern Australia are quite similar to those just described. This is a smaller bird, and males have considerably smaller breast shields, although they have iridescent central tail feathers. Descriptions summarized by Gilliard (1969) and in Cooper and Forshaw (1979) indicate that the wings are extended and raised so that they nearly touch overhead, and the breast shield is expanded. The bill is pointed upward, exposing the green throat, and the head is twisted and turned as the entire body is swayed forward and backward (Figure 56B). The tail is cocked high over the back, and loud, "discordant" notes are uttered, probably in conjunction with gaping. This display is reportedly often performed on high tree branches.

Birds-of-paradise in Which Males Are Solitary but of Unknown Territoriality

The Wilson's bird-of-paradise is fairly closely related to the magnificent bird-of-paradise, described earlier; these two species are also related to the king bird-of-paradise, which is often placed in a separate genus. Though related, the three species appear to follow somewhat different patterns of dispersal. All three are extremely small species; males of the king bird-of-paradise average only about 50 grams and are only about 16 centimeters long, exclusive of their ornamental wirelike central tail feathers. The king is a lowland species occurring over much of New Guinea and its adjoining islands, and is sympatric with the more upland-adapted magnificent. The little-studied Wilson's is confined to two small islands (Waigeo and Batanta) off the west coast of New Guinea, and the king occurs on at least one of these islands (Batanta).

The Wilson's bird-of-paradise is of interest because of its constructed display courts, which are quite similar to those of the magnificent. The male somewhat resembles the magnificent, but its central tail feathers are shorter, wider, and more tightly coiled, it has a shorter yellow "cape," most of its wing surface is red rather than golden yellow, and the top of its head is mostly bare, bright blue skin, interrupted by rows of short black feathers organized in a curious pattern. Females also have a mostly bare blue head with a similar feather pattern on the crown, and otherwise are dark brown above and barred with brown and buff below.

In a zoo situation, one male constructed a court around a sapling and would quickly remove any twigs that were placed within it (Gilliard, 1969). The male's displays have been described from a captive bird by Frith (1974); this male also performed leaf plucking over a fairly large part of the aviary in which it was housed. Other observations on the displays of captive birds have been summarized by Gilliard (1969) and by Cooper and Forshaw (1979). Males occasionally utter loud, single notes that may be audible for about 40–50 meters, while standing on horizontal display perches. In the presence of a female, a male may also quietly "pulse" the green breast shield and somewhat erect his cape and back feathers while tilting his bill slightly upward and standing on a horizontal perch (Figure 59A). However, high-intensity displays are performed from vertical saplings or bamboo stems, less than a meter above ground. The male perches crosswise on such sites, looking upward with his cape slightly raised. If a female is nearby he raises his cape even more, expands his breast shield, and holds his bill close to the perch (Figure 59B). At such times the shield is maximally spread, and the yellow cape is raised to the vertical. The male may thus move up the stem toward the female, while keeping his bill close to the sapling and holding the ornamental tail "wires" out from the stem at right angles to the body, much as was described earlier for the magnificent (Figure 58E). The bird may also gape, exposing his bright yellow-green mouth lining.

Of the solitarily displaying species of unknown territoriality, probably the best described as to its sexual displays is the blue bird-of-paradise. It is a remarkable species for many reasons, one of which is certainly the beauty of the adult male. It occurs in the mountains of eastern New Guinea, in the tall rain forests of midmontane elevations. It is a bird of moderate size (about 30 centimeters in length) and forages on a mixture of insects and fruit that are obtained at canopy level. Males are the only bird-of-paradise species that have opalescent blue wings and deep blue flank plumes. They are otherwise largely black, the slightly iridescent head with contrasting very pale blue to ivory-white bill and incomplete white eye rings. Extending across the lower grayish black breast is an indistinct (except during display), oval black band that is edged below with maroon-red. Below this the feathers become more filamentous and are deep cobalt-blue, this color lightening somewhat toward the tips. Two long, streamerlike central tail feathers are also present. These narrow feathers are almost entirely black but are often tipped with iridescent blue spots. One fully plumaged male weighed about 180 grams, as compared with 165 grams for an immature male (LeCroy, 1981). Female weights are still unavailable, but female measurements are not greatly smaller than those of adult males.

In contrast to all the other species of *Paradisaea,* the blue bird-of-paradise displays solitarily. Reasons for this behavioral anomaly are still uncertain, but Pruett-Jones and Pruett-Jones (1988a) confirmed that the species is promiscuous like the other species of this genus, rather than having evolved pair-bonding and territorial behavior as LeCroy (1981) suggested.

The male advertisement displays of this remarkably beautiful species were first described and illustrated by Crandall (1921, 1932, 1936). At the start of display the male sits on his perch and calls repeatedly. In the wild this occurs on an exposed branch that

59. Displays of male Wilson's bird-of-paradise, including pulsing of breast shield (*A*), and frontal display toward female (*B*). After sketches in Frith (1974). Also those of male king bird-of-paradise, including frontal (dancing) display while "juggling" tail tips (*C*), wing-spreading (wing-cupping) display (*D*), and pendulum display (*E*). Adapted from drawings in Ingram (1907) and photos in Coates (1990).

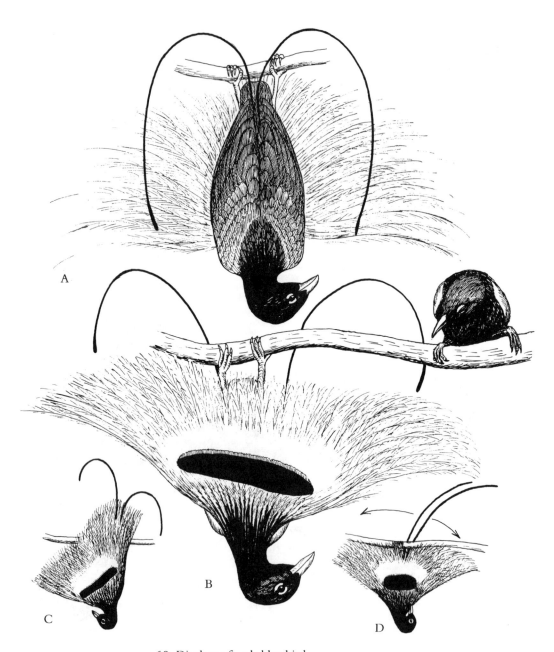

60. Displays of male blue bird-of-paradise, including rear (*A*) and front (*B*) views of inverted display, and swaying of body and central rectrices (*C, D*). After sketches in Cooper and Forshaw (1979) and photos in Coates (1990).

is situated so as to catch the morning sun. He then slowly and cautiously lowers himself backward until his entire body is hanging upside-down. He may remain in this static posture for several minutes without moving his feet, with the wings closed, the head pointing upward, the abdominal feathers spread, and the ornamental tail plumes hanging down on either side of the back. Then he shakes his body, extending his plumes laterally into a filmy blue spray (Figure 60A, B). Using his pelvis as a fulcrum he then rocks or sways his body, causing the blue plumes to flutter and shimmer, and begins to generate a strange buzzing sound of presumably vocal origin (LeCroy, 1981). Additionally, the black oval breast pattern becomes highly conspicuous, rhythmically expanding and contracting in size as the male begins to display in his strange mechanical manner. He also swings his central tail feathers from side to side across a wide arc (Figure 60C, D). During the first part of the display the male's head and bill are turned to one side and are oriented slightly above the horizontal, but later as the tail swinging begins he brings his head forward and turns it more directly upward. This latter phase occurred when a female-plumaged bird stood directly above the displaying male, her feet probably positioned between those of the male. One entire display sequence lasted about three minutes (the first 30 seconds performed in silence), after which the male lifted himself forward and upward to a normal perching posture, displacing the female (Cooper and Forshaw, 1979).

These incredibly beautiful displays have recently been fully described and photographed by Coates (1990). He has observed that during inverted display the male's head may be pointed to either side or sometimes up toward the chest, and the eyes may be almost closed. The associated call is a continuous and fast rhythmic buzzing, with the plumes moved in coordination with the pulsating rhythm of the song. The display perch is a shaded, slim, and gently to steeply sloping branch or stem. When a female is present nearby on the same branch, the male turns his head and twists his body to orient them toward her, and the black breast area becomes less oval and distinctly crescentic in shape. After about a minute of intense display toward a female, the male body is more strongly directed toward her, and his tail is quickly swung from side to side, causing the long wirelike plumes to lash back and forth. The male may also sometimes try to approach the female while in this inverted posture, or may then right himself on the branch.

Birds-of-paradise That Display in Exploded Leks

Gilliard (1969: 69–76) described the distribution of displaying males of numerous bird-of-paradise genera as representing "exploded clans," or organized in "loosely clustered" or "widely spaced" groups. However, their dispersion patterns have been variously reassessed by more recent writers. Thus, Diamond (1986) did not recognize an exploded lek category as such, and Beehler and Pruett-Jones (1983) described only a few species as certainly falling in such a category. These exploded lek species include the king bird-of-paradise, at least two of the four generally recognized *Parotia* or six-wired species, and perhaps one or two forms of *Paradisaea*. The little-studied standardwing may perhaps also fall into the exploded lek category, although it is more often considered (e.g., Diamond, 1986) to be a true lekking form.

As noted earlier, the king bird-of-paradise is perhaps the smallest species of the entire family, and certainly is one of the most attractive. Males are almost entirely a brilliant crimson on the upperparts, head, neck, and breast, the pale eyes being set off by black rounded or exclamation-shaped "eyebrows," and between the red breast and wings are two normally hidden fanlike pectoral plumes that are brown, bordered with emerald green. The remaining underparts are white, the legs and feet are bright blue, and the tail is short and mostly hidden, except for two long and wirelike central rectrices that are lyre-shaped and end in tightly coiled circles of iridescent green barbs. Females are dark brown above and barred with brown and buff below, and strongly resemble females of the closely related magnificent bird-of-paradise.

The king bird-of-paradise is perhaps only questionably a part of the exploded lek group. Beehler and Pruett-Jones (1983) reported that in one area a total of nine males were organized as four pairs and one solitary bird. In another area 22 males were distributed as 8 pairs, one single bird, and one quintet. The mean intermale distances in the two areas were 64 and 67 meters, and the mean interpair distances 219 and 333 meters. Males regularly call and display from specific trees or vine tangles, but there is no evidence that they defend any areas other than their display sites or restrict their movements to areas near the display site. Coates (1990) stated that although males occur in small groups, each male is the sole occupant of his territory, which has a primary display site and two or more nearby secondary sites 25–50 meters away from the primary site. Although male king birds-of-paradise seem frequently to be distributed across their habitat as dispersed clusters or "pairs," there is no indication that these males ever display in concert or as cooperating groups.

Although certainly closely related to the magnificent and Wilson's birds-of-paradise, and though several wild hybrids with the magnificent are known, the king does not establish ground courts and also evidently does not display from vertical saplings, although vertical vines are usually present. It has so far been best observed in captivity (Ingram, 1907; Bergman, 1957a), although notes on birds displaying in the wild have also recently appeared (Coates, 1990). It evidently displays and calls from branches about 10 meters above ground, or just below the canopy in shaded areas, calling with harsh, repeated *wah* notes that are varied with other raspy or buzzing calls. In one case, a captive male began the display with a series of short notes, then spread or flapped his wings in a cuplike manner, nearly hiding his head (Figure 59D). He then puffed out his white abdominal plumage and began a warbling song. The iridescent-tipped fanlike pectoral feathers were alternately opened and closed, the bill was raised, the tail was cocked over the back, and the body was gently swayed, causing the coinlike tips of the central tail feathers to bounce or "juggle" from one side to the other (Figure

59C). The bird then turned around, exhibiting his back, and bent down on his perch in the attitude of a fighting cock, gaped widely, exposing a yellow-green mouth lining, and sang his gurgling song with slowly diminishing movements of his tail and body. In a few cases the male was observed to drop down underneath his perch and walk backward and forward while inverted, and with his wings expanded. He would then close his wings and let his body fall so that his legs were stretched to their full length before stopping, singing all the while (Ingram, 1907).

Some of the same postures seen by Ingram have more recently been described by Bergman (1957a) and Coates (1990). Bergman noted that the female might watch from as close as 5 centimeters during the male's display, sometimes picking at the male's feathers. During singing the male's tail wires were in constant motion, and he typically displayed in the tail-cocked posture while facing away from the female. After this posture (as shown in Figure 59C) had been held for a time, the bird turned around to display in the opposite direction, now facing the female. His pectoral fans were then usually but not always retracted and his bill widely opened. Bergman stated that the first phase of the upside-down display usually lasts 5–10 seconds, and that the second phase, when the bird sways pendulum-like from side to side with closed wings and extended legs, has a similar duration, but on one occasion lasted for nearly a minute. The upside-down display was seen only after a female had been placed in the aviary. She was once seen in a soliciting posture, with her body lowered against a branch, her tail raised, and her head tilted upward in the same manner as the male's during his usual display.

The male's upside-down "pendulum display" (Figure 59E) has also recently been observed in wild birds and photographed by Coates (1990), who noted that in this posture the bird resembles a brightly colored hanging fruit, but swings silently from side to side. Coates further stated that the wing-cupping display phase precedes the singing ("dancing") display, which is typically performed facing away from the female and ended by a tail-swinging phase after the male has suddenly turned around. At times this is then followed by a horizontal open-wings display, with his body held horizontally and rocked from side to side as both wings are spread and vibrated. He may then flip down and perform his upside-down ("inverted") display, with the wings opened, and finally the pendulum display. In one observation of copulation, the male performed dancing and wing quivering prior to mounting.

Perhaps better candidates as exploded lek species are the parotias, a group of four (or five) medium-sized species that occur mainly allopatrically across the mountain ranges of New Guinea. They are thrush- to jay-sized birds that weigh about 150 grams and are about 25–45 centimeters long, depending on tail length. In general, the males are mostly black, with contrasting white, pale-colored, or sometimes iridescent feathers that probably show up well in the sometimes shady areas of forest floor on which they individually perform. Males also have six greatly elongated feathers that extend out from above and behind the eyes like so many flag-tipped antennae, which are the basis for their alternative vernacular name, six-wired birds-of-paradise. Frith and Frith (1981) commented that, in contrast to most birds-of-paradise, male parotias have extravagant shapes rather than extravagant colors, and that the differences in shape and length of the tail cause different male silhouettes during their essentially silent displays. Similarly, Schodde and McKean (1973) noted that male head plumage differences evident during display may be important isolating mechanisms. Frith and Frith suggested that the display of Lawes' parotia is even more complex than that of the red bird-of-paradise, sometimes considered the most complex of any species in the family. They suggest that, by displaying on the ground, male parotias can incorporate complex step routines into a dance, which is further complicated by feather adjustments and changes in basic body postures during their stepping behavior. These authors believed that court distributions of parotias may be either arranged at random or related to topography, and that apparent multiple occupation of a single court is the result of other males appearing during times outside of the breeding season when the court's owner is molting.

Males of apparently all species of parotias (not yet proven for *helenae,* which is extremely similar to and usually considered conspecific with *lawesii*) clear small areas of forest floor as display courts. In one court of Carola's parotia (*P. carolae*) this consisted of a small (1 by 1.5 meter) area of cleared ground that was partly hidden by undergrowth but was exposed to direct light at certain times of the day. Above the court floor, which was seemingly scratched clean, were several low (under 1 meter) branches (Cooper and Forshaw, 1979). In another site, the cleared area was about 2 meters in diameter, and the leaf canopy above was thick. There the aboveground perches were of similar height, and were nearly horizontally oriented. The most prominent perch was a fairly large dead limb about 5 centimeters in diameter that had been stripped of its bark and was well worn from use (Gilliard, 1969). Display courts of Wahnes' parotia (*P. wahnesi*) are similar small arenas with a few saplings stripped of leaves, and different males' courts may be as close as 5 meters apart. Those of Lawes' parotia (*P. lawesii*) may average somewhat larger, and often contain small display objects (chalk, scats, etc.) that are brought in by the males. Courts controlled by different individuals may be as close as 5 meters apart (Coates, 1990).

Observations on the displays of Carola's parotia in captivity were provided by Frith (1968), based on a single pair. The male began display by flying to its perch, turning away from the hen, and crouching, lifting his white flank plumes above the primaries. He then hopped up on a perch near her and began "sideways facing," rapidly turning from one lateral position opposite her to the other, his feet leaving the perch each time he changed position (Figure 61A). A period of preening followed, after which the male began swaying, with rapidly fluttering wings, while facing away from the female and tilting his head upward. His silvery crest feathers were parted, and golden-brown feathers between them were erected and lowered, flashing silver and gold colors in rapid succession. A similar display involved bobbing of the entire body, and

61. Displays of male parotias, including Carola's parotia during sideways facing (*A*) and high-intensity (*B*) display toward female. After sketches in Frith (1968). Also shown are ballerina poses of Lawes' (*C*) and western (*D*) parotias (inset shows rear view) and upright-sleeked postures of western (*E*) and Lawes' (*F*) parotias. After sketches in Frith and Frith (1981) and photos in Coates (1990).

62. Displays of male parotias, including Wahnes' parotia in initial display (*A*) and dancing (*B*) phases, and Lawes' parotia in dancing (*C*) and agonistic (*D*) pose. Also shown (left inset) is dancing sequence of Lawes' parotia. After sketches and photos in Frith and Frith (1981) and Coates (1990).

occasional maximum wing stretching. In the most intense form of display the bird reared up in an upright stance, raised the black breast and white flank feathers over the wings so that a complete "skirt" formed above the back, and erected the six wire feathers and brought them forward in front of his eyes (Figure 61B). The crest feathers were in continuous movement as before, and the iridescent green throat feathers were flickering continuously. In one case the female watched the male from between her legs while on a separate, higher perch (Figure 61B). This display lasted 90 seconds in one instance, during which the two birds didn't move from their perches, and in another case lasted 9–10 seconds. No ground display or ground clearing was observed. Observations of a wild male by C. J. Healey (quoted in Cooper and Forshaw, 1979) were similar to this description, but the male displayed on a partly buried log rather than an aboveground perch. Pulsing of the iridescent breast shield was additionally observed.

Further information on the behavior of the parotias comes from Frith and Frith (1981), who observed two Lawes' parotias in captivity. They observed that court clearing was done by picking up small items with the beak and throwing them away, whereas larger items were rolled off the court. A browse line was also created in the foliage above the court, all leaves being removed that could be reached and plucked from the ground. All observed displays took place on the court. Before displaying, the male increased his rate of apparent court-clearing behavior, but did not actually pick up any items at this time. Each display was clearly part of a ritualized number of stationary postures that preceded or followed a "dance phase." Pseudo–court clearing, alternating with an upright sleeked pose (Figure 61F), ended with an initial head-lowering display bob (or display bow), when the black frontal feathers of the crest were brought forward above the silvery nasal tufts. The bird then flicked his wings, brought the six wirelike plumes forward to a nearly horizontal plane, and shook out the flank plumes, bringing them up to form a complete skirt. He then adopted a ballerina pose (Figure 61C) and, while shaking his head vigorously, began the dance phase. He might dance sideways and then backward, or move back in a semicircle about a half meter from his starting point (Figure 62, inset). While moving backward he would lower the forward edge of the skirt in a bowlike posture (Figure 62C), but when moving forward he held it horizontally. After the backward dance phase, which lasted about 12 seconds, there was a stationary phase (stage 4 in the inset drawing of Figure 62) marked with crouching and forward-backward movements of the ornamental "flags." The male then would crouch, flick the breast shield out in a flash of color, and bob his head in a snakelike manner while the head flags were partly erected. He then adopted an erect, frozen posture, retracted his flags, only to erect them again and begin a final period of strong head pumping. This head pumping was associated with rapid flashing of the iridescent breast shield, which might be erected almost to a right angle from the bird's breast during the low point of the head-pumping cycle. Finally, all feathers were returned to their normal position. Calling often occurred immediately after display, but not during active display. In agonistic interactions between males, a head-lowered posture resembling that of a threatening bull was assumed (Figure 62D) (Coates, 1990).

Of the six major male displays described for this species (Pruett-Jones, 1985; Pruett-Jones and Pruett-Jones, 1990), four (puff, short puff, hopping, and lunging) were performed only by males on the court surface, and two (bouncing and bill pointing) were observed only on the display perch. Dancing on the court floor (called the "puff" display) was directed both toward females and toward males, and occupied the greatest observed duration of display activity. The "short puff" is an abbreviated version of this same display. Hopping across the court in an erect stance also occurred during interactions with both sexes, whereas lunging across the court occurred prior to hopping up to the display perch. Bouncing on the display perch, while flicking the wings and reversing position, occurred as females approached from above. Bill pointing was the most common precopulatory display, with the male bowing and thrusting his bill forward in front of the female in a jerky manner.

In an instructive study by Pruett-Jones and Pruett-Jones (1990), sexual selection pressures on mating traits were estimated by correlating male behavioral and morphometric variables with individual breeding success. Males individually controlled from one to five display courts, but fewer than half of the males under observation were known to mate during any one year. About a third of the male population displayed solitarily (and independently of their ages), and nearest-neighbor distances between male courts ranged from 2 to 500 meters. Females usually visited displaying males singly, but also rather frequently arrived in groups of up to eight birds. Females began visiting male courts as early as six weeks prior to known copulation, and each female evidently visited most if not all of the male courts (which averaged 17) within her own home range. It was determined that a male's probability of displaying to females (as measured by the percentage of females that were displayed to) during their visits during the breeding season was the single factor most clearly and significantly correlated with individual male mating success. Male-male interactions that at times disrupted courtship or copulation attempts did not significantly influence individual male mating success. Neither did a male's ectoparasite or blood parasite levels, the presence or absence of ornamental objects on his court, court dispersal characteristics, or a male's relative court position within clustered-site groupings. Female choice, rather than relative male dominance, appeared to be the primary factor influencing individual male success and thus directing sexual selection in the species. Females showed strong sexual fidelity to particular males, both within and between successive breeding seasons, during a three-year study period. Sixteen males were observed to obtain 103 copulations with 67 different females throughout this period, but there was no evidence of age-related increases in individual male mating success over time.

Displays of the western (Arfak) parotia (*P. sefilata*) are apparently quite similar to those of the Lawes', judging from accounts of Bergman (1957b, 1958). For example, there is a comparable preliminary upright sleeked pose (Figure 61E). The ballerina pose (Figure 61D) is also very similar, except that the rather long tail is held sideways, apparently to prevent it from getting in the way of the dance. In the very long-tailed Wahnes' parotia the dancing phase is preceded by a preliminary head-lowered display-bow posture, with the male's golden forehead tufts extended forward accompanied by repeated tail fanning (Figure 62A). During dancing (Figure 62B), the bird walks forward and to one side, his head bobbing from side to side (Coates, 1990). The elongated tail is also turned to one side during the dancing phase of display in this species, where it seems to play a negligible role in the activity (Crandall, 1940). This species is often regarded as a relatively primitive member of the genus, which perhaps originated from long-tailed ancestors that once displayed in trees and in which the long tail may have been more adaptively significant (Gilliard, 1969; Frith and Frith, 1981).

Goldie's bird-of-paradise, which is isolated on Fergusson and Normanby Islands off southeastern New Guinea, is one of the *Paradisaea* species that have been described by some observers as displaying in exploded leks (LeCroy, Kulupi, and Peckover, 1980; LeCroy, 1981). The displaying males are often distributed 50–100 meters apart within large arenas encompassing several trees, although as many as four fully plumed and five unplumed males may at times be present in a single tree. Plumed males are most frequently distributed two per tree, in which case duetting behavior and close interactions between them are common. Thus it would seem that in many cases visual as well as acoustic interactions are frequent in competing males, and indeed Diamond (1986) considered the Goldie's to be a typical lek species. A similar situation applies in two other classic lekking forms (*guilielmi, raggiana*), in which several males usually assemble in different but nearby trees and then regularly converge and display together in a single large tree (Gilliard, 1969; LeCroy, 1981; Beehler, 1988). Beehler (1988) commented that in the raggiana bird-of-paradise subordinate males are similarly normally dispersed in hidden perches, but that when females arrive at the tree used by dominant males the subordinate males converge on it and form a typical lek. Such examples of seemingly variable male dispersions simply indicate the problem of clearly distinguishing "exploded" leks from typical leks, which might well simply grade into one another. These seemingly variably exploded lek arrangements offer no special interpretational problems nor do they provide any unique insights, and can be most conveniently discussed with the more typical lek-displaying species of *Paradisaea*.

Birds-of-paradise That Display in Arboreal Leks

In this group all of the species of *Paradisaea* can be discussed except for the blue bird-of-paradise, described earlier. The standardwing evidently belongs in this category as well (Goodfellow, 1927; Bishop, 1992). Additionally, at least one species of *Astrapia,* Stephanie's astrapia, is known to perform communally (Healey, 1978), and it is possible that typical lek displays occur in other astrapias as well (LeCroy, 1981).

LeCroy (1981) stated that leklike ("arena") breeding systems often occur in polygynous bird species that typically share a number of traits, including (1) extreme adult sexual dichromatism, (2) considerable adult sexual dimorphism, (3) relatively few adult-plumaged mature males in the total population, relative to younger "unplumed" males and females, (4) loud male advertisement vocalizations, (5) conspicuous displays by males over long seasonal periods, and (6) the included species often part of species-rich genera. To this list could be added such traits as a highly skewed and age-, experience-, or dominance-related male mating success, a prolonged period of male immaturity and associated subordination in competitive display, and the use of traditional mating sites that are often ecologically especially suited for visual or acoustic transmission of signals.

Most of these traits are present in the standardwing, a little-known and geographically isolated bird that occurs in the northern Moluccas rather than in New Guinea. It is a medium-sized species, the adults being about 25 centimeters in length. The plumages of both sexes are mostly a rather nondescript brown, unlike the typically highly dimorphic plumages of many other birds-of-paradise. However, adult males also have a large and iridescent green breast shield, the male's crown is an iridescent lilac color, and two unique white pennantlike feathers emerge from the bend of each wing. An account by Goodfellow (1927) indicates that 30 or more males may aggregate in areas of stunted jungle, where the trees are only saplings about 10 meters high with few branches, and with a few enormous trees also present. The numerous males fluttered constantly about from tree to tree, at times "hanging in all positions," some "turning round and round like a cartwheel," and continuously calling. This description would certainly seem to indicate a typical lekking behavior, and more recent observations by Bishop (1992) confirm that a typical arboreal lekking arrangement exists.

A detailed description of male display was provided by Friedmann (1934), based on observations of a single captive male. In resting posture (Figure 63A) the breast shield and wing pennants are held down along the upper surface of the wing. In early stages of display the bird bends forward, so that his head is lower than his body, and utters several rasping calls. He then flaps or vibrates his partially spread wings rapidly three or four times (Figure 63B). After a few such beats he holds his wings arched horizontally, and his entire body quivers or vibrates briefly. The wings are held so that they nearly touch at the wrists, and the inner pair of pennantlike feathers cross over one another above the body of the bird (Figure 63C, left). In full display the green gorget is fully elevated away from the male's body, and his iridescent crown feathers are ruffled and lowered, causing a rolling beam of light to travel from the forehead to the occiput. Occasionally the male might bend so far forward and downward that he is

63. Displays of male standardwing, including resting position (*A*), beginning of display (*B*), and full display (*C*). Adapted from sketches in Friedmann (1934). Inset (*D*) shows descent phase of aerial display (after Frith, 1992).

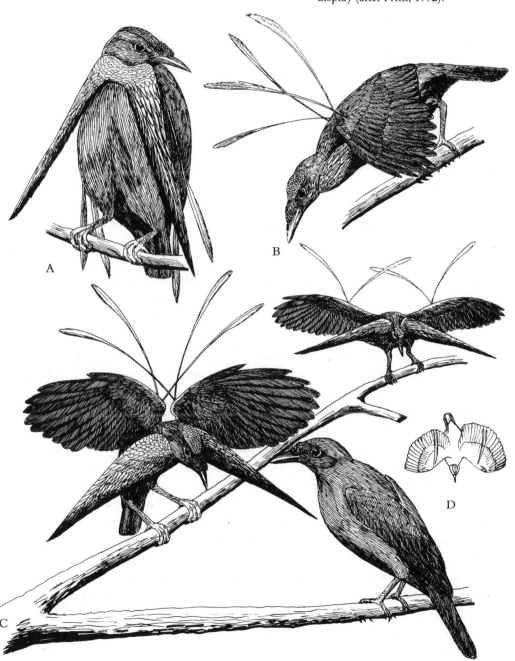

actually upside down. At this point the display might end, or with the bill still pointing downward the wings might suddenly be fully stretched (Figure 63C, right), then
finally folded. Or the bird might then lean back on his perch with his wings half-
arched, presenting the female with a better view of his gorget. He may then release
hold of his perch and tumble over in the air in a backward somersault, finally landing
with closed wings on the ground. In some additional observations, Stephan (1967)
noted that it is the front pennant feather of each wing that is turned away from the
body and somewhat follows the curve of the wing outward and backward during
display, whereas each rear feather is turned inward toward the body and also extends
forward, its tip extending over and in front of the wing and breast shield (Figure 63B).

Recent observations of wild birds (Bishop, 1992) and a later film analysis of their
displays (Frith, 1992) have added several new details. One of these is the discovery of a
vertical aerial display, with a strong ascent followed by a rapid descent (Figure 63D).
This display is unlike any previously known for the family, but may be related to the
inverted display of other species. During the perched "standard presentation" display
the wingtips are vibrated, causing a "rowing effect," and their sudden shifts from a
half-open to a fully open position produce gyrations of the erect standards and evidently generate a cracking sound. Bill-clicking sounds are also produced, as are harsh,
buzzy vocalizations. Males also perform foliage clearing at display sites.

All of LeCroy's typical characteristics of lekking species apply especially well to the
seven species of birds-of-paradise in the genus *Paradisaea*. All of the species in this
genus have males with gloriously colored and ornamental adult plumages. This is
especially true of their anterior flank (pectoral) plumes, which when erected form
beautiful, airy and cascading sprays of color. They also have variously elaborated,
wirelike central tail feathers that extend out well beyond the other rectrices and the
ornamental flank plumes. Most of the species lack iridescent colors except on the
throat and forehead (but extending to the breast on one species), and instead primarily
have richly pigmented plumages. Females are generally dark brown above and are
mostly uniformly white, gray, or brown below, although with some exceptions. The
average weights of fully plumed males range from about 200 to 300 grams, with unplumed males averaging somewhat less and females still smaller. The average
male:female adult mass ratios for several species range from about 1.3:1 to 1.9:1, although available sample sizes are quite small (LeCroy, 1981). The males also produce
loud, rather crowlike vocalizations or mechanical noises, and display for extended
seasonal periods in traditionally used arboreal sites. These sites might perhaps qualify
as "courts" inasmuch as leaf-plucking behavior is present in most species, and such
behavior perhaps serves to create display perches and to keep the area around them
clear (LeCroy, 1981). Most species of *Paradisaea* have subspeciated into several geographic races, and in zones of interspecific contact they also frequently hybridize.

In her description of the male sexual behavior patterns of the *Paradisaea* species,
LeCroy (1981) emphasized the importance of distinguishing male-male behaviors

from heterosexual interactions possibly leading to copulation. She recognized four major male-male *Paradisaea* displays that she considered to be associated with establishing and maintaining male dominance relationships. These include wing posing, charging, zigzagging, and male-male duetting. She also identified four specifically male-female displays, including flower display, inverted display, hopping, and copulation. Other possible male displays or display components, such as bill wiping, pecking-at-perch, ritualized preening, leaf plucking, sun-bathing, seed regurgitation, and the "butterfly dance" (renamed by LeCroy as charging), were also discussed. She implied that females preferentially mate with the most dominant male, after male-male displays have established a stable male dominance hierarchy, with other possible aspects of female choice (such as phenotypic plumage or display comparisons) playing a less significant role.

LeCroy considered that the red bird-of-paradise, based on its rather generalized display, insular distribution, absence of plume spreading during display, and small body size (assuming that *Paradisaea* evolved from smaller ancestors), might be the closest to ancestral stock, reflecting a view that had earlier been suggested by Stonor (1936). However, Frith (1976) argued that the red bird-of-paradise's display is actually one of the most complex of any of the entire family and includes features remarkably similar to those of the king, magnificent, and Wilson's. He also hypothesized that the more elaborate head feathering of the red bird-of-paradise might be a visual compensation for its possible secondary loss of long and ornamental flank feathers, rather than being a primitive feature. Lastly, Frith argued that this species may be intermediate behaviorally between the quite small birds-of-paradise that display solitarily on near-ground vertical perches (such as the magnificent) and the larger *Paradisaea* species that regularly engage in communal arboreal display on horizontal perches.

LeCroy (1981) suggested that the blue bird-of-paradise likewise might be a primitive species in the genus *Paradisaea,* judging from its possible (and presumed by her to be primitive) pair-bonding tendencies and its rather small size. However, she thought it might equally well be a recently derived form, based on its distinctive female plumage pattern and possible male territorial dispersion, and that its relatively short male plumes might also be a derived trait associated with selection for species recognition in areas of sympatry with other more long-plumed *Paradisaea* species. This would seem to be a rather unproductive argument to try to defend or develop, given the present quite limited available information on interactions among these species. Indeed, it seems clear that no fully acceptable scenario of behavioral and structural evolution in this remarkable genus is yet possible, and so these birds-of-paradise will be discussed here in no particular order other than logical convenience.

Frith (1976) has provided the best information on the red bird-of-paradise, which occurs in lowland forests on the islands of Waigeo, Batanta, and Saonek off the western tip of New Guinea. He observed 11 captive birds, including 10 males. These included four fully plumaged and six nearly full-plumaged males that had been in captivity for at least six years, indicating that at least six years are needed to attain full male plumage. Although various display components were observed, most of these were part of a complete sequence Frith called high-intensity display. This is performed on vertical or nearly vertical branches, especially bare perches in an open area. The male begins by perching near the upper part of the display branch, uttering a combination bill click and repeated call while flicking the nearly closed primaries back and forth. While on the vertical branch he perches diagonally, swaying his body from side to side and slightly spreading and quivering his wings (Figure 64A). He then leans progressively to one side (Figure 64B), not only flicking his wings but shaking them, until he has become inverted (Figure 64C). At this point he spreads his wings fully and shakes them so that they seem to vibrate, simultaneously swaying his body laterally, and moves his head about in a stiff, peering manner (Figure 64D). While fully inverted he fully extends his wings and vibrates them rapidly, but only slightly raising his red flank plumes (Figure 64E). His plumage is then suddenly returned to its normal position and the bird hops down the perch, sometimes spiraling until he is almost at its base. He then ascends the perch again, reaching or exceeding his earlier level, and begins a hopping-on-the-spot display (Figure 64F), with his upwardly oriented head and body swaying mechanically and the tail moving pendulum-like back and forth. He then opens his wings again and becomes nearly motionless except for slight but rapid wing vibration and rotary head-peering movements, augmented by occasional bill clicks. He finally begins a series of bill-tapping movements on the perch, and terminates the display by hopping off the display branch. This entire sequence lasts 45–120 seconds. On a few occasions a "butterfly dance" preceded high-intensity display, during which the male hopped quickly from branch to branch, fluttering, flicking, and extending his wings like a giant butterfly.

Although similar inverted displays might occur in all of the *Paradisaea* species (LeCroy, 1981), they are most highly developed and extended in the blue bird-of-paradise (as described earlier) and the emperor bird-of-paradise. The latter species' inverted display has been most fully described and illustrated photographically by Coates (1990). The male begins his display by jumping up and down, with the neck stretched and the iridescent throat area expanded, in an initial upright phase (Figure 65A). He then holds his body almost horizontally, jumping rapidly from one perch to another, while opening and closing his wings in an excitement phase (Figure 65B). At times his body becomes rigid and his wings are spread forward and downward in a wing presentation phase (Figure 65C). He then suddenly swings head-first (or occasionally tail-first) below his perch and hangs there completely inverted for about five minutes. In this pose his flank plumes are fully extended, the open wings are held nearly horizontal, his head is turned sideways and slightly upward, and his tail wires are directed upward, in a posture somewhat like that of the blue bird-of-paradise. During this phase he performs a slow, graceful weaving motion of his body, causing

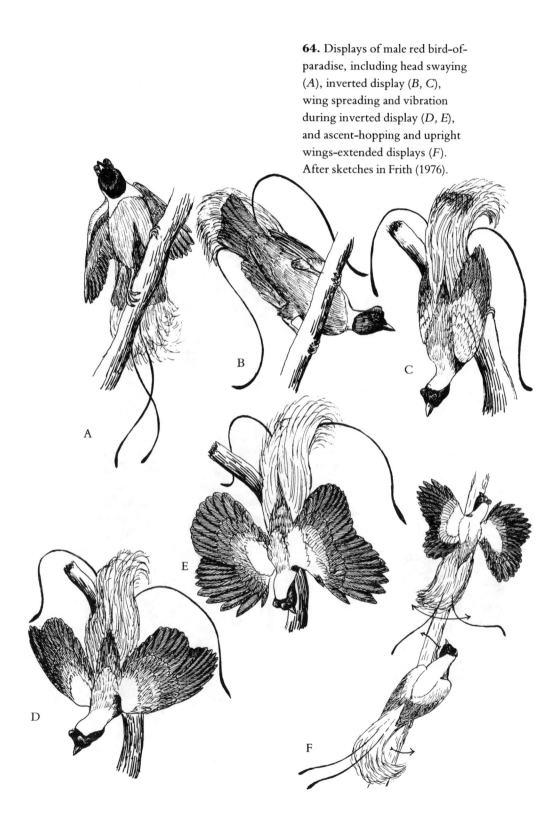

64. Displays of male red bird-of-paradise, including head swaying (*A*), inverted display (*B, C*), wing spreading and vibration during inverted display (*D, E*), and ascent-hopping and upright wings-extended displays (*F*). After sketches in Frith (1976).

his ornamental white flank plumes to swirl and sway, while generally remaining silent but occasionally calling in response to other males (Figure 65D–F).

The displays and male plumages of the greater, lesser, and raggiana birds-of-paradise are all very similar, and these species hybridize whenever they come into geographic

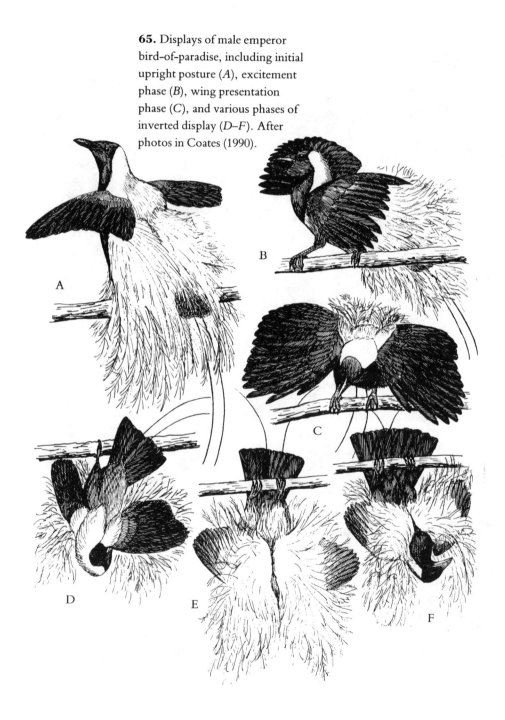

65. Displays of male emperor bird-of-paradise, including initial upright posture (*A*), excitement phase (*B*), wing presentation phase (*C*), and various phases of inverted display (*D–F*). After photos in Coates (1990).

contact. They are thus appropriately regarded as constituting a superspecies both by LeCroy (1981) and by Sibley and Monroe (1990). These authors also included in this superspecies one similar but geographically isolated species (Goldie's bird-of-paradise) that occurs on islands southeast of New Guinea. Its display behavior and male plumage are much like those of the other members of this superspecies, even though there should be no selective pressures on those islands for maintaining distinctive and elaborate social signaling as a reproductive isolating mechanism. The even more distinctively plumaged emperor bird-of-paradise is likewise relatively isolated on the Huon Peninsula of eastern New Guinea. Although not considered part of the *apoda* superspecies, it too hybridizes with the lesser and raggiana species where they are in limited contact. This occurs in spite of the emperor's highly developed inverted display, which LeCroy considered as "undoubtedly" a reproductive isolating mechanism. The highly distinctive blue bird-of-paradise has apparently also hybridized with the very different raggiana. It would seem that male plumage and/or display diversity has either not yet developed as a significant reproductive isolating mechanism (greater, lesser, raggiana), or if present (emperor, blue) has not resulted in fully effective avoidance of hybridization. Conversely, the male plumages and displays of an insular and isolated species (Goldie's) are scarcely less impressive than those of species presently exposed to potential hybridization from several other congeneric forms. One is led to conclude that intraspecific selection, in the form either of female choice or of male dominance signaling devices, must be the major driving factor in the evolution of male plumages and displays within this remarkable genus.

Displays of the greater bird-of-paradise have been well described by Dinsmore (1970), based on studies of four fully plumaged wild males plus additional younger males and at least one female. He observed display at four main sites, three of which were within 63 meters of one another and each had a male in regular attendance. These males could easily hear and probably see each other, and sometimes visited each other's sites. The fourth site, used by a male in incomplete plumage, was 365 meters from the others. The birds displayed on horizontal or slightly sloping and leafless branches, under fairly thick cover. Leaf-tearing behavior was common at or near the display ground, and comparable foliage-clearing behavior has also been observed in the red, lesser, raggiana, and Goldie's birds-of-paradise. Male display consisted of five phases, beginning with preliminary loud *wauk* calling and the wing pose, in which the wings are held in front of the head and body with the flank plumes erect and the tail forward below the perch (Figure 66A). This posture ends with wing flapping and several *wauk* calls. In the second (pump) phase, the male lowers his body and cups his wings around the branch, with his head and bill pointed down. He then hops along the branch in a display often called charging, while uttering a series of rapid *wauk* notes and moving from branch to branch, but usually returning to the display perch. At the end of a pumping phase, the male's body is tilted even farther forward, his back is humped, his plumes are fully erected, and a rigid "bow" posture is assumed for a few seconds or up to a minute (Figure 66B). The display sequence may end at this point, or the male may follow with a "dance" phase in which he slowly shuffles back and forth along the branch with exaggerated vertical movements, accompanied by a distinctive clicking call. On five occasions the male hung upside-down for several seconds after displaying, once holding on with a single foot, and on another occasion a male hung upside-down while another displayed nearby.

Of six copulations observed by Dinsmore, all involved the same male. The male approached the female by dancing toward her with his plumes erect, his wings spread, and uttering the clicking call. Mating was preceded by body-contact "dancing," bill rubbing, and holding the female by extending one wing over her body. After about 20 seconds of such behavior, the male mounted.

The displays of the lesser bird-of-paradise were first described in detail by Ogilvie-Grant (1905). He stated that the first phase of the display is preceded by loud, repeated

66. Displays of male greater bird-of-paradise, including phases in wing pose and bowing display sequence (*A, B*) (after sketches in Dinsmore, 1970); and lesser bird-of-paradise, including phases in the flower display (*C, D*) and inverted display (*E*) (mostly after sketches and photos in LeCroy, 1981, plus photos by author).

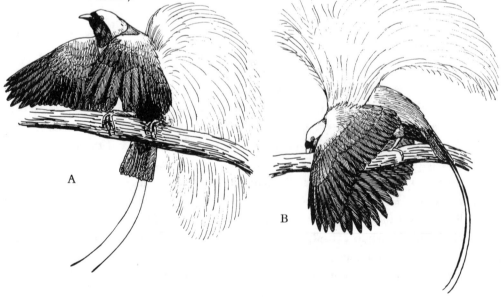

waa calls, with the body held diagonally, the bill opened, and the flank plumes lowered. The male then raises his wings to a semivertical position, bends his tail forward below the perch, and erects his plumes so that they form an arched cascade above his back (Figure 66C). In this posture the bird may stand for 10–20 seconds, slightly quivering his wings and at times "hitching up" his long plumes to above head level. He then begins wildly dancing and hopping ("charging" in LeCroy's terminology) backward and forward along the perch, extending his wings horizontally with his head lowered, while uttering *ca* notes. He then stops, occasionally glances back below his feet, rubs his bill on the branch, and remains for some seconds in a state of apparent ecstasy (Figure 66D). He may then return to the first stage of the display. As with the greater bird-of-paradise, an inverted posture has been seen and even photographed (Figure 66E), but no details of this are yet available.

Beehler (1983d) reported that a group of eight male lesser birds-of-paradise "owned" and daily occupied perches in a single tree, regularly plucking leaves from the immediate vicinity. These perches were as close as a half meter apart, and the four most central perches were occupied by the most active males. Additional subadult-plumaged males were also seen visiting the lek. A single male ("Male 1") performed all but one of 26 observed copulations, and matings only occurred with males at their

C

D

E

67. Displays of male raggiana bird–of–paradise, including phases in high-intensity display with wing beating (*A–C*), ascent dance (*D*), and frontal display to female (*E, F*). After sketches in Frith (1982) except for *E,* which is after a sketch in Cooper and Forshaw (1979).

particular display perches. Only the three most regular attendants at the lek received copulation solicitations from females. Beehler's observations suggested to him that perch ownership is hierarchically organized, and that male-male interactions may sometimes control male access to matings. However, one female solicited copulation from and mated with "Male 2" once. This occurred while Male 1 was only a half meter away, but he did not interfere with the mating, suggesting that female choice (choosing either the "best" male or the most central perch) might also influence mating.

The displays of the raggiana bird-of-paradise have been described in some detail by various writers, of which the accounts by Frith (1982) and Beehler (1988) are especially useful. This species, which has brilliant red flank plumes and a rich blackish maroon breast in adult males, occurs rather commonly in eastern and southern parts of New Guinea, where it is in sympatric contact with the lesser and greater and in limited contact with the emperor bird-of-paradise, which replaces it at higher altitudes on the Huon Peninsula. Beehler (1988) observed ten leks at eight different sites, nine of which consisted of birds using a single tree. The other consisted of two trees 70 meters apart, supporting four and two males in the larger and smaller tree respectively. From as few as one to as many as eight males attended these leks regularly. They were especially active in early morning hours, with a secondary late afternoon peak. All of 35 copulations at one lek were observed by no later than 8:00 A.M., and the sexually most successful male was also the most faithful attendant of the lek.

Frith and Beehler each described a variety of male raggiana vocalizations, of which the advertisement call, a repeated *wau,* is the most frequent male note, especially when females are not present at the lek. When a female arrives, the male begins uttering a display call while performing vertical wing beating (Figure 67A), initially perched on a horizontal or slightly sloping branch. The wing beating produces a thudding sound every 1.4 seconds. In Beehler's terminology, this represents the convergence sequence, since other males are attracted to and converge on the lek. The displaying male then moves to a more inverted position on a sloping branch and performs vigorous and rapid wing beating while uttering rapidly repeated notes (Figure 67B). This is replaced by a phase of fully open wing beating or opened-wing descent with the male in a more static posture, his wings moving only slowly and his head lowered and turned from side to side as his wings are brought forward over it, producing a soft thudding sound as their carpals meet (Figure 67C). Beehler (1988) called this the static display. The male may then perform an "ascent dance" (Figure 67D) as he moves back up the branch toward a female perched above him, at which point he might perform a frontal display to her. In this display the male's wings may be held above his back or in a nearly normal position, and his dark breast feathers may be expanded in a ball-like manner (Figure 67E). With increased excitement his flank feathers are more fully raised and his wings are opened (Figure 67F). He may peck at the female's bill, all the while bouncing and swaying from side to side. This behavior immediately preceded mounting and attempted copulation in Beehler's observations.

Beehler (1988) described the copulation sequence of the raggiana in detail. It includes the male drawing his bill back and forth across the top of the female's head. He simultaneously holds his wings outstretched and flaps downward, so that one of his wings strikes her back, seemingly pummeling her. The female sits quietly on her perch, absorbing these blows for 20–35 seconds, after which the male mounts. Beehler suggested that his observations on this species provided limited support for the male dominance hypothesis of sexual access to females, and strong support for the female choice model.

12

Whydahs and Widowbirds: Tales of African Tails

The enormous diversity of finch- and sparrowlike birds of the world is far too great to be easily comprehended even by most professional ornithologists, but luckily for the limits of this book a discussion can be confined to only two minor groups, each of which is now generally placed taxonomically in a single genus.

The first of these groups consists of the African whydahs of the genus *Vidua*, which are most famous for their entirely "brood-parasitic" mode of reproduction. That is, the females insinuate their eggs into the nests of various "host" species (other African finches), causing the latter to hatch and rear them in company with their own young. The breeding males of some species are quite elaborately adorned with elongated tail feathers having species-specific shapes and colors, whereas other species have rather simple plumage patterns and typically sparrowlike tail shapes. The whydahs have at various times been regarded as comprising a unique passerine family (Viduidae) or have been included as a subfamily (Viduinae) of the waxbill group. At yet other times, they have been regarded as constituting a subfamily of the Old World ploceid finches.

The second group consists of the African bishops and widowbirds of the genus *Euplectes,* which are part of a rather large group of Old World weaver finches (Ploceidae). As a group, weaver finches typically construct surprisingly well made or "woven" nests, which are not only largely constructed by males but often serve as the male's display site. Females are attracted to males displaying at or near their partially completed nests, and presumably choose mates with regard to both the apparent relative quality of the male and the quality of his nest and the associated territory.

Sibley and Monroe (1990) have recently included all of these Old World sparrow- and finchlike groups within a single large family Passeridae. The bishops and widowbirds were included within their proposed tribe Ploceini, and the whydahs alone comprised a separate tribe Viduini. Suffice it to say for our present purposes that, in spite of their distinctly similar vernacular names, the widowbirds and whydahs are not now believed by ornithologists to be very closely related, and that any of the several similarities in their male plumages or mating displays are almost certainly the result of convergent evolution, presumably at least partly under the influence of sexual selection.

The Parasitic Whydahs and Their Remarkable Tails

There are probably about a dozen species of whydahs (the exact number is still in some dispute), of which the social behavior and organization of less than half are yet well known. Inclusion of the whydahs in a book on lekking birds might be questioned by some, inasmuch as none of them is actually known to be a lek-forming species. However, territorial male distributions resembling an exploded lek pattern have been described in some. Additionally, at least some of these species offer especially interesting insights into the effects of sexual selection on the behavior and appearance of males operating within polygynous mating systems.

Although sexual dichromatism is typical of all whydahs and all of them are believed to be polygynous, the well-studied village indigobird (*Vidua chalybeata,* including its several geographic variants) has the least proliferation of male breeding plumage ornamentation (Figure 68A). In common with the other whydah species, as well as with the widowbirds, males of this species exhibit distinctly femalelike juvenal and nonbreeding or "eclipse" plumages. Among the several geographic forms (races or species) of indigobirds, breeding males are rather variable in appearance, but they all lack distinctly elongated tails and associated complex postural displays. The males are highly territorial. They defend rather dispersed singing sites (considered to be exploded leks by Payne and Payne, 1977), from which they vocalize with harsh chattering songs and thus attract females. Evidently most males defend only a single singing site. Singing sites of neighboring males are usually about 300 meters apart, and are neither clearly audible to nor in sight of one another. Based on their relative singing intensity, males are apparently able to attract varying numbers of females, and the more successful males are those having other males singing on sites within hearing range. Individual females usually visit more than one male singing site (based on a sample of 19 marked females, as many as six such sites may be visited), and they may mate with one or more territorial males during the course of a single breeding season (Payne, 1973; Payne and Payne, 1977).

Some of the other widespread and generally considerably more colorful or elaborately plumaged whydahs of eastern and southern Africa are the straw-tailed (*V. fischeri*), steel-blue (*V. hypocherina*), and pin-tailed whydahs (*V. macroura*) (Figure 68B–D). In southern Africa the shaft-tailed or queen whydah (*V. regia*) geographically replaces the closely related straw-tailed whydah. The extraordinarily long-tailed eastern paradise-whydah (*V. paradisaea*) (Figure 69B, F) and the similar broad-tailed paradise-whydah (*V. obtusa*) (Figure 69C–E) also are broadly sympatric with most of these species in eastern and southern Africa. In all these species the females (and probably all

68. Male plumage variations of several whydah species, including village indigobird (*A*), straw-tailed whydah (*B*), steel-blue whydah (*C*), and pin-tailed whydàh (*D*). Also shown is a male pin-tailed whydah in upright posture (*E*), performing wing shaking (*F*), and hovering before a wing-shivering female (*G*). After sketches in Shaw (1984).

69. Male long-tailed widowbird displaying to a female (*A*, after photo by author). Male eastern paradise-whydah crouching (*B*), and broad-tailed paradise-whydah choking (*C*), rustle-curtsy (*D*), and head swinging (*E*). Male eastern paradise-whydah in flight display (*F*), and tail structure, as seen from above (*G*). After sketches in Nicolai (1969) and Koenig (1962).

first-year males) are similar-looking and very inconspicuous, with camouflaged spar-rowlike plumage patterns of drab brown and buff. However, breeding adult males have variably elongated and shaped central tail feathers, which are either golden yel-low (in the straw-tailed) or iridescent blue to black (in all the others). Males also have bright red (in the shaft-tailed and straw-tailed) or black bills, as well as variously col-ored and often contrastingly patterned breeding plumages, which range from mostly white through shades of bright yellows or reds to steel-blue or black. Presumably these substantial differences in male plumage serve at least in part as species recogni-tion signals among these seemingly closely related forms.

Data on the adult mass ratios of the two sexes are still limited for the long-tailed species of whydahs (Maclean, 1985). Nevertheless, the sexual differences among breeding adults would appear to be largely reflected in differential tail development and plumage color, rather than in marked body mass differences between the sexes. Thus, male:female linear measurement ratios for bill, tarsus, wing, and nonornamen-tal rectrices average about 1.0–1.1:1 (Alatalo, Höglund, and Lundberg, 1988), sug-gesting that adult male:female mass ratios are unlikely to be greater than about 1.2:1. This judgment is supported by a small sample of adult weights for the shaft-tailed and eastern paradise-whydahs (Skead, 1974), which suggest that mean male:female mass ratios of only about 1.1:1 may be typical in these highly dichromatic species. How-ever, ornamental tail lengths for breeding males of the long-tailed species of whydahs are highly variable, their coefficients of variability being about three times greater than various body size characters (Alatalo, Höglund, and Lundberg, 1988). This re-markable variability is probably in part attributable to age differences, and is also partly positively correlated with body size characters, suggesting that the lengths of these longest nuptial tail feathers may provide a useful and simple visual index to indi-vidual male fitness.

The whydahs are essentially open-country and grassland-dependent birds, occupy-ing a variety of dry brush, scrub, and savanna habitat types. The males' advertisement vocalizations typically consist of rather simple warbling or twittering songs that are uttered during regular periods from defended call sites. The degree of territorial de-velopment is only poorly documented for most species, and this information is some-times seemingly contradictory. In some cases interspecific as well as intraspecific ter-ritorial defense is known to be present, although in others interspecific sharing of call sites by males has been observed. Territories of some whydahs such as the indi-gobirds appear to function as mating stations only, whereas in others breeding re-sources such as seeding-stage grasses or the presence of water may be important ele-ments of territorial defense. Descriptions of male territoriality, advertisement behavior, and/or sexual displays are now available for three species of indigobirds (Payne and Payne, 1977; Payne and Groschupf, 1984), the pin-tailed whydah (Shaw, 1984; Barnard, 1989), and the shaft-tailed whydah (Barnard, 1989). Sexual advertise-ment displays and some aspects of territoriality have also been described for two species of paradise-whydahs, together with limited corresponding information on the straw-tailed whydah (Nicolai, 1969).

Pin-tailed whydahs were studied in northern Ghana by Shaw (1984), using a sample of 89 wild, color-banded birds. He found that territorial males increased to a maximum of 13 on his study area (of about 12 hectares) during the early breeding season (July). He also found that the mean distance separating favored display sites on neighboring territories was about 100 meters. Only about 2–3 percent of the total area defended by a territorial male was usually used for foraging. The majority of the daylight hours were spent by the male at his most favored display site, and 90 percent of the time was spent within the limits of his territory. Although some males had substantially larger territories than others (ranging from 0.3 to 1.2 hectares, and averaging 0.63), there was no clear correlation between attractiveness to females (as judged by the number of females visiting multiplied by visit durations) and either the male's territory size or the percentage of time he spent on his territory.

Up to as many as nine females were observed by Shaw to visit a single male's territory. The operational sex ratio was judged to be about 2.2 females per breeding male, with nearly half of the total male breeding-season population not in breeding plumage and presumed to be first-year birds. The number of observed copulations was too small to establish the relative mating success of individual males as compared to their territory sizes or territorial positions. However, by removing established males from their territories Shaw found that vacated territories were filled by new males at varying speeds, suggesting that some territorial sites were more favored by males than others, and that some breeding-condition males are regularly nonterritorial and thus excluded from breeding. This would also suggest that a skewed male mating success distribution probably exists in wild populations, as might be expected in any polygynous species.

Male pin-tailed whydah displays toward intruding males differ from those toward females. When confronted by another male, the territorial male and the intruder both adopt an upright posture with the plumage sleeked and wings held close to the body (Figure 68E). In this posture a soft, twittering song is uttered, with the bill open and elevated. By moving progressively closer to the intruder, the resident bird eventually displaces him.

Displays performed by resident male whydahs toward visiting females consist of the male fluffing his plumage, adopting a horizontal stance, and rapidly vibrating his wings and tail (Figure 68F). In this manner the male may lead a female to a particular feeding area, which is usually near a display tree. The other heterosexual display directed by males toward females is a hovering flight. In this display the flying male approaches a perched female and hovers less than a meter from her, while inclining his body diagonally and intermittently vibrating and closing his wings, causing him to jerk up and down in the air. The watching female would usually adopt a horizontal posture, with her wings slightly spread and their tips quivering (Figure 68G). Females visited territories as singles, in pairs, or in groups of up to eight birds.

Barnard and Markus (1989) found that female solicitation intensity in pin-tailed whydahs was strongly correlated with individual differences in male display intensity, suggesting that quantitative male display characteristics may be an important component in stimulating some females to mate. Likewise, Payne and Payne (1977) judged male mating success in village indigobirds to depend on individual male song characteristics as well as on the outcome of male-male competition for favored call sites. However, Barnard (1989) found that neither variations in body size nor length of breeding-plumage rectrices among color-marked male pin-tailed whydahs were correlated with their estimated seasonal mating success, but that a male's capacity for defense of a perennial water source was evidently a significant factor. Individual male mating success was highly variable, ranging from 0 to 50 observed copulations among 11 color-banded males. Evidently females selected specific male call sites primarily on the basis of their variable ecological resources, rather than of evident quality differences of the resident males. Perhaps the lack of apparent correlation between individual male body size and relative mating success helps account for the seemingly low male:female mass ratios that appear to be typical of this genus.

In a counterpart field study of shaft-tailed whydahs, Barnard (1989) found that females were attracted to those territorial males having call sites containing a high density of seeding-stage grasses, and that male territorial intrusion rates were higher at clumped mating sites than at isolated ones. Interestingly, territorial males tended to gravitate away from call sites of lower sexual activity and toward those of higher activity, suggesting that a tendency may exist in this species for male aggregations to develop around relatively sexually successful males.

Much more information is needed on possible hotshot-related clustering of male territories in *Vidua,* and on the differential mating success of competing males based on their physical or behavioral attributes; both questions may offer fruitful lines of future study. One recent aviary study by Barnard (1990) illustrates how this relationship might be expected to operate in wild shaft-tailed whydahs. She found that females, when presented with two males of widely differing tail lengths, preferentially performed solicitation displays to those males having artificially lengthened tails (ca. 40 percent longer) rather than to normal-tailed controls. Females also favored wild-type control males to those with experimentally shortened tails (ca. 40 percent shorter). Remarkably, the males with the experimentally lengthened tails displayed and vocalized significantly more than did their control counterparts, and those with shortened tails did so significantly less often. Potential explanations for these interesting results include possible differences in male "self-awareness" and, more believably, differences in sensory feedback to such males from interacting females.

Wild populations of the two sympatric species of paradise-whydahs were studied by Nicolai (1969) in Tanzania. Breeding males of these two species are quite similar to one another except for relatively minor differences in their variably elongated and highly ornamental central tail feathers. However, males of the broad-tailed paradise-whydah evidently lack flight displays, whereas in the eastern paradise-whydah males have a highly conspicuous territorial flight display (Figure 69F). The male of this species takes off and with fast wingbeats ascends 20–100 meters. He then makes a series of straight sweeps above his territory, flying very slowly, the longest tail feathers dropping and the shorter and very wide central pair raised well above body level. He finally drops back down to land on another perch. Should a rival male enter his territorial air space, the resident male approaches and maintains a position under the intruder, thereby escorting him back to the edge of the territory.

The stationary form of male advertisement display is similar in the two species. In the eastern paradise-whydah the male imitates, in highly ritualized form, the begging response of a nestling bird (Figure 69B). In the broad-tailed species one of the male's stationary displays mimics a feeding response toward the female (Figure 69C). He also performs a so-called "rustle-curtsy," making apparent stridulation noises with the ornamental tail feathers (Figure 69D). This may be done by scraping the shafts and vanes of the vertically oriented central tail feathers against the inner sides of the more diagonally oriented upper vanes of the adjoining, highly elongated feathers. Males also perform a lateral head-swinging display (Figure 69E). Curiously, just before sunset both sexes of the local whydah population assemble and fly as a group out to forage. Thus, the male's territory seemingly serves largely or entirely for self-advertisement rather than as a food-rich resource that might attract females. Since all the species of this genus are brood parasites, the male's territory obviously need not include any possible nesting sites or specific brood-rearing resources.

Curious Tales of Widows and Bishops

A more clear trend toward arena behavior among males can be detected within the African genus *Euplectes,* a group of some 15 species of sparrowlike birds. In several species (the "bishops") the breeding males have large and brilliant red plumage patches on the head and neck, which sometimes extend over the rump to nearly cover their short tails. However, in many other species the breeding plumages of males are predominantly black and their tails usually elongated. These species are called widows or widowbirds. Breeding plumages of male widowbirds range from entirely brown or black, as in the Jackson's widowbird (*E. jacksoni*), to species in which the back, neck, and head plumage may be a colorful orange, yellow, or red. The "epaulets" on the upper wing coverts may also be uniformly yellow or distinctly bicolored, with varied combinations of yellow or red and buff or white present. The longest tails by far are found in the long-tailed widowbird (*E. progne*), in which males' tails may exceed half a meter in length and the males' overall length may be up to five times longer than the females'.

Widowbirds are frequently and easily confused with the outwardly similar-appearing whydahs, and in some classifications they have been placed fairly close together. The resulting confusion is compounded because, although the whydahs' English name is reportedly a variant of Ouidah (a coastal locality in Benin), their generic name *Vidua* derives from Latin and likewise means "widow." Also in common with the whydahs, adult male widowbirds are distinctively plumaged during breeding but assume femalelike eclipse plumages outside the breeding season. Likewise, in some if not all species of both groups the first-year males are both sexually immature and femalelike in plumage. However, in contrast to the whydahs, none of the widowbirds are brood parasites.

In all widowbirds the adult males defend territories during the breeding season, often against males of other species of widowbirds as well as their own species. It is believed that all 16 species in the genus are polygynous, although data on this point are still lacking for four. Males display to all femalelike birds that enter their territories, including eclipse plumage males, and may even direct their courtship toward females of other widowbird species. Perhaps as a result, several interspecific hybrid combinations have been reported under captive conditions. In no case is the male's breeding territory known to be related to available food supplies, and, in common with the whydahs, males may join in flocks to feed elsewhere than within their territories during the breeding season (Craig, 1980).

Comprehensive comparative adult male:female mass data for widowbirds are limited, but Craig (1978) has documented significant differences in adult mass ratios for several species. In this group, male:female wing length ratios collectively average about 1.25:1, suggesting that an adult mass ratio as high as about 1.5:1 might exist (assuming volume and mass are related to the square of the linear wing length measurement). In general agreement with this, the long-tailed widowbird's mean adult male:female mass ratio is 1.32:1 (Skead, 1977), and in the lek-forming Jackson's widowbird (*E. jacksoni*) the adult ratio is about 1.45:1 (Andersson, 1992).

So far there is no clear evidence in *Euplectes* that an individual male's physical size is directly related to his mating success, and Craig was unable to demonstrate any specific effects of individual male behavior on mate attraction in the red bishop (*E. orix*). In his view, the primary function of the male territory (beyond its seemingly obvious role in mate attraction) is to provide a nest site. However, in the long-tailed widowbird the nest (or nests) might not always be located within the territory. This is certainly also the case with the Jackson's widowbird; Andersson (1989) found that all of 26 nests were beyond but within 300 meters of a lek area. In the Jackson's widowbird the male evidently plays no role whatever in nest building, but in the long-tailed widowbird the male might start building a nest that is later to be completed by the female. It is probable that in the other species of *Euplectes* the male likewise constructs the basic nest structure, with its lining provided by the female. Evidently males of all species play no subsequent reproductive role, such as in incubation and feeding of the young, and thus become emancipated by the time nest building is completed (Craig, 1980).

The sizes of the males' territories are not clearly related to body size, but may be more closely related to the type of vegetation present and to population density. The largest species, the long-tailed widowbird, does not correspondingly defend the largest territory, and the lek-breeding Jackson's widowbird's court or "dancing ground" is no larger than 10 square meters. Males of many species advertise their territories by overhead aerial displays (Emlen, 1957; Crook, 1964). In some species such as the fan-tailed widowbird (*E. axillaris*) the male may spend at least 10 percent of his time flying above his territory, whereas in others such as the red bishop the flying time might be as little as about 2 percent. In this species territories are very small, and the breeding and roosting sites may be identical (Craig, 1980).

In the long-tailed widowbird, adult males have extraordinarily long central tail feathers that in some individuals reach a length of a half meter. This tail is expanded vertically into a deep inverted V as the male displays above his 0.5–3-hectare territory, around which he flies with slow wingstrokes at heights of no more than about 2 meters above ground. Such displaying birds may be visible from as far as a kilometer away. Females are attracted to such males, and are believed to mate with the male on whose territory they nest (Andersson, 1982). Males also display from perches directly toward receptive females, by spreading their wings, raising their red epaulets, and calling loudly as they bend forward (Figure 69A), in a display somewhat similar to the song-spread posture of North American red-winged blackbirds (*Agelaius phoeniceus*).

Andersson (1982) initially found no significant correlation between male tail length and the numbers of nests on each male long-tailed widowbird's territory (among seven males the coefficient of variation in tail length was 9.4 percent). He then experimentally shortened the tails of nine males, lengthened (by gluing on additions) the tails of nine others, and left the tail lengths alone in two control groups. The males with lengthened tails had a significantly higher mating success, as measured by new active nests on their territories, than did males having normal or reduced tails. Males with reduced tail lengths increased their mean rate of flight displays by about 10 percent (perhaps because flying was now easier), whereas males with longer tails and control males showed decreases in flight display rates of about 40–50 percent. Changes in the intensity of male display behavior were therefore probably not responsible for increased breeding success. Males with longer tails apparently also did not increase their territory sizes nor their territorial tenacity. Thus their improved mating success was probably due to female choice factors rather than to changes in relative male dominance (Andersson, 1982). Interestingly, more than a century previously Charles Darwin had related an anecdotal account to the effect that the female of this species "disowns" the male when it is "robbed of the long tail-feathers with which he is ornamented during the breeding season."

Similar studies carried out on the yellow-shouldered widowbird (*Euplectes macrourus*) by Savalli (1992) have produced somewhat comparable results; males with experimentally shortened tails were more likely to lose or not be able to establish territories than were control and long-tailed males. However, males that remained on their territories did not vary in their reproductive success. These results suggest that male tail length may not only be an important heterosexual signal in widowbirds, but may also serve as an index of male fitness during male-male interactions.

The Jackson's widowbird provides the final example of sexual selection and arena behavior in this group. Indeed, it is the only true lek-forming species of the entire group, and the only one to construct a display court. Some of the first observations on the court-building and lek-displaying behavior of this species were provided by van Someren (1945, 1956). He stated that in the Nairobi area of Kenya males display between March and August, although annual variations occur, probably in conjunction with the time of ripening of grass seeds (Andersson, 1992). Van Someren described the "dancing ring" constructed by the male as initially an ill-defined circle or oval in the grass with a tuft or hummock in the middle, typically somewhat less than a meter in diameter. This ring is enlarged by the male trimming the grass with his bill; as the ring expands, the central tuft is also trimmed and the grassy interior of the ring is sloped off. The usually oval-shaped ring may then be somewhat over a meter in diameter and the central tuft about 60 centimeters in diameter, the latter with typically two shallow recesses at its base at each end of the longer axis. These recesses are cup-shaped, with one usually more distinct and better worked than the other (Andersson, 1991). When the resident male arrives he first inspects the court and may snip off bits of grass (Figure 70A). He then walks around the track; crouches and quivers, shaking his body from side to side; and enters a slight recess in the central tuft, ruffles his head feathers while his head is held well back, and cocks his tail. He thus sways and moves about with jerky movements, making wheezy and rattling noises. He then jumps upward a highly variable distance, with his head held back, the body plumage ruffled, and the wings quivering (Figure 70B), and utters a tinkling call. The tail feathers are stacked into an upright V-like shape with the central pair dangling, just the opposite of the condition typical of displaying long-tailed widowbirds (S. Andersson, pers. comm.). The male may thus jump five or six times before resting. He might perform in a single place on the ring, or may gradually circle the tuft in the course of his display.

Females evidently are selectively attracted to individual displaying males. Usually they sit in front of one of the recesses, watching as the male begins a ground display of flaunting and quivering his body plumage while vertically cocking and fluttering his tail. During this time the male stands in the ring opposite the female, keeping the grass tuft between them and seemingly hiding from her direct continuous sight (Figure 70C), but at times peeking to see if she is still there. After about a minute of such

70. Displays of Jackson's
widowbird, including a male
standing in his court (*A*), display-
jumping (*B*), and displaying to a
female (*C*). Insets show shape of
court and positions of male and
female as seen from above. After
photos by van Someren (1945)
and S. Andersson.

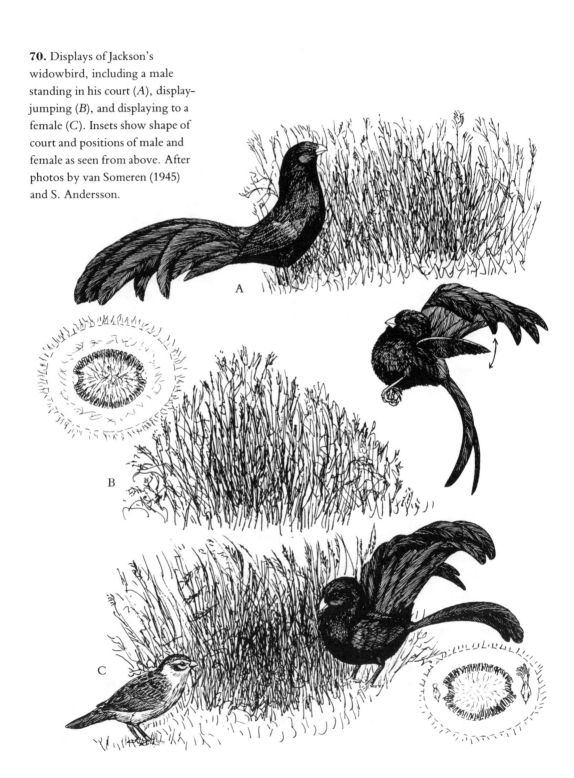

ground display, the male approaches the female on stiff legs and flaunting his raised tail from side to side, simultaneously producing a distinctive gurgling sound. She then may fly, move away from the male while remaining in the ring, or crouch and solicit copulation, which occurs within about ten seconds (Andersson, 1989, 1991). Males have not been observed near nests, even during nest construction phases, so male reproductive involvement probably ceases with copulation. However, the cup-shaped recesses in the display court resemble the early stages of widowbird nests, and visits to them by females may represent a symbolic nest inspection in Andersson's view.

The lekking behavior and aspects of female choice in this species have been studied in detail by Andersson (1989, 1991, 1992). In one study (1989) four leks were observed, with a maximum of from 9 to 17 males observed in attendance. Average distances between neighboring males in 13 cases was 31 meters (Andersson, 1992). Overall mating success did not differ among the four leks, based on a sample of 1,672 female visits and 29 copulations. However, of the 36 males individually tracked, one male obtained five matings, one obtained four, two obtained three, four obtained two, six obtained one, and 22 males did not mate at all. Statistically, the six factors judged to be most important in causing females to choose among males were, in descending significance: male tail length, male display rate (jumps per unit time), male bill length, relative lek attendance, male wing length, male tarsus length, and estimated "recess quality" (evidence of male shaping of the court's central tuft). The mating success rate of individual males seemed to depend on two separate components. Display rate and lek attendance affected the number of female visits, whereas the male's tail length seemed to primarily influence his chances of copulating with a particular visiting female (Andersson, 1989). These results suggest that, within leks, mating success among males seems to be the result of female choice factors. However, Andersson indicated that male-male aggression is common during lek formation, and to the degree that some males may have been excluded from participation, intrasexual selection cannot be fully excluded from possible sexual selection effects. However, established males do not attempt to damage one another's courts. Little male interference with the courting activities of other males is apparent, and there is no apparent preference by females for the most dominant males or those in the most centrally located courts.

Andersson also considered the possible influence of the display court on female preferences. He concluded (1991) that the apparent quality of the display court is only a weak predictor of male mating success, although it does correlate positively with male tail length and relative lek attendance. Court quality may, nevertheless, have to meet a certain minimum construction standard for the female initially to visit a particular male. Female visits to courts with damaged or destroyed recesses averaged shorter than those to intact courts, and their visits to experimentally improved courts (with deeper than normal recesses) lasted longer.

Andersson finally (1992) experimentally tested the preference of females for mating with long-tailed males. He used 26 males (13 matched pairs) in the experiment, either trimming their tail feathers to 14 centimeters (experimentals), or adjusting them to the species' average normal tail length of 18 centimeters (controls). Male display rates and lek attendance rates were not measurably affected by these tail manipulations. Andersson observed 631 visits by females to these 26 males, of which the longer-tailed males had a mean of 41 female visits and the tail-shortened males only 23 visits. He also videorecorded an additional 178 visits to 16 males. Within this latter group, four copulations were determined to be obtained by the longer-tailed males, and only one by a tail-shortened male. These results strongly suggest that females are sufficiently selective as to impose sexual selection pressures favoring longer male tails. Since there was a positive relationship established between male tail length and body condition, females choosing males with longer tails are also thereby selecting males in good physical condition (Andersson, 1992).

With a discussion of these beautiful finches, we have reached the end of our long tale. It has been not only an extended but at times an almost unbelievable story. Sexual selection has repeatedly sent various polygynous bird groups down evolutionary pathways leading convergently to magnificent male plumages and wonderful behaviors that conform to some of our highest human ideals of aesthetic beauty. Some readers might wonder how such simple biological processes could possibly produce such similarly wondrous end products in a group of animals seemingly lacking refined aesthetic tastes. On this it is only fair to let Charles Darwin (1871) have the final word.

> Everyone who admits the principle of evolution, and yet feels great difficulty in admitting that female mammals, birds, reptiles, and fish, could have acquired the high taste implied by the beauty of the males, and which generally coincides with our own standard, should reflect that the nerve-cells of the brain in the highest as well as the lowest members of the Vertebrate series, are derived from those of the common progenitor of this great Kingdom. For we can thus see how it has come to pass that certain mental faculties, in various and widely distinct groups of animals, have developed in nearly the same manner and to nearly the same degree.
>
> He who admits the principle of sexual selection will be led to the remarkable conclusion that the nervous system not only regulates most of the existing functions of the body, but has indirectly influenced the progressive development of various bodily structures and of certain mental qualities. Courage, pugnacity, perseverance, strength and size of body, weapons of all kinds, musical instruments, both vocal and instrumental, bright colours and ornamental appendages, have all been indirectly gained by the one sex or the other, through the exertion of choice, the influence of love and jealousy, and the appreciation of the beautiful in sound, color or form; and these powers of the mind manifestly depend on the development of the brain.

Glossary

The following definitions provide explanations for most of the technical and some nontechnical terms that might be encountered in this book, as well as for relevant terms in the cited readings. Considerable assistance in constructing this glossary was obtained from those of Mayr (1963), Ricklefs (1973), Wilson (1975), and Wittenberger (1981).

Aberrant. Deviating from the normal type.

Active mate choice. Nonrandom mating behavior, in which mates are chosen for particular phenotypic traits rather than because of simple proximity or for other nonselective ("passive") reasons. *See also* Passive mate choice.

Adaptation. A structural, behavioral, or physiological trait that increases an organism's Darwinian fitness.

Adaptive value. Equivalent to Darwinian fitness.

Advertising behavior. The social behaviors ("displays") of an animal of either sex that serve to identify and broadcast (visually and/or acoustically) its species and sex, and may also tend to announce or reveal its relative reproductive capabilities, social status, and general vigor. Additional specific functions may also exist, such as territorial proclamation and defense, communication between pair or group members, etc.

Age-graded dominance. A dominance hierarchy in which social status is directly correlated with the age and associated experience of the participants.

Age indicator hypothesis. A hypothesis that female preferences for mating with older males can be linked to their favoring of male traits that are age-correlated; a variant of the good genes hypothesis. *See also* Female choice behavior; Good genes hypothesis; Total viability hypothesis.

Aggregation. A temporary assemblage of animals, sometimes of more than one species, that cluster because of their attraction to certain environmental (biotic or nonbiotic) factors, but are not engaged in cooperative behavior nor exhibit internal organization. *See also* Assemblage; Congregation.

Aggression. Hostile interaction that is associated with threat, fighting, and determining social dominance; the complement to aggression is submissive behavior that leads to social subordination.

Agonistic behavior. Hostile interactions between individuals that range from overt attack and fighting to submission and escape, including both interspecific (often survival-related) and intraspecific (often competition-related) interactions. *See also* Aggression.

Agonistic sexual selection. *See* Intrasexual selection.

Air sac. As used here, a structure (usually the anterior esophagus) in the neck region of some grouse, bustards, shorebirds, and ducks that by air inflation visibly expands that area and may also help to modulate or resonate vocal sound production. In male grouse, paired areas of unfeathered and colorful skin may also be exposed and enlarged during esophageal inflation. Such "air sacs" are not directly associated with the functional air sacs of the avian respiratory system, but in some stiff-tailed ducks one or more tracheal air sacs are present and serve similar visual or sound-related display functions.

Allele. Any of the alternative states of a gene occupying a particular chromosomal locus.

Allopatry. The relationship of populations or species occupying mutually exclusive but sometimes adjacent geographic ranges. Parapatric populations are those in actual extended geographic contact at their boundaries. *See also* Sympatry.

Alpha. Pertaining to the first rank in a social dominance hierarchy; alpha males are often called "master cocks." *See also* Dominance hierarchy.

Altricial. Refers to birds or mammals that are physically and physiologically very poorly developed at hatching or birth, and thus remain dependent on parental care over an extended period.

Altruism. Selfless or self-sacrificing behavior by one individual that increases another's fitness at the expense of its own. The opposite of selfish behavior, and unlikely to evolve through known means of individual selection. *See also* Kin selection.

Anthropomorphism. The application of human attributes to nonhuman animals.

Arboreal arenas (*or* leks). Elevated arenas in trees, saplings, or vines.

Arena. A fixed or mobile site at which males of polygynous species congregate for epigamic purposes (male dominance determination and female attraction competition), including both strongly clustered groups (typical arenas) and more dispersed groups (exploded arenas). Arenas typically are spatially localized, include no resources useful to females beyond mating opportunities, and have the participating males structurally organized as to social rank. *See also* Lek.

Arena behavior. The variable clustering and social display interactions of sexually active males of a polygynous species at a localized and often traditional site for reproductive purposes. Males contribute nothing beyond their genes to the attracted females, who nevertheless often exhibit active mate choice from among the assembled males. *See also* Lekking.

Assemblage. Any group of organisms, potentially comprising a congregation, an aggregation, or even a random accumulation of individuals.

Assortative mating. Mating choices in which mates tend to be more similar in phenotype than randomly chosen mates would be. *See also* Disassortative mating; Mate choice behavior.

Attenuated. Tapering to a sharp point.

Avenue bower. A type of bower constructed by bowerbirds, in which a central pathway is surrounded by two or more wall-like structures. *See also* Maypole bower.

Bateman's principle. A. J. Bateman's concept that, because of the marked differences in the energetic investment of each sex in its gametes, a male's reproductive success is limited only by his success in fertilizing females, whereas a female's fitness is limited by her ability to produce gametes. This addition to basic Darwinian sexual selection theory largely accounts for the observable sex differences in the mating behavior of many animals, namely the eagerness of most males to mate relatively indiscriminately, as compared with the females' stronger tendencies toward individual discrimination among potential mates.

Benefits. In ecological game theory, those consequences of an individual's phenotypic attributes, such as its behavior, that increase its fitness. *See also* Costs; Risks.

Beta. Pertaining to a second-level rank in a dominance hierarchy; beta males are immediately subordinate to alpha males but dominant to all others, and are sometimes called "subcocks." *See also* Alpha; Dominance hierarchy.

Bimodal. *See* Mode.

Bisexual selection. Selection in monogamous species that affects both sexes equally as a result of mutual intersexual selection.

Bishops. A vernacular name for a group of relatively carotenoid-rich (in breeding males) African finches in the genus *Euplectes*. *See also* Widowbirds.

Bower. A structure individually constructed, maintained, advertised, and defended by male bowerbirds, which serves to attract females for mating purposes. *See also* Display court.

Brood parasitism. Describes those birds whose females deposit their eggs in the nests of another (rarely the same) species, leaving them to be incubated and reared by the latter. Sometimes also called nest or egg parasitism. *See also* Social parasitism.

Bulla. A bony inflation of the syringeal region that may generate, modulate, or resonate vocalizations. *See also* Syrinx.

Call. A term used in ornithology to describe avian vocalizations that are acoustically simple, are often innately uttered and perceived, and the production of which is usually not limited by sex, age, or season. *See also* Song; Vocalizations.

Capercaillie. A vernacular English name (meaning "horse of the woods") for two very large species of Eurasian woodland grouse in the genus *Tetrao*. *See also* Grouse.

Carotenoid. A group of brilliant red to yellow organic pigments (biochromes) occurring widely in plants and also in the plumages of some arena-breeding birds, such as cotingids, manakins, and whydahs.

Catherine wheel. Descriptive of a revolving cartwheel.

Character. Generally used in biology as a synonym for a trait or a trait component, as in the term "derived characters" used in cladistics. *See also* Trait.

Character displacement. The selection-driven divergence of behavioral or structural traits in two sympatric populations, resulting either in increased niche segregation, thereby reducing competition (ecologic character displacement), or in increased reproductive isolation, thereby reducing hybridization probabilities (reproductive character displacement).

Clade. A group of organisms linked by common evolutionary descent; also a branch of a cladogram.

Cladistics. A procedure for arranging taxa by comparing their shared derived character traits (synaptomorphies) in such a way that the resulting arrangement (or cladogram) reflects their apparent recency of common ancestry.

Cladogram. A diagrammatic representation of branching phylogeny, showing the apparent splitting sequences of taxonomic groups through time, based on character state changes. Unlike typical dendrograms or treelike representations of phylogeny, cladograms do not normally attempt to represent actual time scales, although cladistic distances (the number of branching points between any two points) are of primary phyletic significance.

Clustered (*or* clumped) arena. An arena in which the display sites of the participating males are very closely spaced, as contrasted with an exploded (or dispersed) arena. *See also* Arena behavior.

Clutch. The total number of eggs laid during an egg-laying cycle by a single female, and usually (except in social parasites) incubated collectively.

Coefficient of relatedness. A measure of genetic relationship, ranging from 0 (unrelated) to 1.0 (genetically identical), and representing the proportion of identical alleles that are shared by two individuals by common descent.

Coefficient of variation. A statistical measure of relative variation occurring around the mean of a sample, expressed as a fraction or percentage of the mean.

Coevolution. Evolution in which reciprocally induced evolutionary changes occur in each of two or more species over time, potentially including either antagonistic responses (prey-predator or parasite-host adaptations) or mutually desirable ones (mutualism). *See also* Red Queen's rule.

Communal display. Sexual advertising display that regularly involves two or more participating males displaying in close proximity, but not interacting cooperatively. *See also* Arena behavior; Cooperative display.

Communication. The adaptive transmission of information from one individual to another by various means, including visual, acoustic, tactile, and possibly others, altering the behavior of the information recipient in an adaptive manner. Two-way transmission of information between individuals is sometimes (teleologically) called "purposive" communication; two-way message transfers that have adaptive feedback effects on both the transmitter and the recipient are sometimes called "semantic" communication.

Competition. Supply-and-demand-related interactions between organisms over limited resources, including those between individuals or groups (intraspecific competition) and those between species (interspecific competition). It may occur as contest competition (resource access controlled by individual aggression) or scramble competition (access controlled by individual speed and efficiency).

Condition. A convenient, often subjective, term for an individual's general health and vigor; condition-based characteristics are those features of an animal that reflect such phenotypic aspects, rather than its overall viability-based traits. *See also* Viability.

Congeneric. Belonging to the same genus.

Congregation. An assemblage of animals that interact temporarily (1) as a result of social interactions, and that often (2) exhibit internal social organization and sometimes also (3) perform cooperative behavior. *See also* Aggregation; Assemblage.

Conspecific. Belonging to the same species.

Convergent evolution. A pattern of evolution during which two or more relatively unrelated lineages progressively approach one another in various phenotypic attributes, because of similar ecological adaptations.

Cooperative behavior. Noncompetitive behavior in social groups that is of mutual benefit to all participants (such as mutual defensive or cooperative foraging behavior) or to some of them, including possible altruistic interactions. *See also* Altruism; Cooperative breeding.

Cooperative breeding. Cooperative behavior specifically associated with breeding. In monogamous species this rare behavior most often occurs during the nesting/brooding phases (building, sanitation, and defense of nests, incubation, and care of young), and is related to kin selection. In promiscuous species it is even rarer (reported only in manakins, ruffs, and wild turkeys), and involves such prenesting activities as mate attraction, courtship facilitation, and possibly protection of the courting pair from disruptive influences.

Cooperative display. Social displays in which two or more males participate in a seemingly noncompetitive manner when displaying sexually toward potential mates.

Costs. In ecological game theory, the consequences of producing and/or maintaining a phenotypic trait, in terms of diminished Darwinian fitness. Examples might include increased visibility and associated predation risks, or increased energy expenditure associated with performing sexual displays. *See also* Benefits; Risks.

Cotingids. Members of the New World passerine family Cotingidae, such as bellbirds and cocks-of-the-rock.

Court. *See* Display court.

Courtship. A nontechnical term referring collectively to various heterosexual premating behaviors of animals. In monogamous species these usually include diverse "pairing" interactions that are broadly related to forming and maintaining pair bonds, but in polygynous or promiscuous species they often involve only heterosexual attraction and precopulatory ("mating") signals. *See also* Mating; Pair bonding.

Cracids. A collective vernacular name for members of the New World galliform family Cracidae, such as chachalacas.

Crepuscular. Pertaining to dawn and twilight periods. *See also* Diurnal; Nocturnal.

Crypsis. Biological camouflage, including both reduced overall visual conspicuousness (countershading and disruptive patterning) and also special protective visual resemblances (background mimicry and interspecific mimicry). *See also* Mimicry.

Darwinian fitness. The tendency for a particular allele to change in frequency relative to other genes at that locus as a result of natural selection.

Darwinism. An approach to biology using evolutionary principles first propounded by Charles Darwin, and the application of these principles to natural populations. Neo-Darwinism incorporates modern concepts of genetics and population biology. *See also* Evolution; Natural selection.

Deme. *See* Population.

Detached arena. *See* Mobile arena.

Dialects. Local intraspecific variations in songs or other communication signals resulting from learning and cultural transmission effects.

Dichromatism. The occurrence of two discrete color phenotypes within one taxon, such as in sexual dichromatism, where the adult sexes differ in color (hue or brightness) or pattern. Intrasexual dichromatism (or polychromatism) may also exist, such as occurs in genetically controlled plumage morphs (so-called "color phases") of male ruffs. *See also* Dimorphism; Morphism; Polymorphism.

Diethism. The occurrence of two discrete behavioral phenotypes within one taxon, such as in sexual diethism, where the sexes differ markedly in certain aspects of behavior, at least among adults. *See also* Dimorphism; Morphism; Polymorphism.

Differential access hypothesis. A hypothesis stating that an individual's own relative mating attractiveness or social status affects its access to potential mates.

Differential allocation hypothesis. A hypothesis stating than an individual's own relative mating attractiveness affects the amount of parental investment it can secure from its mate.

Dimorphism. The occurrence of two discrete structural phenotypes within one taxon. In this book "dimorphism" normally is used specifically to refer to grouping based on physically defined distinctions, such as linear mensural differences (size dimorphism) or weight differences (mass dimorphism). However, "dimorphism" is often used in a broader sense by biologists to encompass dichromatism and diethism as well. *See also* Dichromatism; Diethism; Morphism; Polymorphism; Reversed sexual dimorphism; Sexual dimorphism.

Direct (*or* individual) fitness. A measure of an individual's ability to produce and raise direct descendants in future generations relative to others in that same population. *See also* Fitness; Indirect fitness; Viability.

Directional selection. A process of selection by which a trait is shifted away from its mean condition in either direction, by selection operating against one end of the trait's range of variation. *See also* Disruptive selection; Selection pressure; Stabilizing selection.

Direct selection. A type of natural or sexual selection in which changes in gene frequencies in subsequent generations result from individuals producing direct descendants at differing rates. *See also* Direct fitness; Indirect fitness; Indirect selection.

Disassortative mating. Mating choices in which mates tend to be less similar in phenotype than randomly chosen mates would be. *See also* Assortative mating; Mate choice behavior.

Dispersed arena. *See* Exploded arena.

Display. A convenient nontechnical term for a social signal (usually visual or acoustic in birds) that transmits specific information between one individual and another, the latter often but not necessarily of the same species. Many avian displays are also believed to be genetically transmitted between generations, and therefore to be innately performed and perceived. *See also* Instinctive; Sign stimulus.

Display court. As used here, a surface-level or elevated site that is regularly used for sexual display by one or more individuals, and has been adaptively modified to facilitate that function. Such modifications often include vegetation manipulation, debris removal, or soil excavation. Avian display courts that have been constructed using brought-in materials are usually called bowers.

Display flight. A variably extended flight, often above a territory, marked by distinctive vocalizations, aerial posturing, or other signaling behavior. *See also* Flutter-jump.

Display mound. A terrestrial site that is higher than surrounding areas, and on which one or more individuals regularly display. If the mound is piled up by the participants themselves it is by definition a "court"; otherwise it is simply a "stage." Dug-out sites are sometimes called display bowls.

Display perch. An elevated site, such as a branch or vine, on which one or more birds regularly display. If the site has been adaptively modified for that function, by vegetational manipulation or removal, it is by definition a display "court."

Display stage. A terrestrial site that is regularly used for display and is in some respect especially suitable for it, but has not been adaptively modified for display functions by the participants. *See also* Display court.

Disruptive behavior. Direct interference in the behavior of one individual by another individual of the same sex during sexual competition, such as disruption of courtship displays and copulation attempts. *See also* Mate competition.

Disruptive selection. A process of selection that operates against the middle of a population's range of phenotypic variation, thus tending to split the population genetically. *See also* Directional selection; Stabilizing selection.

Diurnal. Pertaining to the hours of daylight. *See also* Crepuscular; Nocturnal.

Dominance-fitness gradient. A hierarchical social structuring in which the most dominant individual in the social unit also exhibits the greatest inclusive fitness, and the most subordinate one has the least. *See also* Dominance hierarchy.

Dominance hierarchy. A structured social grouping that is organized from the most dominant (alpha) individuals to the least dominant or most subordinate ones, either temporarily or permanently, and with the individuals stratified either linearly ("peck order") or in some more complex arrangement. Leks are typically site-related dominance hierarchies. Such hierarchies often strongly influence individual mating success, and presumably reflect dominance-fitness gradients in the population. *See also* Dominance-fitness gradient.

Dominant. Pertaining to the social superiority of one individual over another (social dominance); or, with reference to a genetic allele, determining the phenotypic appearance of a heterozygotic individual (genetic dominance). Social dominance may be site-dependent (territorial dominance) or group-related (stable social hierarchies); leks uniquely exhibit both characteristics.

Duetting. Coordinated and (1) overlapping or (2) alternating vocalizations by two birds, performed either cooperatively (e.g., between mates) or as competitive countersinging contests (e.g., between competing males).

Eclipse plumage. Dull, often femalelike plumages carried by nonbreeding males of some bird species (such as ducks and widowbirds), and which alternate with their brighter breeding ("nuptial") plumages.

Ecology. The study of the interrelationships between organisms and their biotic and physical environments.

Egg parasitism. *See* Brood parasitism.

Emancipation. Freedom from responsibility for the brooding, defending, and feeding of dependent young; obtained by males (rarely females) of some groups of birds, usually those birds having precocial young or that at least can be easily tended by the females alone.

Epigamic. Pertaining to sexual reproduction, especially structures (secondary sexual characteristics) or behaviors (advertising displays) that serve to facilitate such reproduction.

Epigamic selection. *See* Intersexual selection.

Ethological characteristics. Behavioral traits, especially species-specific behaviors that are at least in part genetically controlled, such as those operating in ethological isolation. *See also* Instinctive; Isolating mechanisms; Species-specific.

Ethology. The study of behavior from an evolutionary, often naturalistic or comparative, biological viewpoint.

Evolution. Any gradual change; biological evolution involves change as a result of altered gene frequencies between generations.

Evolutionarily stable strategy (ESS). A behavioral strategy that, when it predominates in a population, leads to higher inclusive fitness than any other. *See also* Strategy.

Evolutionary biology. A branch of biology concerned with the interpretation of contemporary populations and their evolved characteristics using principles of natural selection, along with evolutionary aspects of ecology and ethology.

Exploded arena (*or* lek). A social display ground in which the participating males display at greater distances from one another than in clustered arenas; often out of site of one another but within hearing range. Sometimes also called a dispersed arena or lek.

Extant. Not extinct.

Extinct. No longer surviving anywhere.

Extirpated. No longer surviving in a particular location.

Extrapair copulations. Additional copulations obtained by an already paired individual with a conspecific other than its own mate. Sometimes termed "sneaky" copulations. *See also* Forced copulations.

Facultative polygyny/promiscuity. Used here to describe species that are normally monogamous, but the males of which under certain conditions are able to fertilize additional females in a single breeding season. *See also* Extrapair copulations.

Fecundity. A term for describing relative reproductive potential, especially in females. *See also* Fertility.

Female choice behavior. The variable attraction of females to individual or grouped males for mating, thus providing a basis for intersexual selection. Passive attraction (indirect mate choice) involves an attraction to potential mates because of their convenient location, sheer numbers, or through copying the behavior of other females. Active (direct) mate choice requires discrimination among potential mates having varied phenotypic attributes, based on the female's relative costs (time and energy spent) and her potential benefits (direct resources or high-quality genes received). *See also* Active mate choice; Mate choice behavior; Passive mate choice.

Female copying. Mate choice behavior in which the probability of a female mating with a particular male increases if other females have already mated with that male, and decreases if they have not. *See also* Active mate choice; Passive mate choice.

Female defense polygyny. A mating system in which some males are able to gain and control sexual access to two or more breeding females simultaneously, usually owing to female tendencies toward intrasexual gregariousness. May be "economically" or "noneconomically" based; the former if sexually successful males also defend resource-rich territories. Sometimes also called harem polygyny. *See also* Male dominance polygyny; Noneconomic polygyny; Resource defense polygyny.

Female preference model. A hypothesis stating that male clustering (lekking) behavior is promoted by female preferences for mating with grouped rather than solitary males, and that clustered males facilitate efficient phenotypic comparison of potential mates, thus promoting rapid mating. This model predicts that larger leks should be favored by females over smaller leks, and that mate choice by females, rather than the ability by a few dominant males to control or influence female mating, should determine mating patterns. *See also* Hotshot model; Hot spot model.

Feral. Living in a secondarily wild state, following a period of captivity or domestication.

Fertility. A measure of reproductive potential, especially with regard to males. *See also* Fecundity.

Fertilization. The process by which two gametes unite to form a zygote; in birds, internal fertilization occurs and is achieved by copulation.

Fidelity. *See* Site fidelity.

Fisherian runaway process. *See* Runaway process.

Fitness. The genetic contributions of an individual (individual or direct fitness) as well as that of its close genetic relatives (indirect fitness) toward subsequent generations of a population, as measured by some form of reproductive rate. The combination of direct and indirect fitness is inclusive fitness. Population fitness is the tendency of one population to replace another within a group of populations over time. *See also* Inclusive fitness; Indirect fitness; Individual fitness.

Floater. A term used to describe a nonterritorial, mobile male, as opposed to a territorial resident.

Flutter-jump. A descriptive term for brief jumping-and-fluttering bird displays that are shorter and lower than either distinctly vertically oriented "rocket flight" displays or the more extended and horizontally oriented display flights. *See also* Display flight.

Forced copulations. Rapelike behavior in birds and other nonhumans. *See also* Extrapair copulations.

Form. A deliberately loose or neutral term for a population (i.e., a species or species subdivision) of unstated taxonomic rank.

Frugivorous. Fruit eating.

Fundamental frequency. The lowest sound frequency in a group of harmonically related sounds. *See also* Harmonic.

Galliform. A member of the order (Galliformes) of fowllike birds, such as pheasants and grouse.

Gamete. A haploid germ cell (egg or sperm). *See also* Zygote.

Gene. The functional unit of heredity, as represented by the genetic material (DNA). Also generally agreed to be the fundamental unit of evolution, through selection for traits at the individual phenotype level, rather than at the level of the gene (selfish gene hypothesis) or population (group selection hypothesis). Thus, genes mutate, individual phenotypes (and genotypes) are selected, and populations evolve through time-dependent changes in gene frequencies.

Generic. Pertaining to a genus.

Genic selection. The genetic consequence of natural selection on phenotypes, producing genetic changes in population gene pools over time.

Genome. All the genetic factors characteristic of a cell or an organism.

Genotype. The genetic identity of a gene locus of an individual; more broadly speaking, the gene constitution of an individual.

Genus (*plural* genera). The taxonomic category immediately above the species level, and also the first component of an organism's binomial name.

Good genes hypothesis. A hypothesis stating that the male traits that breeding females use to discriminate among potential mates are directly correlated with fitness differences among the available males (thus being "honest advertising" traits), and that these traits are heritable, so that their offspring directly benefit by the mating choices made by such females. *See also* Age indicator hypothesis; Female choice behavior; Handicap hypothesis; Total viability hypothesis.

Group (*or* interdemic) selection. Selection based on differences in survival and reproduction among groups (populations or population subunits) rather than among individuals.

Grouse. A collective vernacular term for members of the galliform tribe Tetraonini, including the typical grouse, ptarmigans, and capercaillies, all of which exhibit variably feathered tarsi. *See also* Capercaillie; Galliform.

Habitat. The specific environment within which a particular organism lives, including both its physical and biotic components. *See also* Niche.

Handicap. A trait the production or maintenance of which is costly (in terms of natural selection) to the individual carrying it.

Handicap hypothesis. A hypothesis stating that female mate choice is favored when it selects for sexual characters that actually handicap the survival of males carrying them. Costly morphological or behavioral traits can thus evolve in a population because they are reliable indicators of the genetic superiority of the individuals that carry and exhibit them. Females choosing to mate with males having such traits will have offspring of higher average viability because of the superior genes with which these traits are genetically linked. A variant of the good genes hypothesis.

Harem. A group of females simultaneously under the sexual control of a single male, as in harem polygyny. *See also* Female defense polygyny.

Harmonic. One in a series of consecutive multiple frequencies, or overtones, variably resonated above a fundamental sound frequency. *See also* Fundamental frequency; Resonation.

Hectare. A metrically defined area (10,000 square meters) that is approximately equal to 2.47 acres.

Heritability. The genetic component of phenotypic variability, which may theoretically range from 1.0 (all variation is genetically controlled) to 0 (all variation is environmentally controlled).

Hermit. A collective vernacular name for a large group of relatively dull-colored and tropical hummingbirds of the genus *Phaethornis*.

Hertz (*abbreviation* Hz). A measure of sound frequency (pitch), measured in cycles per second. *See also* Kilohertz.

Heterosexual. Pertaining to between-sex (intersexual) interactions or differences.

Heterozygote. An individual carrying two different alleles at corresponding loci of its two parental chromosomes.

Hierarchy. A structured social status in animal groups. Hierarchical promiscuity is a mating system in which a male's access to females is influenced by his relative social dominance status.

Hill. A traditional (and inappropriate) name for the lekking site of ruffs.

Hitchhiking principle. The concept that an allele lacking in positive selective value can increase in frequency within a population because it is genetically linked to another allele having positive selective value.

Home range. The area within which an individual ranges over a stated period of time (e.g., daily home range, annual home range, lifetime home range).

Homologous. Structural, physiological, or behavioral traits that are shared by two or more groups and reflect their common evolutionary ancestry.

Homozygote. An individual carrying identical alleles at corresponding loci of its two parental chromosomes.

Honest advertising. *See* Good genes hypothesis.

Hotshot model. A hypothesis stating that, since some males are socially more dominant and sexually successful, other less dominant males should congregate around them to gain possible sexual access to females. This hypothesis predicts that the dominant ("hotshot") males are the nuclei around which leks may form, and that mating opportunities are largely determined by relative male dominance relationships (and related experience-based or copying-based mate choice decisions by females) rather than by site-based characteristics. *See also* Female preference model; Hot spot model.

Hot spot model. A hypothesis stating that areas used by females for various nonreproductive reasons (such as for foraging) should be ones to which males may also gravitate for display purposes. This hypothesis predicts that such sites are the nuclei around which leks may form, and that females may make individual mating choices within such groups without specific mating constraints related to male-male competitive and dominance interactions. *See also* Female preference model; Hotshot model.

Hybridization. The interbreeding of individuals from two genetically unlike natural populations, such as different races (intraspecific hybridization), different species (interspecific hybridization), different genera, etc.

Hybridization paradox. Used here to observe that although lekking bird species usually exhibit a very high degree of species-specific male plumage and display signals, they are paradoxically more prone to hybridizing than are many less colorful and less specifically distinctive monogamous relatives. *See also* Lek paradox.

Hypothesis. A testable proposition or conclusion established by inductive reasoning to help explain some specific phenomenon. A generally accepted and widely applicable hypothesis is often called a principle, and a collection of related principles may comprise a general theory. *See also* Model; Prediction.

Inbreeding. Mating choices in which chosen mates tend to be more closely related genetically than are randomly chosen individuals. *See also* Outbreeding.

Inciting. A mate choice and pair-bonding behavior typical of many duck species, during which females seemingly encourage their mates or prospective mates to threaten or attack other individuals, often other conspecific males.

Inclusive fitness. The total of direct fitness and indirect fitness, namely the sum of an individual's direct fitness plus its influence on the fitness of its relatives other than direct descendants, based on their relative degree of relatedness. *See also* Coefficient of relatedness; Fitness.

Indirect fitness. A measure of the degree to which an individual increases the proportion of its genes in future generations by increasing the direct fitness of genetically similar relatives (the similarity measured by their coefficients of relatedness to that individual). *See also* Direct fitness.

Indirect selection. A form of selection in which (in the case of runaway sexual selection for mate preferences) the mating choices made by females do not directly affect their own survival or fecundity, and thus do not improve their direct fitness, but instead only influence male traits.

Individual distance. The distance within which one individual will not tolerate the presence of another; such spacing behavior is especially evident in flocking or colonial species that do not use territoriality for spacing out.

Individual (*or* direct) fitness. The genetic contribution of an individual's direct descendants to future generations of that species' population. *See also* Direct fitness; Direct selection; Fitness.

Individual selection. Selection based on individual differences in survival and reproduction within a population. *See also* Group selection; Kin selection.

Innate. Pertaining to genetically transmitted traits.

Instinctive. Refers to those genetically inherent (innate) behavioral traits that appear to be more complex than various simpler innate responses (such as reflexes) and whose performance seemingly depends on specific and transitory internal states ("tendencies" or "drives").

Intergeneric. Pertaining to interactions between genera.

Intersexual. Pertaining to heterosexual interactions.

Intersexual selection. Sexual selection caused by varied heterosexual attraction abilities producing differential individual reproductive success rates, and resulting in the evolution of secondary sexual characteristics typical of one sex. Sometimes called epigamic sexual selection. *See also* Female choice behavior; Intrasexual selection.

Interspecific. Pertaining to interactions between species.

Intrageneric. Pertaining to interactions between species of a single genus.

Intrasexual selection. Sexual selection caused by the outcome of male-male (rarely female-

female, as in polyandry) dominance contests, producing differential individual reproductive success rates within one sex, and resulting in the evolution of secondary sexual characteristics typical of that sex. Sometimes called agonistic sexual selection. Potentially includes both mate competition and sperm competition among interacting males of nonmonogamous birds. *See also* Intersexual selection.

Intraspecific. Pertaining to interactions within a species.

Invertebrate. Any nonvertebrate animal; an animal lacking a backbone.

Iridescent. Producing rainbowlike hues through light refraction and reflection, as occurs in most hummingbird feathers.

Isolating mechanisms. Inherent ("intrinsic") characteristics of individuals of a species that tend to prevent successful interbreeding with other sympatric species. Environmental constraints that prevent interbreeding by maintaining allopatry are often called extrinsic isolating factors. *See also* Reproductive isolation.

Kilohertz (*abbreviation* kHz). A measure of sound frequency (pitch) equal to 1,000 hertz (cycles per second).

Kin selection. Variously defined as (1) the combined effects of direct and indirect selection, or (2) selection resulting from individuals influencing the survival and reproduction of relatives (other than direct offspring) who possess the same genes by common descent; thus limited to indirect selection operating through nondescendant relatives.

Kinship. The relative sharing of a common recent ancestor, as measured by the coefficient of relatedness.

Learned behaviors. Experience-dependent responses resulting from an individual's past history, as contrasted to its innate (experience-independent) attributes.

Least costly male hypothesis. A hypothesis accounting for lek evolution by assuming that, whenever a male's presence on a female's breeding range may interfere with her reproductive success, it is a better strategy for males to wait for females to find them rather than to search actively for them. Such strategies are only likely to evolve in species having high population densities and relatively mobile females.

Lek. A traditional Scandinavian name for a site used by males for epigamic (intersexual and intrasexual) display. Originally specifically used in Sweden to describe black grouse and ruff arenas, but now much more widely applied to birds, other vertebrates, and even some invertebrate mating assemblages. *See also* Arena; Clustered arena; Exploded arena; Mobile arena.

Lekking. A category of epigamic behavior in which males gather in variably clustered arenas (leks) for reproductive purposes. *See also* Arena; Lek.

Lek paradox. The paradox that poses the question of why females of lek-breeding species, which obtain almost nothing more from males than their sperm, should have evolved such strong mating preferences for particular males within assembled mating clusters. Contrariwise, such strong preferences are not nearly so evident in more clearly "economic" mating systems, in which male involvement in female breeding success is much greater and thus effective mating choice should be more directly important to females. *See also* Hybridization paradox.

Linkage. A term used (1) ethologically, to describe sequentially associated behaviors, and (2) genetically, to describe the condition of genes that are located on the same chromosome.

Lyrate. Lyre-shaped.

Male buffet model. *See* Female preference model.

Male dominance polygyny. A "noneconomic" polygynous mating system in which male reproductive success is related to individual social dominance and associated sexual access to females, rather than to defense of a resource-rich territory; cases of male dominance polygyny where males are strongly clustered (or "clumped") represent typical arena behavior (or "lekking"). In polyandrous systems a comparable male-access dominance system might occur, based on female dominance interactions that control the sexual access of other females to adult males. *See also* Noneconomic polygyny; Resource defense polygyny.

Manakins. Members of the New World avian family Pipridae (not to be confused with mannikins, a group of Old World finches).

Mass ratio. Used here to indicate the average adult male-to-female weight ratio, expressed relative to unity (e.g., 1.5:1).

Mate choice behavior. The intersexual aspect of sexual selection; namely, choosing mating partners based on individual heterosexual attraction attributes. Mating choices may be passive or active. Active mating choices may tend to increase population homozygosity (through assortative mating and inbreeding choices), or contrariwise may increase heterozygosity (through disassortative mating and outbreeding choices). *See also* Active mate choice; Assortative mating; Disassortative mating; Female choice behavior; Inbreeding; Outbreeding; Passive mate choice.

Mate competition. The direct and indirect competition among members of one sex for sexual access to the other sex. *See also* Disruptive behavior; Sperm competition.

Mating. A term variously used with reference either (1) to initial pair bonding (pairing behavior), or more specifically (2) to copulation (mating behavior per se). *See also* Pair bonding.

Mating strategy. A species' evolved behavioral basis for engaging in a particular mating system. *See also* Mating system; Strategy.

Mating success. A measure of an individual's capacity for transmitting its genes to the next generation.

Mating system. A general descriptive term for various kinds of mating strategies, such as monogamy, polygyny, polyandry, and promiscuity. Definitions sometimes also include additional mating attributes, such as the duration of pair bonding (if any), the degree of random or selective mating, and possible differential parental investments. *See also* Mating.

Maypole bower. One of the types of bowers constructed by bowerbirds, in which a vertical sapling provides the axis around which the entire structure is built. *See also* Avenue bower.

Mean. A numerical average. *See also* Mode.

Mechanical sounds. Nonvocal (nonsyringeal) noises produced by percussion, stridulation, feather vibration, or other similar means.

Megapodes. A collective vernacular name for members of the galliform family Megapodidae, or mound builders, of Australia and New Guinea.

Mimesis. *See* Mimicry.

Mimicry. Imitative of another species or environmental attribute, such as vocal mimicry (as in lyrebirds), or visual mimicry, including protective visual resemblance to another species (as in Batesian mimicry) or to the species' environment (background mimicry). Aggressive mimicry directly exploits or harms another species, as when brood parasite finches lay eggs and have young that visually mimic those of their hosts, reducing the latter's reproductive success. *See also* Crypsis.

Mobile arena (*or* lek). An arena- or leklike assemblage of males competing sexually for females and whose interactions are not dependent upon specific territories nor restricted to fixed display sites. Sometimes also called a "detached" arena.

Mode. The most frequently occurring ("modal") group in a frequency distribution. Unimodal distributions have a single peak; bimodal distributions have two.

Model. A working concept used to test a hypothesis by generating predictions that can be applied to some natural phenomenon. A "robust" model is one that offers unusually rich predictions and opportunities for testing. *See also* Hypothesis; Prediction.

Modulation. Controlled temporal variations in sound frequencies or amplitudes, especially in vocalizations. *See also* Resonation.

Molting. In birds, the sequential loss and replacement of individual feathers or entire plumages.

Monogamy. A mating system in which adults of both sexes establish pair bonds that persist through at least one breeding cycle (single-brood monogamy), or for a longer, sometimes lifelong period (permanent or indefinite monogamy). Single-brood monogamous species sometimes become facultatively polygamous if the breeding season is long enough.

Monomorphism. The occurrence of a single morphological phenotype ("morph") within a population, as in a monomorphic species, where the sexes appear phenotypically identical as adults. *See also* Dimorphism; Morphism; Polymorphism.

Monotypic. Referring to a taxonomic group having a single representative, such as a monotypic genus with a single included species.

Morph. Any of the phenotypic variants that occur in cases of sexual or nonsexual morphism. Nonsexual morphs are sometimes called phases.

Morphism. Here used to refer to structural or behavioral phenotypic variations in a population's traits that result in dimorphism (*sensu lato*) or polymorphism. *See also* Dimorphism; Monomorphism; Polymorphism; Sexual dimorphism.

Morphological. Pertaining to structures of an organism, especially external or generally apparent structures as opposed to internal (anatomical) structures.

Mortality rate. The rate at which individuals are removed from a population by death; often calculated on an annual basis. Usually expressed as a percentage (e.g., 15 percent) and reciprocally related to the population's survival rate, the sum of the two being 100 percent (e.g., M = 15 percent, S = 85 percent). *See also* Turnover rate.

Mosaic evolution. *See* Primitive.

Mutation. A potentially permanent alteration in a gene or chromosome that results in changes in the cell's genotype.

Natural selection. In a contemporary and broad sense, the changes in gene frequencies and associated traits resulting from differential survival and reproduction of individuals within an interbreeding population. Also used by Darwin in a stricter sense to refer only to traits associated with differential survival in nature, with "sexual selection" used to designate differential reproduction effects influencing only a single sex separately. *See also* Sexual selection.

Nearctic region. A zoogeographic region encompassing nontropical portions of North America.

Nest parasitism. *See* Brood parasitism.

Niche. The behavioral, morphological, and physiological adaptations of a species to its habitat. Also sometimes defined from an environmental standpoint, namely as the range of ecological conditions under which a species potentially exists (fundamental niche), in which it survives best (preferred niche), or in which it actually survives (realized niche). Part of the overall niche is the foraging niche, including not only a species' foods but also behavioral characteristics affecting individual foraging efficiency.

Niche segregation. The ecological separation of two populations (or subpopulations) so as to reduce ecological competition between them.

Nocturnal. Pertaining to the hours of darkness. *See also* Crepuscular; Diurnal.

Noneconomic polygyny. A type of polygynous mating system in which a male contributes only his genes to the females with whom he mates; thus distinct from resource-based polygyny, where other resources are also provided. *See also* Female defense polygyny; Male dominance polygyny.

Normalizing selection. *See* Stabilizing selection.

Null hypothesis. A hypothesis stated in such a manner as to be the converse of the expected results, in order to avoid accepting a false conclusion without first disproving (falsifying) the opposite position. *See also* Hypothesis.

Nuptial. Pertaining to breeding, as in nuptial plumages or nuptial displays.

Ocelli. Eyelike structures on the feathers of some birds, especially peacock-pheasants, argus pheasants, and peacocks.

Ontogeny. The temporal development of an individual, from zygote to adult. *See also* Phylogeny.

Operational sex ratio. The ratio of fertilizable females available to sexually active males at any given time. *See also* Sex ratio; Tertiary sex ratio.

Optimal traits. In evolutionary biology, those characteristics whose collective benefit-to-cost ratio is greater than that associated with alternate phenotypes that have arisen in the species' past history. *See also* Benefits; Costs.

Oscines. The subdivision of passerine birds that includes the true "songbirds," defined anatomically by their relatively great syringeal complexity. *See also* Suboscines; Syrinx.

Outbreeding. Mating choices in which chosen mates tend to be less closely related genetically than are randomly chosen individuals. *See also* Inbreeding.

Pair bonding. The establishment of a variably prolonged and individualized heterosexual relationship. Pair-bonding (or pairing) behavior normally occurs adaptively in conjunction with reproduction, although some homosexual pairings have been reported between females of various nondimorphic bird species, such as geese and gulls. *See also* Mating.

Palearctic region. A zoogeographic region encompassing all of Europe, Asia south to the Himalayas, and Africa north of the Sahara.

Parapatry. The relationship of populations having extensively abutting but not overlapping ranges. *See also* Allopatry; Sympatry.

Parental investment. Any investment (of food, energy, time, etc.) made by the parent in an individual offspring that increases the offspring's chance of surviving at the cost of the parent's ability to invest in other activities. *See also* Costs; Reproductive effort.

Parotia. A collective vernacular name for species of the genus *Parotia,* also known as six-wired birds-of-paradise.

Passerine. Pertaining to typical perching birds (members of the order Passeriformes), and including both the anatomically more primitive suboscines and the true songbirds, or oscines.

Passive mate choice. Mating behavior in which mating choices are made for reasons other than individual selection of phenotypic traits, such as local proximity or temporal convenience. *See also* Active mate choice.

Peck order. *See* Dominance hierarchy; Hierarchy.

Pectoral. Pertaining to the breast and shoulder area.

Phase. *See* Morph.

Phenotype. The collective characteristics (behavioral, physiological, morphological) comprising the attributes of an individual organism, and resulting from interactions between its genotype and the environment. *See also* Genotype.

Phylogenetic inertia. The limits to which evolutionary rates can be altered because of fundamental population characteristics; often used to account for persistent traits of a species or lineage that are seemingly no longer adaptive.

Phylogenetic order. A linear sequence of taxonomic groups that best reflects their known phyletic affinities.

Phylogeny. The evolutionary history of an organism or related group of organisms. *See also* Ontogeny.

Pleiotropy. The capacity of a gene or linkage group to influence two or more different phenotypic characteristics of an individual carrying it.

Plumage. A group of feathers associated with a single molting period and thus representing a single feather generation.

Pochards. A collective vernacular name for species of the diving duck genus *Aythya*.

Polyandry. A mating system in which females mate concurrently or sequentially with two or more males during a single reproductive cycle, either by monopolizing critical resources (resource defense polyandry) or by limiting sexual access by other females to males (female access polyandry). Polyandrous mating systems occur in several nonpasserine bird families (Turnicidae, Jacanidae, Rostratulidae, Rallidae, and Charadriidae) with precocial young.

Polygamy. A mating system in which either individual males gain sexual access to more than one female (polygyny), or individual females gain sexual access to more than one male (polyandry). Rapid multiple-clutch polygamy involves individuals breeding several times in rapid succession with varied but briefly monogamous mates, with each of the participants incubating the resulting clutches separately. *See also* Polyandry; Polygyny; Promiscuity.

Polygenic inheritance. The inheritance of traits that are controlled by more than a single gene locus or linkage group.

Polygyny. A mating system in which males mate with two or more females during a single reproductive cycle, either establishing pair bonds with them simultaneously or pairing with several different females in rapid sequential, sometimes overlapping, succession. *See also* Female defense polygyny; Male dominance polygyny; Polygamy; Promiscuity; Resource defense polygyny.

Polygyny threshold. A hypothetical stage at which a greater reproductive benefit accrues to a female by joining a polygynous male holding a good territory than by pairing monogamously with a male holding a poor territory.

Polymorphism. The occurrence of several discrete phenotypes (behavioral or structural polymorphism) or associated genes (genetic polymorphism) within a population simultaneously. Genetic polymorphism is caused by the maintenance of two or more alternative alleles in a gene pool above the frequencies that can be maintained by mutations and gene flow alone, often because of heterozygotic superiority. Behavioral polymorphism may have either a genetic (innate) or acquired (experience-dependent) basis.

Population. A group of individuals belonging to the same species and occupying the same general area simultaneously. Local population subunits are often called demes; groups of related populations may be termed metapopulations.

Precocial. Refers to those animals (here defined only for birds) that at or shortly after hatching are feathered sufficiently (ptilopaedic) to retain body heat, that often leave the nest site well before fledging (are nidifugous), and that are often sufficiently developed to begin foraging soon for themselves. *See also* Altricial.

Predator deflection hypothesis. A hypothesis stating that bright colors or conspicuous patterns of some birds may be signals to predators that they represent unprofitable (e.g., hard to catch) prey. Also called the unprofitable prey hypothesis. In a broader sense, predator deflection signals may also include eyelike or other conspicuous markings on the body or tail that perhaps confuse or startle a potential predator.

Prediction. In evolutionary biology, the logical outcome of a working hypothesis, preferably one that can be measured as a test of the hypothesis.

Primaries. Major flight feathers of birds that are supported by the hand and finger bones. *See also* Secondaries.

Primitive. Pertains to generalized traits that appeared earlier in a group's phylogeny but later gave rise to derived (but not necessarily more complex) traits. The retention of both primitive and derived traits in a single phyletic line is called mosaic evolution.

Principle. *See* Hypothesis.

Promiscuity. A polygamous mating system in which breeding adults form no heterosexual pair bonds, but instead associate with the opposite sex only for reproductive purposes. Not always easily distinguished from polyandry and polygyny, which (at least in birds) imply the existence of temporary pair bonding.

Proximate causation. Pertaining to those environmental factors that are currently operating, and that produce or influence some biological phenomenon. *See also* Ultimate causation.

Quality. A term used in evolutionary biology to describe the extent to which a particular attribute may enhance an individual's inclusive fitness.

Race. *See* Subspecies.

Rape. *See* Forced copulations.

Recessive. Refers to the condition of those genes in a heterozygotic individual whose effects are not expressed. *See also* Dominant; Heterozygote.

Reciprocal altruism. The trading of altruistic acts by individuals at different times. *See also* Altruism.

Rectrices. The tail feathers of birds. *See also* Tail coverts.

Red Queen's rule. The concept that in cases of antagonistic coevolution, one species must constantly be evolving simply to keep up with evolutionary adaptations occurring in an adversarial species, as in host-parasite relationships. Suggested as an influence on genetic diversity among males of lekking species whose reproductive success rates are strongly affected by their parasite levels. *See also* Coevolution.

Releaser. *See* Sign stimulus.

Reproductive effort. The proportion of an organism's total available energy that is used in reproduction, including both its mating effort (effort spent in trying to obtain mates) and its parental effort (effort spent in parental investment). Sometimes also stated in terms of an organism's relatively diminished ability to reproduce later. *See also* Parental investment.

Reproductive isolation. The genetically based isolation of one species from another, by various devices or "isolating mechanisms" such as behavioral (ethological mechanisms), time-dependent (temporal mechanisms), structural (mechanical mechanisms), and habitat-related differences (ecological mechanisms) occurring between these species. Selection for traits improving reproductive isolation is sometimes called "reproductive character displacement" (as opposed to ecological character displacement, which improves ecological isolation between species). *See also* Isolating mechanisms.

Reproductive success. The number of surviving offspring of an individual, a measure of direct fitness.

Resonation. The transmission and amplification by nonsyringeal structures of vocalizations, especially those having particular frequency characteristics. *See also* Air sac; Tracheal air sac.

Resource. An environmental attribute needed for an organism's survival and reproduction, including but not necessarily restricted to those that are in limited supply and over which competition might occur.

Resource defense polygyny. A polygynous mating system in which the male's territory includes environmental resources of potential use to the females with whom he breeds, thus comprising an "economic" mating system. *See also* Noneconomic polygyny.

Resource-holding potential. The factors that influence an individual's relative dominance or fighting ability during competition for limited resources.

Resource-inclusive territories. Territories that include within their boundaries various resources (nest sites, feeding areas, etc.) of potential importance for successful breeding.

Reversed sexual dimorphism. A condition of size and/or mass dimorphism in which females are the larger and/or heavier sex. Especially typical of polyandrous species, but also found among many polygynous species in which selection for male aerial agility occurs. Reversed sexual dichromatism (females brighter) and diethism (females more aggressive) may also occur, especially in polyandrous mating systems. *See also* Dimorphism.

Risks. A term from game theory used to describe the expected potential costs, especially possible direct bodily harm, of engaging in a particular activity. *See also* Costs.

Ritualized. In ethology, refers to behaviors that have gradually acquired signal functions (become displays) through evolution. *See also* Display; Signaling behavior.

Runaway process. R. A. Fisher's hypothetical process of sexual selection, which proposes that females may increasingly tend to prefer a male secondary sexual trait that becomes progressively enhanced through directional selection associated with this female mate choice behavior. Although it might initially have provided the females' offspring with some general selective advantage, the trait may eventually become so magnified in males as to be disadvantageous in its effect on overall viability, and the process is thus gradually brought to a halt. *See also* Good genes hypothesis; Sexy son hypothesis.

Satellite. A descriptive term for males in some lekking species, especially ruffs, that do not establish fixed territorial residences within the lek but may temporarily occupy display positions near resident males.

Scenario. A hypothetical explanation or description of a natural phenomenon, based on evolutionary principles and currently available information.

Secondaries. Major flight feathers that are associated with the forearm (radius and ulna) of birds. *See also* Primaries.

Secondary sexual characteristic. Any sex-related trait (usually limited to or best developed in adults of only one sex) that is associated with obtaining a mate or facilitating reproduction, but is usually not directly required for such reproduction or for parental care (these latter structures being primary sexual characteristics).

Selection pressure. Any factor tending to cause natural selection to operate in some consistent manner. *See also* Directional selection; Disruptive selection; Stabilizing selection.

Sex ratio. The ratio of males to females in a population, ranging from the ratio existing at fertilization (primary sex ratio) to that of mature adults (tertiary sex ratio), and shown either as a direct male-to-female ratio (e.g., 1.5:1) or as a proportional percentage (e.g., 60:40). *See also* Operational sex ratio.

Sex role reversal. A condition, typical of polyandrous species, in which the more usual asymmetries in the distribution of secondary sex characteristics are reversed; for example, reversed sexual dimorphism (females are brighter and larger) and reversed sexual diethism (females compete aggressively for mates; males select mates and perform most or all of the parental care).

Sexual bimaturism. Differential sexual maturation rates in a species, males usually being slower to mature than females.

Sexual competition. Competition to achieve fertilization, as expressed through mate competition and/or sperm competition.

Sexual dimorphism. As used broadly, the condition by which the adults of a bird species variously exhibit measurably different secondary sex characteristics, including general external morphology or linear dimensions (size dimorphism), overall adult body weights (mass dimorphism), and differences in hue, brightness, or patterns of feather and/or softpart phenotypes (sexual dichromatism). *See also* Morphism.

Sexual selection. A type of natural selection (*sensu lato*) in which the evolution and maintenance of traits of one sex result from the social interactions that produce differential individual reproductive success, including both interactions between members of the same sex (intrasexual or agonistic selection) and those between the sexes (intersexual or epigamic selection). *See also* Intersexual selection; Intrasexual selection; Natural selection.

Sexy son hypothesis. A hypothesis stating that female mating preferences for unusually attractive males can evolve even if mating with these males results (because of reduced male investment in parental care) in lower immediate female fecundity, inasmuch as these females' sons will inherit the traits that are especially attractive to females and thereby eventually leave more descendants. Compare with the related Runaway process, which applies more directly to polygynous or promiscuous species.

Shelducks. A collective vernacular name for ducklike species (*Tadorna* spp.) of the waterfowl tribe Tadornini. Shelducks and the more gooselike sheldgeese represent a structural and ethological transition between true geese (Anserinae) and more typical ducks (Anatinae).

Shorebirds. A collective vernacular name (in North America) for species of the avian suborder Charadrii, such as plovers, sandpipers, and snipes. Called waders in Britain.

Siblings. Offspring of the same parents; assemblages of brothers and/or sisters may comprise sibling groups.

Sibling species. Populations that are morphologically similar or identical to, but reproductively isolated from, one another and thus represent separate species.

Signaling behavior. Communication behavior, during which information is transmitted from one individual to another through the production of social signals ("displays") such as posturing, vocalizations, or other channels of transmission. *See also* Display; Signaling device.

Signaling device. A structure (such as feathers, softparts, etc.) that facilitates the production and broadcasting of social signals ("displays" or sign stimuli) by an individual. *See also* Display; Sign stimulus.

Sign stimulus. In traditional ethological terminology, a stimulus signal (visual, acoustic, etc.) produced by one individual that "releases" an instinctive response ("fixed action pattern") in another individual who is already in an appropriate internal state conducive to performing such a response (i.e., having an adequate "specific action potential"). The "releaser" is that part of the overall sign stimulus that is most important in transmitting the signal. *See also* Display.

Single-brood monogamy. Monogamous mating systems in which a pair bond persists through the completion of a single nesting cycle, and in which both sexes typically participate to some degree in parental responsibilities. *See also* Monogamy.

Site fidelity (*or* tenacity). Describes the tendency of individuals to return to a previously occupied site, such as a prior year's nest (nest site tenacity), a particular territory or lek site (territorial or lek fidelity), or some more general environmental location.

Skewed distributions. Frequency distributions that are shifted away from a normal (bell-shaped) distribution in either direction.

Social dominance. *See* Dominance hierarchy; Hierarchy.

Social parasitism. Broadly speaking, exploitative intraspecific or interspecific behavior such as brood parasitism, as well as various kinds of piracy including food stealing (kleptoparasitism), copulation stealing (cuckoldry), and stealing of nest sites. *See also* Brood parasitism; Extrapair copulations.

Song. A term used in ornithology to describe vocalizations that are usually more complex acoustically than are calls. Songs usually exhibit species, sex, and individual variations, these variations resulting from genetic, experience-dependent, and internal-state differences. *See also* Call; Vocalizations.

Songbirds. *See* Passerine.

Sonogram. A visual depiction of sound, typically showing variations in frequency and amplitude over time; also called sound spectrograms.

Speciation. The development of reproductive (intrinsic) isolation by a population or group of populations over time; also called species proliferation or species formation.

Species. A population or populations of actually or potentially interbreeding individuals, which are reproductively isolated from all other populations. The term species (unchanged in the plural) also refers to the taxonomic category between the genus and the subspecies. It is written as the second and subsidiary (uncapitalized) component of a two-parted (binomial) name, with the generic component taking precedence and being capitalized.

Species recognition behavior. The exchange of signals between individuals that serve to transmit information concerning species identity. Such signals are especially important during intraspecific courtship (epigamic species recognition), but they sometimes also function interspecifically, between competing species (agonistic species recognition). Interspecific mimicry is a related phenomenon, as are also various cryptic adaptations that result in species nonrecognition.

Species-specific. Pertaining to those traits typical of and also unique to a species, especially ones concerned with reproductive isolation.

Species-typical. Pertaining to traits that are typical of but not necessarily unique to a species. Other taxon-typical traits may be characteristic of a genus, family, etc.

Speculum. A brightly patterned, often iridescent area of feathers on the wings of some birds, especially ducks.

Sperm competition. The hypothetical "competition" between sperm from two or more males to fertilize the eggs of a single female. Although essentially a physiological process, it has behavioral implications in nonmonogamous species, such as influencing copulation rates or stimulating copulation disruption behavior toward other males. *See also* Mate competition.

Stabilizing selection. A process of natural selection in which the mean condition of a particular trait remains unchanged since selection is operating at both ends of the ranges of variation equally. *See also* Directional selection.

Stage. *See* Display stage.

Stereotyped. Refers to behavior patterns having marked rigidity in form, duration, and/or amplitudinal aspects, presumably reflecting their innate bases.

Stifftails. A collective vernacular name for various long-tailed diving ducks of the tribe Oxyurini.

Stimulus pooling. The magnification effect that clustered males produce by merging their display activities, thus forming a more effective collective visual and/or acoustic mate attraction stimulus. *See also* Superoptimal signal.

Strategy. A term describing a set of behaviors a species or individual may follow in order to maximize its fitness, and which may be either-or choices or choices along a continuum of possibilities. Evolutionarily stable strategies (ESS) are those adopted by most members of a population and that cannot be displaced by the spread of an alternate strategy. Default strategies are "standard" responses that are performed in the absence of special conditions requiring the use of alternate ones. *See also* Evolutionarily stable strategy.

Stridulation. Mechanically produced noises that are usually generated by rhythmically rubbing the surfaces of body parts (such as feathers) against one another.

Subordinate. A socially inferior or submissive individual in a dominance hierarchy. *See also* Beta; Dominance hierarchy; Dominant.

Suboscines. A subdivision of passerine birds including those groups (such as lyrebirds, manakins, and cotingids) that have less complex syringeal anatomy than the oscines or true "songbirds." *See also* Passerine.

Subspecies. One or more populations of a species that can be geographically and morphologically distinguished from other such populations, and that have gene and genotype frequencies differing from other such subspecies. Subspecies are also often called races.

Superoptimal (*or* supernormal) signal. A stimulus that, by increased size or number, releases stronger-than-normal responses from signal recipients. *See also* Stimulus pooling.

Superspecies. A group of closely related but entirely or essentially allopatric populations that are too distinct taxonomically to be considered a single species.

Supplanting. The eviction of a subordinate individual from its territory or display site, and occupation by the more dominant bird.

Sympatry. The simultaneous occurrence of two or more populations in the same area, especially during the breeding season; often used as a criterion for biologically determining reproductive isolation and defining completed speciation. *See also* Allopatry; Parapatry.

Syndrome. A collection of related traits that are adaptively associated with the exploitation of a specific niche or habitat, such as the frugivorous syndrome.

Syrinx (*plural* syringes). The vocal organ of birds, functionally comparable to the larynx of mammals. The syrinx is responsible for all avian vocalizations, which however may be modulated or resonated by other structures such as the trachea and anterior esophagus. *See also* Air sac; Bulla; Vocalizations.

Tail coverts. Shorter feathers covering the dorsal (upper tail coverts) and ventral (under tail coverts) bases of the tail feathers (rectrices).

Taxon (*plural* taxa). A group of organisms representing a particular category of biological classification, such as the genus *Anas*.

Taxonomy. A system of naming and describing groups of organisms (taxa), and organizing such groups in ways that reflect established phyletic relationships.

Teleology. Describing behavior of animals in such a manner as to imply conscious purposiveness to their acts, or a foreknowledge of the possible results of their behavior.

Temporal. Pertaining to time.

Tenacity. *See* Site fidelity.

Territory. An area defended by an animal against others of its species, and occasionally also of other species, and within which it tends to be socially dominant. Territories may be defended only during the breeding season (breeding territories), continuously (permanent territories), or for other seasonal periods (such as winter territories), and may include part or all of the individual's home range. Moving or mobile territories (those centering around mates or families) sometimes also exist.

Tertiary sex ratio. The ratio of adult males to adult females in a population. *See also* Operational sex ratio.

Tidbitting. Food presentation behavior; sometimes ritualized and without actual food being present.

Total viability hypothesis. A hypothesis stating that females will choose to mate with males carrying the fewest deleterious genes and thus having the highest viability; a variant of the good genes hypothesis. *See also* Age indicator hypothesis; Female choice behavior; Good genes hypothesis.

Tracheal air sac. An inflatable structure that is directly attached to the tracheal tube (windpipe) rather than being an outgrowth of the lungs. *See also* Air sac.

Trait. A measurable phenotypic attribute (behavioral, structural, or physiological), especially one that is at least in part genetically controlled. Traits may have several character components. *See also* Character.

Transferral hypothesis. A hypothesis stating that species- and sex-specific characteristics of male birds may sometimes be transferred to species-specific male-controlled environmental objects or structures (such as bowers), thereby reducing the need for these males to exhibit more elaborate and possibly costly male secondary sexual characteristics.

Truth-in-advertising hypothesis. *See* Good genes hypothesis.

Turnover rate. The interval between the time a new age class (cohort) enters a population and the time its last survivor dies. *See also* Mortality.

Tyrannid. Pertaining to members of the New World flycatcher family Tyrannidae; tyrannoid birds include a larger taxonomic group of flycatcher-like birds.

Ultimate causation. Those selective factors that have resulted in adaptive traits in the past history of a species and that are still evident in present-day populations. *See also* Proximate causation.

Unprofitable prey hypothesis. A controversial hypothesis that the presence of bright or showy coloration may be a signal to predators that the animals bearing such coloration represent elusive or unprofitable prey. The hypothesis predicts that predators should therefore prefer dull-colored over brightly colored prey organisms. *See also* Predator deflection hypothesis.

Viability. The overall capacity of an individual animal to survive in nature; viability-based traits are those innate attributes that influence such viability. *See also* Condition.

Vocalizations. Sounds generated (at least among birds) by the syrinx, including both calls and songs, and sometimes modulated or resonated by nonsyringeal structures. *See also* Fundamental frequency; Harmonic; Modulation; Resonation; Syrinx.

Waterfowl. A collective vernacular name for ducks, geese, and swans of the family Anatidae. Called wildfowl in Britain.

Whydahs. A collective vernacular name for several African brood-parasitic finches of the genus *Vidua*.

Widowbirds. A collective vernacular name for certain mostly dull-colored African finches of the genus *Euplectes*. *See also* Bishops.

Wing coverts. Feathers of a bird's wings other than its large flight feathers. *See also* Primaries; Secondaries.

Zygote. A fertilized (diploid) cell produced by the fusion of two (haploid) gametes.

Literature

Alatalo, R. V., J. Höglund, and A. Lundberg. 1988. Pattern of variation in tail ornament size in birds. *Biol. J. Linn. Soc.* 34:363–74.

Alatalo, R. V., J. Höglund, and A. Lundberg. 1991. Lekking in the black grouse: a test of male viability. *Nature* 352 (6331):155–56.

Alcock. J. 1987. Leks and hilltopping in insects. *J. Nat. Hist.* 21:319–28.

Ali, S., and A. R. Rahmani (eds.). 1984. Study of ecology of certain endangered species of wildlife and their habitats. The great Indian bustard. Bombay Nat. His. Soc., Bombay.

Andersson, M. 1982. Female choice selects for extreme tail length in a widowbird. *Nature* 299:818–20.

Andersson, M. 1986. Evolution of condition-dependent sex ornaments based on viability differences. *Evol.* 40:804–16.

Andersson, S. 1989. Sexual selection and cues for female choice in leks of Jackson's widowbird (*Euplectes jacksoni*). *Behav. Ecol. & Sociobiol.* 25:403–10.

Andersson, S. 1991. Bowers on the savanna: display courts and mate choice in a lekking bird. *Behav. Ecol.* 2:210–18.

Andersson, S. 1992. Female preference for long tails in lekking Jackson's widowbirds: experimental evidence. *Anim. Behav.* 43:379–88.

Andreev, A. V. 1979. Reproductive behaviour in black-billed capercaillie compared to capercaillie. Pp. 135–39, *in* Behaviour of woodland grouse, woodland grouse symposium, T. W. I. Lovel (ed.). World Pheasant Assn., Bures, Suffolk, England.

Armstrong, A. A. 1947. Bird display and behaviour: an introduction to the study of bird psychology. Lindsay Drummond Ltd., London.

Attenborough, D. 1990. The trials of life: a natural history of animal behavior. Little, Brown, Boston.

Atwood, J. L., V. L. Fitz, and J. E. Bamesberger. 1991. Temporal patterns of singing activity at leks of the white-bellied emerald. *Wils. Bull.* 103:373–86.

Aubin, A. E. 1970. Territory and territorial behavior of male ruffed grouse in southeastern Alberta. M.S. thesis, U. of Alberta, Edmonton.

Avery, M. I. 1984. Lekking in birds: choice, competition and reproductive constraints. *Ibis* 126:177–87.

Avery, M. I., and G. Sherwood. 1982. The lekking behavior of great snipe. *Ornis. Scand.* 13:72–78.

Baker, R. R., and G. A. Parker. 1979. The evolution of bird colouration. *Phil. Trans. Royal Soc. London* B. 287:63–130.

Ballard, W. R., and R. J. Robel. 1974. Reproductive importance of dominant male greater prairie chickens. *Auk* 91:75–85.

Balmford, A. 1992. Social dispersion and lekking in Uganda kob. *Behaviour* 120:177–89.

Balph, D. F., G. S. Innis, and M. H. Balph. 1980. Kin selection in Rio Grande turkeys: a critical assessment. *Auk* 97:854–60.

Bancke, P., and H. Meesenburg. 1958. A study of the display of the ruff *Philomachus pugnax* (L.). *Dansk. Orn. Foren. Tiddskr.* 52:118–41.

Barash, D. P. 1972. Lek behavior in the broad-tailed hummingbird. *Wilson Bull.* 84:202–3.

Barnard, P. 1989. Territory and the determinants of male mating success in the southern African whydahs (*Vidua*). *Ostrich* 60:103–17.

Barnard, P. 1990. Male tail length and female sexual response in a parasitic African finch. *Anim. Beh.* 39:652–56.

Barnard, P., and M. B. Markus. 1989. Male copulation frequency and female competition for fertilization in a promiscuous brood parasite, the pin-tailed whydah (*Vidua macroura*). *Ibis* 131:421–25.

Bateman, A. J. 1948. Intrasexual selection in *Drosophila*. *Heredity* 2:349–68.

Bateson, P. (ed.). 1983. Mate choice. Cambridge Univ. Press, Cambridge.

Beehler, B. M. 1983a. The behavioral ecology of four birds of paradise. Ph.D. diss., Princeton Univ.

Beehler, B. M. 1983b. Frugivory and polygamy in birds of paradise. *Auk* 100:1–12.

Beehler, B. M. 1983c. For gaudy display, it's hard to beat birds of paradise. *Smithsonian* 13(11):90–97.

Beehler, B. M. 1983d. Lek behavior of the lesser bird of paradise. *Auk* 100:992–95.

Beehler, B. M. 1987a. Birds of paradise and mating system theory—predictions and observations. *Emu* 87:78–99.

Beehler, B. M. 1987b. Ecology and behavior of the buff-tailed sicklebill (Paradisaeidae: *Epimachus albertsi*). *Auk* 104:48–55.

Beehler, B. M. 1988. Lek behavior of the raggiana bird of paradise. *Natl. Geo. Res.* 4:343–58.

Beehler, B. M., and C. H. Beehler. 1986. Observations on the ecology and behavior of the pale-billed sicklebill. *Wils. Bull.* 98:505–15.

Beehler, B. M., and M. S. Foster. 1988. Hotshots, hotspots and female preference in the organization of lek mating systems. *Am. Nat.* 131:203–19.

Beehler, B. M., and S. G. Pruett-Jones. 1983. Display dispersion and diet of birds of paradise: a comparison of nine species. *Beh. Ecol. & Sociobiol.* 13:229–38.

Belt, T. 1874. The naturalist in Nicaragua. J. Murray, London.

Benalcazar, C. E., and F. S. de Benalcazar. 1984. Historia natural del gallo de roca andino (*Rupicola peruviana sanguinolenta*). *Cespedesia* 8 (47–48):59–92. (English summary.)

Bent, A. C. 1940. Life histories of North American cuckoos, goatsuckers, hummingbirds and their allies. *U. S. Natl. Mus. Bull.* 176:1–506.

Bergman, S. 1957a. On the display and breeding of the king bird of paradise, *Cicinnurus regius rex* (Scop.) in captivity. *Avic. Mag.* 63:115–24.

Bergman, S. 1957b. Through primitive New Guinea. Robert Hale, London.

Bergman, S. 1958. On the display of the six plumed bird of paradise (*Parotia sefilata*) (Pennant). *Avic. Mag.* 64:3–8.

Bierregaard, R. O., Jr., F. Stotz, L. H. Harper, and G. V. N. Powell. 1984. Observations on the occurrence and behaviour of the crimson fruitcrow, *Haematoderus militaris,* in central Amazonia. *Bull. Brit. Orn. Club* 107:134–37.

Bishop, K. D. 1992. The standardwing bird of paradise *Semioptera wallacii* (Paradisaeidae), its ecology, behaviour, status and conservation. *Emu* 92:72–78.

Blackford, J. L. 1958. Territoriality and breeding behavior in a population of blue grouse in Montana. *Condor* 60:145–58.

Bleiweiss, R. 1985. Iridescent polychromatism in a female hummingbird: is it related to feeding strategies? *Auk* 102:701–13.

Bluhm, C. K. 1985. Mate preferences and mating patterns of canvasbacks (*Aythya valisineria*). *Orn. Monogr.* 37:45–56.

Borgia, G. 1979. Sexual selection and the evolution of mating systems. Pp. 19–80, *in* Sexual selection and reproductive competition, M. Blum and A. Blum (eds.). Academic Press, New York.

Borgia, G. 1985a. Bower destruction and sexual competition in the satin bowerbird (*Ptilonorhynchus violaceus*). *Behav. Ecol. & Sociobiol.* 18:91–100.

Borgia, G. 1985b. Bower quality, number of decorations, and mating success of male satin bowerbirds (*Ptilonorhynchus violaceus*): an experimental analysis. *Anim. Behav.* 33:266–71.

Borgia, G. 1986. Sexual selection in bowerbirds. *Sci. Amer.* 254 (6):92–100.

Borgia, G., and K. Collis. 1989. Female choice for parasite-free male satin bowerbirds and the evolution of bright male plumage. *Behav. Ecol. & Sociobiol.* 25:445–54.

Borgia, G., and K. Collis. 1990. Parasites and bright male plumage in the satin bowerbird (*Ptilonorhynchus violaceus*). *Am. Zool.* 30:279–85.

Borgia, G., and M. A. Gore. 1986. Feather stealing in the satin bowerbird (*Ptilonorhynchus violaceus*): male competition and the quality of display. *Anim. Behav.* 34:727–38.

Borgia, G., and U. Mueller. 1992. Bower destruction, decoration stealing and female choice in the spotted bowerbird (*Chlamydera maculata*). *Emu* 92:11–18.

Borgia, G., S. Pruett-Jones, and M. Pruett-Jones. 1985. Bowers as markers of male quality. *Z. Tierpsychol.* 67:225–36.

Bossema, I., and J. P. Kruijt. 1982. Male activity and female acceptance in the mallard (*Anas platyrhynchos*). *Behaviour* 79:313–24.

Bossema, I., and E. Raemers. 1985. Mating strategy, including mate choice, in mallards. *Ardea* 73:147–57.

Boyce, M. S. 1990. The Red Queen visits sage grouse leks. *Am. Zool.* 30:263–70.

Bradbury, J. W. 1977. Lek mating behavior of the hammer-headed bat. *Z. Tierpsychol.* 45:225–55.

Bradbury, J. W. 1981. The evolution of leks. Pp. 138–69, *in* Natural selection and social behavior, R. D. Alexander and D. W. Tinkle (eds.). Chiron Press, New York.

Bradbury, J. W. 1985. Contrasts between insects and vertebrates in the evolution of male display, female choice and lek mating. *Forts. Zool.* 31:273–92.

Bradbury, J. W., and M. Andersson (eds.). 1987. Sexual selection: testing the alternatives. J. Wiley, New York.

Bradbury, J. W., and R. Gibson. 1983. Leks and mate choice. Pp. 109–38, *in* Mate choice, P. P. G. Bateson (ed.). Cambridge Univ. Press, Cambridge.

Brodsky, L. M. 1988. Ornament size influences mating success in male rock ptarmigan. *Anim. Behav.* 36:662–67.

Brodsky, L. M., C. D. Ankney, and D. G. Dennis. 1988. The influence of male dominance on social interactions in black ducks and mallards. *Anim. Behav.* 36:1371–78.

Brosset, A. 1982. The social life of the African forest yellow-whiskered greenbul *Andropadus latirostris. Z. Tierpsychol.* 60:239–55.

Brosset, A. 1983. Parade et chants collectifs chez les courocous du genre *Apaloderma* [*Apaloderma*]. *Alauda* 51:1–10.

Burley, N. 1986. Sexual selection for aesthetic traits in species with biparental care. *Am. Nat.* 127:415–45.

Byrkjedal, I. 1990. Song flight of the pintail snipe *Gallinago stenura* on the breeding ground. *Ornis. Scand.* 21:239–47.

Carbonell, M. 1983. Comparative studies of stiff-tailed ducks (tribe Oxyurini, Anatidae). Ph.D. diss., Cardiff Univ., Cardiff.

Carranza, J., S. J. Hidalgo de Trucios, and V. Ena. 1989. Mating system flexibility in the great bustard: a comparative study. *Bird Study* 36:192–98.

Cartar, R. V., and B. E. Lyon. 1988. The mating system of the buff-breasted sandpiper: lekking and resource defense polygyny. *Ornis. Scand.* 19:74–76.

Cemmick, D., and D. Veitch. 1987. Kakapo country: the story of the world's most unusual bird. Hodder & Stoughton, Auckland, New Zealand.

Chaffer, N. 1959. Bower-building and the display of the satin bower-bird. *Aust. Zool.* 12:295–305.

Chapman, F. M. 1935. The courtship of Gould's manakin (*Manacus vitellinus vitellinus*) on Barro Colorado Island, Canal Zone. *Bull. Amer. Mus. Nat. Hist.* 68:471–525.

Cheng, K. M., R. N. Shoffner, R. E. Phillips, and F. L. Lee. 1978. Mate preference in wild and domesticated (game farm) mallards (*Anas platyrhynchos*). Initial preference. *Anim. Behav.* 26:996–1003.

Cheng, K. M., R. N. Shoffner, R. E. Phillips, and F. L. Lee. 1979. Mate preference in wild and domesticated (game farm) mallards (*Anas platyrhynchos*). Pairing success. *Anim. Behav.* 27:417–25.

Cherry, M. I. 1990. Tail length and female choice. *Trends Ecol. & Evol.* 5:359–60.

Cicero, J. M. 1983. Lek assembly and flash synchrony in the Arizona firefly *Photinus knulli* Green (Coleoptera, Lampyridae). *Coleopt. Bull.* 37:318–42.

Clayton, D. H. 1991. The influence of parasites on host sexual selection. *Parasit. Today* 7:329–35.

Clutton-Brock, T. 1991. Lords of the lek. *Nat. Hist.* 100 (10):34–41.

Clutton-Brock, T., D. Green, and M. H. Hasegawa. 1988. Passing the buck: resource defense, lek breeding and mate choice in fallow deer. *Behav. Ecol. & Sociobiol.* 23:281–91.

Coates, B. J. 1990. The birds of Papua New Guinea. Vol. II. Passerines. Dove Publications, Alderley, Queensland, Australia.

Cockburn, A., and K. A. Lazenby-Cohen. 1992. Use of nest trees by *Antechinus stuarti*, a semelparous lekking marsupial. *J. Zool.* 226:657–80.

Collins, D. R. 1984. A study of the Canarian houbara (*Chlamydotis undulata fuertaventurae*), with species reference to its behaviour and ecology. M.Phil. thesis, Univ. of London, London, England.

Cooper, W. T., and J. M. Forshaw. 1979. The birds of paradise and bower birds. D. R. Godine, Boston.

Cordier, C. 1943. The umbrellabird comes to the zoo. *Anim. Kingdom* 46:3–10.

Cox, C. R., and B. J. LeBoeuf. 1977. Female incitation of male competition: a mechanism of sexual selection. *Am. Nat.* 111:317–35.

Craig, A. J. F. K. 1978. Seasonal variation in the weight of red bishops, redcollared widows and redshouldered widows. *Ostrich* 49:153–57.

Craig, A. J. F. K. 1980. Behaviour and evolution in the genus *Euplectes*. *J. Ornith.* 121:144–61.

Craig, A. J. F. K. 1988. Tail length and sexual selection in the polygynous longtailed widow (*Euplectes progne*): a cautionary tale. *S. Afr. J. Sci.* 85:523–24.

Cramp, S., and K. E. L. Simmons (eds.). 1980. The birds of the western Palearctic. Vol. II. Oxford Univ. Press, Oxford.

Cramp, S., and K. E. L. Simmons (eds.). 1983. The birds of the western Palearctic. Vol. III. Oxford Univ. Press, Oxford.

Crandall, L. S. 1921. The blue bird of paradise. *Bull. N. Y. Zool. Soc.* 24:111–13.

Crandall, L. S. 1932. Notes on certain birds of paradise. *Zoologica* 17:77–87.

Crandall, L. S. 1936. Birds of paradise on display. *Bull. N. Y. Zool. Soc.* 39:87–103.

Crandall, L. S. 1937a. Further notes on certain birds of paradise. *Zoologica* 22:193–95.

Crandall, L. S. 1937b. Position of wires in display of the twelve-wired bird of paradise. *Zoologica* 22:307–10.

Crandall, L. S. 1940. Notes on the display forms of Wahnes' six-plumed bird of paradise. *Zoologica* 25:257–59.

Crandall, L. S. 1945a. A brilliant flash—that's the manakin's display. *Animal Kingdom* 48:67–69.

Crandall, L. S. 1945b. The umbrella bird is not a dull fellow any more. *Animal Kingdom* 48:109–12.

Crandall, L. S. 1946. Further notes on display forms of the long-tailed bird of paradise. *Zoologica* 31:9–10.

Crandall, L. S. 1948. Notes on the display of the three-wattled bell-bird (*Procnias tricarunculata*). *Zoologica* 33:113–14.

Crandall, L. S., and C. W. Leister. 1937. Display of the magnificent riflebird. *Zoologica* 22:311–14.

Crook, J. H. 1964. The evolution of social organization and visual communication in the weaver birds (Ploceinae). *Behaviour,* suppl. 10:1–178.

Curtis, H. S. 1972. The Albert lyrebird in display. *Emu* 72:81–84.

Dane, B., and W. G. van der Kloot. 1964. An analysis of the display of the goldeneye duck (*Bucephala clangula* (L.)). *Behaviour* 22:282–328.

Darnton, I. 1958. The display of the manakin *M. manacus*. *Ibis* 100:52–58.

Darwin, C. 1871. The descent of man and selection in relation to sex. J. Murray, London.

Davis, T. A. W. 1949. Display of the white-throated manakin, *Corapipio gutturalis*. *Ibis* 91:146–47.

Davis, T. A. W. 1958. The displays and nests of three forest hummingbirds in British Guiana. *Ibis* 100:31–39.

Davis, T. A. W. 1982. A flight-song display of the white-throated manakin. *Wils. Bull.* 94:594–95.

Davison, G. W. H. 1978. Studies of the crested argus. II. Gunong Rabon 1976. *Wld. Pheasant Ass. J.* 3:46–53.

Davison, G. W. H. 1981. Sexual selection and the mating system of *Argusianus argus* (Aves: Phasianidae). *Biol. J. Linn. Soc.* 15:91–104.

Davison, G. W. H. 1982. Sexual displays of the great argus pheasant *Argusianus argus*. *Z. Tierpsychol.* 58:185–202.

Davison, G. W. H. 1983a. Behaviour of Malay peacock pheasant, *Polyplectron malacense*. *J. Zool.* 201:57–66.

Davison, G. W. H. 1983b. The eyes have it: ocelli in a rain-forest pheasant. *Anim. Behav.* 31:1037–42.

De Vos, G. J. 1979. Adaptiveness of arena behavior in black grouse (*Tetrao tetrix*) and other grouse species (Tetraonidae). *Behaviour* 68:277–314.

De Vos, G. J. 1983. Social behavior of black grouse: an observational and experimental field study. *Ardea* 71:1–103.

Dharmakumarsihnji, K. S. 1950. The lesser florican (*Sypheotis indica* (Miller)): its courtship display, behavior and habits. *J. Bombay Nat. Hist. Soc.* 49:201–16.

Diamond, J. 1981. Birds of paradise and the theory of sexual selection. *Nature* 293:257–58.

Diamond, J. 1986. Biology of birds of paradise and bowerbirds. *Ann. Rev. Ecol. & Syst.* 178:17–37.

Diamond, J. 1987. Bower building and decoration by the bowerbird *Amblyornis inornatus*. *Ethology* 74:177–204.

Dilger, W. C., and P. A. Johnsgard. 1959. Comments on "species recognition," with special reference to the wood duck and the mandarin duck. *Wils. Bull.* 71:46–53.

Dinsmore, J. J. 1970. Courtship behavior of the greater bird of paradise. *Auk* 87:305–21.

Draffan, R. D. W. 1978. Group display of the Emperor of Germany bird-of-paradise *Paradisaea guilielmi* in the wild. *Emu* 78:157–59.

Emlen, J. T. 1957. Display and mate selection in the whydahs and bishop birds. *Ostrich* 28:202–13.

Emlen, S. T., and L. W. Oring. 1977. Ecology, sexual selection and the evolution of mating systems. *Science* 197:215–23.

Endler, J. A. 1983. Natural and sexual selection on color patterns in poecilid fishes. *Eviron. Biol. Fishes* 9:173–90.

Fay, F. H., C. G. Ray, and A. A. Kibalchick. 1984. Time and location of mating and associated behavior of the Pacific walrus *Odobenus rosmarus divergens*. *In* NOAA Tech. Rep. NMFS 12.

Ferdinand, L. 1966. Display of the great snipe (*Gallinago media* Latham). *Dansk Orn. Foren. Tidssk.* 60:14–34.

Ffrench, R. 1980. A guide to the birds of Trinidad and Tobago. Harrowood Books, Newton Square, Pennsylvania.

Fisher, R. A. 1930. The genetical theory of natural selection. Clarendon Press, Oxford.

Fitzherbert, K. 1978. Observations on breeding and display in a colony of captive Australian bustards (*Ardeotis australis*). B.S. thesis, Monash Univ., Clayton, Victoria, Australia.

Fitzherbert, K. 1983. Seasonal weight changes and display in captivity of Australian bustard. Pp. 210–26, *in* Bustards in decline, P. D. Goriup and H. Vardhan (eds.). Tourism & Wildlife Soc. India, Jaipur.

Fjeldså, J., and J. Krabbe. 1990. Birds of the high Andes. Apollo, Denmark.

Forshaw, J. M. 1989. Parrots of the world. 3d ed. Blandford, London.

Foster, M. S. 1977. Odd couples in manakins: a study of social organization and cooperative breeding in *Chiroxiphia linearis*. *Am. Nat.* 111:845–53.

Foster, M. S. 1978. Social organization and behavior of the swallow-tailed manakin, *Chiroxiphia caudata*. *Natl. Geog. Soc. Res. Rpt.,* 1978, pp. 314–19.

Foster, M. S. 1981. Cooperative behavior and social organization of the swallow-tailed manakin (*Chiroxiphia caudata*). *Behav. Ecol. & Sociobiol.* 9:167–77.

Foster, M. S. 1983. Disruption, dispersion and dominance in lek-breeding birds. *Am. Nat.* 122:53–72.

Foster, M. S. 1984. Jewel bird jamboree. *Nat. Hist.* 93(7):55–59.

Foster, M. S. 1987. Delayed maturation, neoteny and social system differences in two manakins of the genus *Chiroxiphia*. *Evolution* 41:547–58.

Friedmann, H. 1934. The display of Wallace's standard-wing bird of paradise in captivity. *Scient. Monthly* 39:52–55.

Frith, C. B. 1968. Some displays of Queen Carola's parotia. *Avic. Mag.* 74:85–90.

Frith, C. B. 1974. Observations on Wilson's bird of paradise. *Avic. Mag.* 80:207–12.

Frith, C. B. 1976. Displays of the red bird of paradise *Paradisaea rubra* and their significance, with a discussion on displays and systematics of other Paradisaeidae. *Emu* 76:69–78.

Frith, C. B. 1982. Displays of Count Raggi's bird of paradise *Paradisaea raggiana* and congeneric species. *Emu* 82:193–201.

Frith, C. B. 1992. Standardwing bird of paradise *Semioptera wallacii* display and relationships with comparative observations on displays of other Paradisaeidae. *Emu* 92:79–86.

Frith, C. B., and D. W. Frith. 1981. Displays of Lawes' parotia *Parotia lawesi* (Paradisaeidae), with reference to those of congeneric species and their evolution. *Emu* 81:227–38.

Frith, D. W., and C. B. Frith. 1988. Courtship display and mating of the superb bird of paradise *Lophorina superba*. *Emu* 88:183–88.

Fryxell, J. M. 1987. Lek breeding and territorial aggressiveness in white-eared kob. *Ethology* 75:211–20.

Fullagar, P., and M. Carbonell. 1986. The display postures of the male musk duck. *Wildfowl* 37:142–50.

Fuller, E. 1979. Hybridization amongst the Paradisaeidae. *Bull. Brit. Orn. Club* 99:145–52.

Garrod, A. H. 1876. On some anatomical characters which bear upon the major sub-divisions of the passerine birds. Pt.1. *Proc. Zool. Soc. London,* pp. 506–19.

Gewalt, W. 1959. Die Grosstrappe. Neue Brehm-Bücherei, Wittenberg-Lutherstadt, Germany.

Gibson, R. M. 1990. Relationships between blood parasites, mating success and phenotypic cues in male sage grouse *Centrocercus urophasianus*. *Am. Zool.* 30:270–78.

Gibson, R. M. 1992. Lek formation in sage grouse: the effect of female choice on male territory settlement. *Anim. Behav.* 43:443–50.

Gibson, R. M., and G. C. Bachman. 1992. The costs of female choice in a lekking bird. *Behavioral Ecol.* 3:300–309.

Gibson, R. M., and J. W. Bradbury. 1985. Sexual selection in lekking sage grouse: phenotypic correlates of male mating success. *Behav. Ecol. & Sociobiol.* 18:117–23.

Gibson, R. M., J. W. Bradbury, and S. L. Vehrencamp. 1992. Mate choice in lekking sage grouse revisited: the roles of vocal display, female site fidelity and copying. *Behav. Ecol.* 2:165–80.

Gilliard, E. T. 1959. The courtship behavior of Sanford's bowerbird (*Archboldia sanfordi*). *Am. Mus. Novitates* 1935:1–18.

Gilliard, E. T. 1962. On the breeding behavior of the cock-of-the-rock (Aves, *Rupicola rupicola*). *Am. Mus. Nat. Hist.* 124:31–68.

Gilliard, E. T. 1969. Birds of paradise and bower birds. Natural History Press, Garden City, N.Y.

Goodfellow, W. 1927. Wallace's bird of paradise *Semioptera wallacei*. *Avic. Mag.* ser. 4, no. 6:57–65.

Göransson, G., T. von Schantz, I. Fröberg, A. Helgée, and H. Wittzell. 1990. Male characteristics, viability and harem size in the pheasant, *Phasianus colchicus*. *Anim. Behav.* 40:89–104.

Gorman, M. L. 1974. The endocrine basis of pair-formation in the male eider *Somateria mollissima*. *Ibis* 116:451–67.

Gosling, L. M. 1986. The evolution of mating strategies in male antelopes. Pp. 244–81, *in* Ecological aspects of social evolution, D. I. Rubenstein and R. W. Wrangham (eds.). Princeton Univ. Press, Princeton.

Gosling, L. M., and M. Petrie. 1990. Lekking in topi: a consequence of satellite behavior in small males at hotspots. *Anim. Behav.* 40:272–87.

Götmark, F. 1992. Anti-predator effect of conspicuous plumage in a male bird. *Anim. Behav.* 44:51–56.

Gratson, M. W., G. K. Gratson, and A. T. Bergerud. 1991. Male dominance and copulation disruption does not explain variance in male mating success of sharp-tailed grouse (*Tympanuchus phasianellus*) leks. *Behaviour* 118:187–213.

Gray, A. P. 1958. Bird hybrids: a check-list with bibliography. Comm. Agr. Bureaux, Farnham Royal, England.

Gray, B. J. 1980. Reproduction, energetics, and social structure in the ruddy duck. Ph.D. diss., U. of Calif. Davis, Davis, Calif.

Gregory, P. T. 1974. Patterns of spring emergence of the red-sided garter snake (*Thamnophis striatus parietalis*) in the interlake region of Manitoba. *Can. J. Zool.* 52:1063–69.

Guangmei, Z., Y. Ronglun, and Z. Zhengwang. 1989. Courtship display behavior of Cabot's tragopan. *Acta Zool. Sinica* 35:332. (In Chinese, English summary.)

Gullion, G. W. 1967. Selection and use of drumming logs by male ruffed grouse. *Auk* 84:87–112.

Gullion, G. W. 1981. Non-drumming in a ruffed grouse population. *Wils. Bull.* 93:372–93.

Hamilton, D. W., and M. Zuk. 1982. Heritable true fitness and bright birds: a role for parasites? *Science* 218:384–87.

Harger, M., and D. Lyon. 1980. Further observations on lek behavior of the green hermit hummingbird *Phaethornis guy* at Monteverde, Costa Rica. *Ibis* 122:525–30.

Hartzler, J. E., and D. A. Jenni. 1988. Mate choice by female sage grouse. Pp. 240–69, *in* Adaptive strategies and population ecology of northern grouse, A. F. Bergerus and M. W. Gratson (eds.). U. of Minn. Press, Minneapolis.

Healey, C. 1975. A further note on the display and mating of *Pteridophora alberti*. *New Guinea Bird Soc. Newsl.*, no. 110:6–7.

Healey, C. 1978. Communal display of Princess Stephanie's astrapia *Astrapia stephaniae* (Paradisaeidae). *Emu* 78:197–200.

Hellmich, J. 1988. (On the mating behavior of the kori bustard.) *Zool. Gart.* 58:345–52. (In German.)

Hidalgo, S. J., and J. Carranza. 1991. Timing, structure and function of the courtship display in male great bustard. *Ornis. Scand.* 22:360–66.

Hill, G. E. 1990. Female house finches prefer colorful males: sexual selection for a condition-dependent trait. *Anim. Behav.* 40:563–72.

Hill, W. L. 1991. Correlates of male mating success in the ruff *Philomachus pugnax*, a lekking shorebird. *Behav. Ecol. Sociobiol.* 29:367–72.

Hillgarth, N. 1984. Social organization of wild peafowl in India. *World Pheasant Assoc. J.* 9:47–56.

Hillgarth, N. 1990a. Parasites and female choice in the ring-necked pheasant. *Amer. Zool.* 30:227–33.

Hillgarth, N. 1990b. Pheasant spurs out of fashion. *Nature* 345:119–20.

Hirons, G. 1980. The significance of roding by woodcock *Scolopax rusticola:* an alternative explanation based on marked birds. *Ibis* 122:350–54.

Hirons, G., and R. B. Owens, Jr. 1982. Comparative breeding behavior of European and American woodcock. *Wildl. Res. Rept.* 14:179–86.

Hjorth, I. 1970. Reproductive behaviour in Tetraonidae, with special reference to males. *Viltrevy* 7:184–596.

Hjorth, I. 1982. Attributes of capercaillie display grounds and the influence of forestry: a progress report. Pp. 26–35, *in* Proc. Second Int. Symp. on Grouse, 16–20 March 1981. T. W. I. Lovel (ed.). Lamarch: World Pheasant Assoc.

Hochbaum, H. A. 1944. The canvasback on a prairie marsh. Amer. Wild. Inst., Washington, D.C.

Hogan-Warburg, A. J. 1966. Social behavior of the ruff, *Philomachus pugnax* (L.). *Ardea* 54:109–229.

Hogan-Warburg, A. J. 1993. Female choice and the evolution of mating strategies in the ruff *Philomachus pugnax* (L.). *Ardea* 80:395–403.

Höglund, J. 1989. Size and plumage dimorphism in lek-breeding birds: a comparative analysis. *Am. Nat.* 134:72–87.

Höglund, J., R. V. Alatalo, and A. Lundberg. 1989. Copying the mate choice of others? Observations on female black grouse. *Behaviour* 114:221–31.

Höglund, J., M. Eriksson, and L. E. Lindell. 1990. Females of the lek-breeding great snipe, *Gallinago media,* prefer males with white tails. *Anim. Behav.* 40:23–32.

Höglund, J., J. A. Kålås, and L. Løfaldi. 1990. Sexual dimorphism in the lekking great snipe. *Ornis. Scand.* 21:1–6.

Höglund, J., and A. Lundberg. 1987. Sexual selection in a monomorphic lek-breeding bird: correlates of male mating success in the great snipe *Gallinago media. Behav. Ecol. & Sociobiol.* 21:211–16.

Höglund, J., and A. Lundberg. 1989. Plumage color correlates with body size in the ruff (*Philomachus pugnax*). *Auk* 106:336–38.

Höglund, J., and J. G. M. Robertson. 1990. Female preferences, male decision roles and the evolution of leks in the great snipe. *Anim. Behav.* 40:15–22.

Holmberg, K., L. Edsman, and T. Klint. 1989. Female mate preferences and male attributes in mallard ducks *Anas platyrhynchos. Anim. Behav.* 38:1–7.

Howard, R. D. 1978. The evolution of mating strategies in bullfrogs (*Rana catesbiana*). *Evolution* 32:850–71.

Huxley, J. S. 1914. The courtship habits of the great crested grebe, *Podiceps cristatus,* with an addition to the theory of sexual selection. *Proc. Zool. Soc. London,* pp. 491–562.

Huxley, J. S. 1938. The present study of the theory of sexual selection. Pp. 11–42, *in* Evolution; essays on aspects of evolutionary biology, G. de Beer (ed.). Clarendon Press, Oxford.

Ingram, W. J. 1907. On the display of the king bird of paradise. *Ibis,* ser. 9, vol. 1:225–29.

Inskipp, C., and T. P. Inskipp. 1983. Report on a survey of Bengal floricans *Houbaropsis bengalensis* in Nepal and India, 1982. Study Rpt. No. 2, ICBP, Cambridge, England.

Islam, K. 1991. Evolutionary history and speciation of the genus *Tragopan*. Ph.D. diss., Oregon State Univ., Corvallis.

Iwasa, Y., A. Pomiankowski, and S. Nee. 1991. The evolution of costly mate preferences. II. The "handicap" principle. *Evol.* 45:1431–42.

Jehl, J. R., and B. G. Murray. 1986. The evolution of normal and reverse sexual dimorphism in shorebirds and other birds. Pp. 1–86, *in* Current Ornithology, R. F. Johnson (ed.), vol. 3. Plenum Press, New York.

Johnsgard, P. A. 1961a. Evolutionary relationships among the North American mallards. *Auk* 78:1–43.

Johnsgard, P. A. 1961b. The sexual behavior and systematic position of the hooded merganser. *Wils. Bull.* 73:477–84.

Johnsgard, P. A. 1965. Handbook of waterfowl behavior. Cornell Univ. Press, Ithaca.

Johnsgard, P. A. 1966. Behavior of the Australian musk duck and blue-billed duck. *Auk* 83:98–110.

Johnsgard, P. A. 1981. The plovers, sandpipers and snipes of the world. Univ. of Nebr. Press, Lincoln.

Johnsgard, P. A. 1983a. The grouse of the world. Univ. of Nebr. Press, Lincoln.

Johnsgard, P. A. 1983b. The hummingbirds of North America. Smithsonian Inst. Press, Washington, D.C.

Johnsgard, P. A. 1986. The pheasants of the world. Oxford Univ. Press, Oxford.

Johnsgard, P. A. 1991. Bustards, hemipodes and sandgrouse: Birds of dry places. Oxford Univ. Press, Oxford.

Johnsgard, P. A., and C. Nordeen. 1981. Display behavior and relationships of the Argentine blue-billed duck. *Wildfowl* 32:5–11.

Johnston, G. W. 1969. Ecology, dispersion and arena behaviour of black grouse *Lyrurus tetrix* (L.) in Glen Dye, N.E. Scotland. Ph.D. diss., Aberdeen Univ., Scotland.

Jones, D. N. 1988. Construction and maintenance of the incubation mounds of the Australian brush-turkey, *Alectura lathami*. *Emu* 88:210–18.

Jones, D. N. 1990. Social organization and sexual interactions in Australian brush-turkeys (*Alectura lathami*): implications of promiscuity in a mound-building megapode. *Ethology* 84:89–104.

Kemp, A., and W. Tarboton. 1976. Small South African bustards. *Bokmakierie* 28:40–43.

Kenyon, R. F. 1972. Polygyny among superb lyrebirds in Sherbrooke Forest Park, Kallista, Victoria. *Emu* 72:70–76.

Kermott, L. H., III. 1982. Breeding behavior in the sharp-tailed grouse. Ph.D. diss., U. of Minn., Minneapolis.

Kirkpatrick, M. 1982. Sexual selection and the evolution of female choice. *Evolution* 36:1–12.

Kirkpatrick, M. 1987. Sexual selection by female choice in polygynous animals. *Ann. Rev. Ecol. & Syst.* 18:43–70.

Kirkpatrick, M., T. Price, and J. Arnold. 1990. The Darwin-Fisher theory of natural selection in monogamous birds. *Evol.* 44:180–91.

Kirkpatrick, M., and M. J. Ryan. 1991. The evolution of mating preferences and the paradox of the lek. *Nature* 350:33–38.

Klint, T. 1980. Influence of male nuptial plumage on mate selection in the female mallard (*Anas platyrhynchos*). *Anim. Behav.* 28:1230–38.

Knapton, R. W. 1985. Lek structure and territoriality in the chryxus arctic butterfly, *Oeneis chryxus* (Satyridae). *Behav. Ecol. & Sociobiol.* 17:389–95.

Kodric-Brown, A., and J. H. Brown. 1984. Truth in advertising: the kinds of traits favored by sexual selection. *Am. Nat.* 124:309–23.

Koenig, O. 1962. Der Schrillapparat der Paradieswitwi *Steganura paradisea. J. Ornith.* 103:86–91.

Kruijt, J. P., and J. A. Hogan. 1967. Social behaviour on the lek in black grouse, *Lyrurus tetrix tetrix* (L.). *Ardea* 55:203–40.

Kruse, K. C. 1981. Mating success, fertilization potential and male body size in the American toad (*Bufo americanus*). *Herpetologica* 37:228–33.

Landell, H. F. 1989. A study of female and male mating behavior and female mate choice in the sharp-tailed grouse. Ph.D. diss., Purdue Univ.

Lank, D. B., and C. M. Smith. 1987. Conditional lekking in ruff (*Philomachus pugnax*). *Behav. Ecol. & Sociogiol.* 20:137–47.

Lank, D. B., and C. M. Smith. 1992. Females prefer larger leks: field experiments with ruffs. *Behav. Ecol. & Sociobiol.* 30:323–29.

LeCroy, M. K. 1981. The genus *Paradisaea:* display and evolution. *Amer. Mus. Novitates* 2714:1–52.

LeCroy, M. K., A. Kulupi, and W. S. Peckover. 1980. Goldie's bird of paradise: display, natural history and traditional relationships of people to the bird. *Wils. Bull.* 92:289–301.

Lemnell, P. A. 1978. Social behavior of the great snipe, *Capella media,* at the arena display. *Ornis. Scand.* 9:146–63.

Ligon, J. D., R. Thornhill, M. Zuk, and K. Johnson. 1990. Male-male competition, ornamentation and the role of testosterone in sexual selection in red junglefowl. *Anim. Behav.* 40:367–73.

Lill, A. 1966. Some observations on social organization and non-random mating in captive Burmese red jungle fowl (*Gallus gallus spadiceus*). *Behaviour* 26:228–42.

Lill, A. 1974a. Sexual behavior of the lek-forming white-bearded manakin (*M. manacus trinitatis* Hartert)). *Z. Tierpsychol.* 36:1–36.

Lill, A. 1974b. Social organization and space utilization in the lek-forming white-bearded manakin *M. manacus trinitatis* (Hartert). *Z. Tierpsychol.* 36:513–30.

Lill, A. 1976. Lek behavior in the golden-headed manakin *Pipra erythrocephala* in Trinidad (West Indies). *Z. Tierpsychol.,* Suppl. 18:1–84.

Lill, A. 1979. An assessment of male parental investment and pair bonding in the polygamous superb lyrebird. *Auk* 96:489–98.

Loffredo, C. A., and G. Borgia. 1986. Sexual selection, mating systems and the evolution of avian acoustical displays. *Am. Nat.* 128:773–94.

Loiselle, P. W., and G. W. Barlow. 1978. Do fish lek like birds? Pp. 31–75, *in* Contrasts in behavior: adaptations in the aquatic and terrestrial environments, E. S. Reese and F. J. Lighter (eds.). J. Wiley & Sons, New York.

Lorenz, K. Z. 1951–53. Comparative studies on the behaviour of Anatinae. *Avic. Mag.* 57:157–82; 58:8–17, 61–72, 86–94, 172–84; 59:24–34, 80–91.

Lovvorn, J. R. 1990. Courtship and aggression in canvasbacks: influence of sex and pair-bonding. *Condor* 92:369–78.

Lowe, P. R. 1942. The anatomy of Gould's manakin (*Manacus vitellinus*) in relation to its display. *Ibis* ser. 14, vol. 6:50–83.

Loye, J. E., and M. Zuk (eds.). 1991. Bird-parasite interactions: ecology, evolution and behaviour. Oxford Univ. Press, Oxford.

Lumsden, H. G. 1965. Displays of the sharptail grouse. Ont. Dept. Lands & Forest tech. series, res. rept. no. 66.

Lumsden, H. G. 1968. The displays of the sage grouse. Ont. Dept. Lands & Forest tech. series, res. report (wildlife) no. 83.

McDonald, D. B. 1989a. Cooperation under sexual selection: age-graded changes in a lekking bird. *Am. Nat.* 134:709–30.

McDonald, D. B. 1989b. Correlates of male mating success in a lekking bird with male-male competition. *Anim. Behav.* 37:1007–22.

MacDonald, S. D. 1968. The courtship and territorial behavior of Franklin's race of the spruce grouse. *Living Bird* 7:4–25.

Mackey, R. D. 1989. The bower of the fire-maned bowerbird, *Sericulus bakeri. Aust. Bird Watcher* 13:62–64.

McKinney, D. F. 1961. An analysis of the displays of the European eider *Somateria mollissima mollissima* (Linnaeus) and the Pacific eider *Somateria mollissima v-nigra* (Bonaparte). *Behaviour* suppl. no. 7.

Maclean, G. L. 1985. Roberts Birds of Southern Africa. 5th ed. John Voelcker Bird Book Fund, Cape Town.

McWilliam, A. N. 1990. Mating system in the bat *Minopterus minor* (Chiroptera, Verpertilionidae) in Kenya, East Africa: a lek? *Ethology* 35:302–12.

Manakadam, R., and A. R. Rahmani (eds.). 1986. Study of ecology of certain endangered species of wildlife and their habitats. Great Indian bustard. Bombay Nat. Hist. Soc., Bombay.

Manning, J. T. 1984. Males and the advantage of sex. *J. Theor. Biol.* 108:215–20.

Manning, J. T. 1985. Choosy females and correlates of male age. *J. Theor. Biol.* 116:349–54.

Manning, J. T. 1987. The peacock's train and the age-dependency model of female choice. *World Pheasant Assoc. J.* 12:44–56.

Manning, J. T. 1989. Age advertisement and the evolution of the peacock's train. *J. Evol. Biol.* 2:299–313.

Manning, J. T., and M. A. Hartley. 1991. Symmetry and ornamentation are correlated in the peacock's train. *Anim. Behav.* 42:1020–21.

Manson-Bahr, P. H. 1935. Remarks on displays of birds of paradise. *Bull. Brit. Orn. Club* 56:63–68.

Marshall, A. J. 1954. Bower-birds, their displays and breeding cycles: a preliminary statement. Oxford Univ. Press, Oxford.

Marshall, A. J. 1955. We are beginning to understand the bower-birds. *Animal Kingdom* 58:34–43.

Marshall, A. J. 1970. Bower-building and decorating by the regent bowerbird in captivity. *Emu* 70:28–29.

Matthews, G. V. T., and M. E. Evans. 1974. On the behaviour of the white-headed duck with especial reference to breeding. *Wildfowl* 23:56–66.

Maynard Smith, J. M. 1987. Sexual selection—a classification of models. Pp. 9–20, *in* Sexual selection: testing the alternatives, J. W. Bradbury and M. N. Anderson (eds.). John Wiley, New York.

Maynard Smith, J. M. 1991. Theories of sexual selection. *Trends in Ecol. & Evol.* 6:146–51.

Mayr, E. 1935. Bernard Altum and the territory theory. *Proc. Linn. Soc. N.Y.* 45–46:1–15.

Mayr, E. 1963. Animal species and evolution. Oxford University Press, Oxford.

Mayr, E. 1982. The growth of biological thought. Belknap Press, Cambridge, Mass.

Merton, D. V., R. B. Morris, and I. A. E. Atkinson. 1984. Lek behaviour in a parrot: the kakapo *Strigops habroptilus* of New Zealand. *Ibis* 126:277–83.

Mjelstad, H. 1991. Display intensity and sperm quality in the capercaillie *Tetrao urogallus*. *Fauna Norv.* (ser. C) 14:93–94.

Möller, A. P. 1988. Female choice selects for male sexual adornments in the monogamous swallow. *Nature* 332:640–42.

Möller, A. P. 1991. Sexual selection in the monogamous barn swallow (*Hirundo rustica*). I. Determinants of tail ornament size. *Evolution* 45:1823–36.

Morony, J. J., Jr., W. J. Bock, and J. Ferrand, Jr. Reference list of the birds of the world. American Mus. of Nat. History, New York.

Morris, R., and H. Smith, 1988. Wild south: saving New Zealand's endangered birds. TVNZ & Century Hutchinson, Auckland.

Morrison-Scott, T. 1936. Display of *Lophorina superba minor*. *Proc. Zool. Soc. London*, p. 809.

Moss, R. 1980. Why are capercaillie cocks so big? *Br. Birds* 73:440–47.

Müller, F. 1974. (Territorial behavior and distribution structure of a capercaillie population.) Ph.D. dissertation, Phillips Univ. of Marburg, Germany. In German.

Müller, F. 1979. A 13-year study of a capercaillie lek in the western Rhon-mountains (W. Germany). Pp. 120–30, *in* Behaviour of woodland grouse, woodland grouse symposium, T. W. I. Lovel (ed.). World Pheasant Assn., Bures, England.

Myers, J. P. 1979. Leks, sex, and buff-breasted sandpipers. *Am. Birds* 33:823–25.

Myers, J. P. 1983. The promiscuous pectoral sandpiper. *Am. Birds* 36:119–22.

Myers, J. P. 1989. Making sense of sexual nonsense. *Audubon* 91 (7):41–46.

Narayan, G., and L. Rosalind. 1988. Ecology of Bengal florican in Manas Wildlife Sanctuary. Pp. 9–42, *in* The Bengal florican: status and ecology, A. R. Rahmani, G. Narayan, R. Sankaran, and L. Rosalind (eds.). Bombay Nat. Hist. Soc., Bombay.

Newton, A. (ed.). 1896. A dictionary of birds. Adam & Chas. Black, London.

Nicolai, J. 1969. (Observations on the paradise finches *Steganura paradisaea, S. obtusa* and *T. fischeri* in East Africa.) *J. Ornith.* 110:421–47. (German, English summary.)

Nilsson. 1969. (The behavior of the goldeneye *Bucephala clangula* in winter.) *Vår Fågelvärld* 28:199–210. (In Swedish, English summary.)

Nitchuk, W. M., and R. M. Evans. 1978. A volumetric analysis of sharp-tailed grouse sperm in relation to dancing ground size and organization. *Wils. Bull.* 90:460–62.

Ogilvie-Grant, W. R. 1905. On the display of the lesser bird-of-paradise (*Paradisaea minor*). *Ibis* ser. 8, vol. 5:429–40.

Oring, L. W. 1982. Avian mating systems. Pp. 1–92, *in* Avian biology, vol. 6, D. S. Farner and J. R. Kings (eds.). Academic Press, New York.

Parker, G. A. 1978. Evolution of competitive mate searching. *Ann. Rev. Entomol.* 23:173–96.

Parkes, K. C. 1961. Intergeneric hybrids in the family Pipridae. *Condor* 63:345–50.

Patterson, I. J. 1983. The shelduck: a study in behavioral ecology. Cambridge Univ. Press, Cambridge, England.

Payne, R. B. 1973. Behavior, mimetic songs and song dialects, and the relationships of the parasitic indigobirds (*Vidua*) of Africa. *Ornith. Monogr.* 11.

Payne, R. B. 1984. Sexual selection, lek and arena behavior and sexual size dimorphism in birds. *Ornith. Monogr.* 33.

Payne, R. B., and K. D. Groschupf. 1984. Sexual selection and interspecific competition: a field experiment on territorial behavior of nonparental finches (*Vidua* spp.). *Auk* 101:140–45.

Payne, R. B., and K. Payne. 1977. Social organization and mating success in local song populations of village indigobirds *Vidua chalybeata*. *Z. Tierpsychol.* 45:113–73.

Petrie, M., T. Halliday, and C. Sanders. 1991. Peahens prefer peacocks with elaborate trains. *Anim. Behav.* 41:323–31.

Phillips, J. B. 1990. Lek behavior in birds: do displaying males reduce nest predation? *Anim. Behav.* 39:555–65.

Pitelka, F. A., R. T. Holmes, and S. F. MacLean, Jr. 1974. Ecology and evolution of social organization in arctic sandpipers. *Am. Zool.* 14:184–204.

Pomiankowski, A. 1989. Mating success in female pheasants. *Nature* 337:696.

Pomiankowski, A., Y. Iwasa, and S. Nee. 1991. The evolution of costly mate preferences. I. Fisher and biased mutation. *Evol.* 45:1422–30.

Powlesland, R. G., B. D. Lloyd, H. A. Best, and D. V. Merton. 1992. Breeding biology of the kakapo *Strigops habroptilus* on Stewart Island, New Zealand. *Ibis* 134:361–73.

Pruett-Jones, M. A., and S. G. Pruett-Jones. 1982. Spacing and distribution of bowers in MacGregor's bowerbird (*Amblyornis macgregoriae*). *Behav. Ecol. & Sociobiol.* 11:25–32.

Pruett-Jones, M. A., and S. G. Pruett-Jones. 1983. The bowerbird's labor of love. *Nat. Hist.* 92(9):49–55.

Pruett-Jones, S. G. 1985. The evolution of lek mating behavior in Lawes' parotia (Aves: *Parotia lawesii*) Ph.D. diss., U. of Calif., Berkeley.

Pruett-Jones, S. G. 1988. Lekking versus solitary display: temporal variations in dispersion in the buff-breasted sandpiper. *Anim. Behav.* 36:1740–52.

Pruett-Jones, S. G. 1992. Independent versus nonindependent mate choice: do females copy each other? *Amer. Nat.* 140:1000–09.

Pruett-Jones, S. G., and M. A. Pruett-Jones. 1988a. A promiscuous mating system in the blue bird of paradise *Paradisaea [Paradisaea] rudolphi*. *Ibis* 130:373–77.

Pruett-Jones, S. G., and M. A. Pruett-Jones. 1988b. The use of court objects by Lawes' parotia. *Condor* 90:538–45.

Pruett-Jones, S. G., and M. A. Pruett-Jones. 1990. Sexual selection through female choice in Lawes' parotia, a lek-mating bird of paradise. *Evolution* 44:486–501.

Prum, R. O. 1985. Observations of the white-fronted manakin (*Pipra serena*) in Suriname. *Auk* 102:384–87.

Prum, R. O. 1986. The displays of the white-throated manakin *Corapipio gutturalis* in Suriname. *Ibis* 128:91–102.

Prum, R. O. 1989. Phylogenetic analysis of the morphological and behavioral evolution of the Neotropical manakins. Ph.D. diss., U. of Mich., Ann Arbor.

Prum, R. O., and A. E. Johnson. 1987. Display behavior, foraging ecology, and systematics of the golden-winged manakin (*Masius chrysopterus*). *Wils. Bull.* 99:521–39.

Queller, D. 1987. The evolution of leks through sexual selection. *Anim. Behav.* 35:1425–32.

Rahmani, A. R. (ed.). 1990. Status and ecology of the lesser and Bengal floricans. Bombay Nat. Hist. Soc., Bombay.

Rand, A. L. 1940. Results of the Archbold Expedition No. 26. Breeding habits of the birds of paradise: *Macgregoria* and *Diphyllodes*. *Am. Mus. Novitates* 1073:1–14.

Rands, M. R. W., M. W. Ridley, and A. D. Lelliott. 1984. The social organization of feral peafowl. *Anim. Behav.* 32:830–35.

Reynolds, J. D., and M. T. Gross. 1990. Costs and benefits of female mate choice: is there a lek paradox? *Amer. Nat.* 136:230–43.

Ricklefs, R. E. 1973. Ecology. Chiron Press, Portland, Oregon.

Ridgway, R. 1907. The birds of North and Middle America. Part IV. *Bull. U.S. Natl. Mus.* 50:1–973.

Ridley, M. W., and D. A. Hill. 1987. Social organization in the pheasant (*Phasianus colchicus*): harem formation, mate selection and the role of mate guarding. *J. Zool.* 211:619–30.

Ridley, M. W., A. D. Lelliott, and M. R. W. Rands. 1984. The courtship display of feral peafowl. *J. World Pheasant Assn.* 9:57–68.

Rimlinger, D. S. 1984. Display behavior of Temminck's tragopan. *J. World Pheasant Assn.* 9:19–32.

Rimlinger, D. S. 1985. Observations on the display behaviour of the Bulwer's pheasant. *J. World Pheasant Assn.* 10:15–26.

Ripley, D. 1950. Strange courtship of birds of paradise. *Nat. Geog.* 97:247–78.

Robbins, M. B. 1983. The display repertoire of the band-tailed manakin. *Wils. Bull.* 95:321–42.

Robbins, M. B. 1985. Social organization of the band-tailed manakin (*Pipra fascicauda*). *Condor* 87:449–56.

Robel, R. J. 1966. Booming territory size and mating success of the greater prairie chicken (*Tympanuchus cupido pinnatus*). *Anim. Behav.* 14:328–31.

Robel, R. J., and W. B. Ballard. 1974. Lek social organization and reproductive success in the greater prairie chicken. *Am. Zool.* 14:121–28.

Robinson, F. N., and H. J. Frith. 1981. The superb lyrebird *Menura novaehollandiae* at Tidbinbilla, ACT. *Emu* 81:145–57.

Sankarin, R., and A. R. Rahmani. 1986. Study of ecology of certain endangered species of wildlife and their habitats. The lesser florican. Bombay Nat. Hist. Soc., Bombay.

Savalli, U. M. 1992. Tail length functions in male-male competition in the yellow-shouldered widowbird. Abstracts of proceedings, Am. Ornith. Union annual meeting, Ames, Iowa, 24–30 June.

Schodde, R. 1976. Evolution in the birds-of-paradise and bowerbirds, a resynthesis. Proc. 16th Intern. Ornithol. Congr., pp. 138–49.

Schodde, R., and J. L. McKean. 1973. The species of the genus *Parotia* (Paradisaeidae) and their relationships. *Emu* 73:145–56.

Schulz, H. 1985. Grundlagenforschung zur Biologie der Zwergtrappe. 401 pp. Staatliches Nat. Mus. Braunschweig, Braunschweig, Germany.

Schulz, H. 1986. Agonistisches Verhalten, Territorialverhalten und Balz der Zwergtrappe (*Tetrax tetrax*). *J. Ornith.* 127:125–204.

Schwartz, P., and D. W. Snow. 1978. Display and related behavior of the wire-tailed manakin. *Living Bird* 17:51–78.

Scott, J. W. 1942. Mating behavior of the sage grouse. *Auk* 59:472–98.

Searcy, W. A., and M. Andersson. 1986. Sexual selection and the evolution of song. *Ann. Rev. Ecol. & Syst.* 17:507–33.

Selander, R. K. 1972. Sexual selection and dimorphism in birds. Pp. 180–230, *in* Sexual selection and the descent of man, B. G. Campbell (ed.). Aldine, Chicago.

Selous, E. 1909–10. An observational diary on the nuptial habits of the black cock (*Tetrao tetrix*) in Scandinavia and England. *Zoologist* ser. 4, 13:401–13; 14:23–29, 51–56, 176–82, 248–65.

Seth-Smith, D. 1923. On the display of the magnificent bird-of-paradise. *Proc. Zool. Soc. London,* pp. 609–13.

Shaw, P. 1984. The social behavior of the pin-tailed whydah *Vidua macroura* in northern Ghana. *Ibis* 126:463–73.

Shelly, T. E. 1989. Waiting for mates: variations in female encounter rates within and between leks of *Drosophila conformis. Behaviour* 111:34–48.

Sibley, C. G. 1957. The evolutionary and taxonomic significance of sexual dimorphism and hybridization in birds. *Condor* 59:166–91.

Sibley, C. G., and B. L. Monroe, Jr. 1990. Distribution and taxonomy of birds of the world. Yale Univ. Press, New Haven.

Sick, H. 1954. Zur Biologie des amazonischen Schirmvogels *Cephalopterus ornatus. J. Ornith.* 95:233–42.

Sick, H. 1959. Der Balz der Schmuckvogel (Pipridae). *J. Ornith.* 100:269–302.

Sick, H. 1967. Courtship behavior in the manakins (Pipridae): a review. *Living Bird* 6:5–22.

Siegfried, W. R. 1976. Social organization in ruddy and maccoa ducks. *Auk* 93:560–70.

Sigurjønsdøttir, H. 1981. The evolution of sexual size dimorphism in gamebirds, waterfowl and raptors. *Ornis. Scand.* 12:249–60.

Skead, D. M. 1974. Bird weights from the central Transvaal bushveld. *Ostrich* 45:189–92.

Skead, D. M. 1977. Weights of birds handled at Barberspan. *Ostrich,* suppl. 12:117–31.

Skutch, A. F. 1960. Life histories of Central American birds. II. *Pacific Coast Avifauna* 34.

Skutch, A. F. 1967. Life histories of Central American highland birds. *Publ. Nuttall Club* 7:1–213.

Skutch, A. F. 1969. Life histories of North American birds. III. *Pacific Coast Avifauna* 35.

Skutch, A. F. 1972. Studies of tropical American birds. Nuttall Ornith. Club. Pub. 10.

Skutch, A. F. 1989. Courtship of the rufous piha *Lipaugus unirufus. Ibis* 131:303–4.

Smith, L. H. 1968. The lyrebird. Lansdowne Press, Melbourne, Australia.

Snow, B. K. 1961. Notes on the behavior of three Cotingidae. *Auk* 78:150–61.

Snow, B. K. 1970. A field study of the bearded bellbird in Trinidad. *Ibis* 112:299–329.

Snow, B. K. 1972. A field study of the calfbird *Perrisocephalus tricolor. Ibis* 114:139–62.

Snow, B. K. 1973a. The behavior and ecology of hermit hummingbirds in the Kanaku Mountains, Guyana. *Wils. Bull.* 85:163–77.

Snow, B. K. 1973b. Notes on the behavior of the white bellbird. *Auk* 90:743–51.

Snow, B. K. 1974. Lek behaviour and breeding of Guy's hermit hummingbird *Phaethornis guy. Ibis* 116:278–97.

Snow, B. K. 1977a. Comparison of the leks of Guy's hermit hummingbird *Phaethornis guy* in Costa Rica and Trinidad. *Ibis* 119:211–14.

Snow, B. K. 1977b. Territorial behavior and courtship of the three-wattled bellbird. *Auk* 94:623–45.

Snow, B. K. 1978. Calls and displays of the male bare-throated bellbird. *Avic. Mag.* 84:157–61.

Snow, B. K., and D. W. Snow. 1979. The ochre-bellied flycatcher and the evolution of lek behavior. *Condor* 81:286–92.

Snow, B. K., and D. W. Snow. 1985. Display and related behavior in male pin-tailed manakins. *Wils. Bull.* 97:273–82.

Snow, D. W. 1961. The displays of the manakins *Pipra pipra* and *Tyranneutes virescens. Ibis* 103:110–13.

Snow, D. W. 1962. A field study of the black and white manakin, *M. manacus,* in Trinidad. *Zoologica* 47:65–104.

Snow, D. W. 1963a. The display of the black-backed manakin *Chiroxiphia pareola* in Tobago. *Zoologica* 48:167–76.

Snow, D. W. 1963b. The display of the orange-headed manakin. *Condor* 65:44–48.

Snow, D. W. 1963c. The evolution of manakin displays. Proc. 13th Intern. Ornith. Congr., pp. 553–61.

Snow, D. W. 1968. The singing assemblies of little hermits. *Living Bird* 7:47–56.

Snow, D. W. 1971. Notes on the biology of the cock-of-the rock. *J. Ornith.* 112:323–33.

Snow, D. W. 1976. The web of adaptation, bird studies in the American tropics. Quadrangle/New York Times Book Co., New York.

Snow, D. W. 1977a. The display of the scarlet-horned manakin *Pipra cornuta. Bull. Brit. Orn. Club* 97:23–27.

Snow, D. W. 1977b. Duetting and other synchronized displays of the blue-backed manakins *Chiroxiphia* spp. Pp. 239–51, *in* Evolutionary ecology, B. Stonehouse and C. Perrins (eds.). Univ. Park Press, Baltimore.

Snow, D. W. 1982. The cotingas. British Mus. (Nat. Hist.), London.

Snow, D. W. 1985. Lek. Pp. 326–28, *in* A dictionary of birds, B. Campbell and E. Lack (eds.). T. & A. D. Poyser, Carlton, England.

Steadman, D. W., J. Stull, and S. W. Eaton. 1979. Natural history of the ocellated turkey. *World Pheasant Assn. J.* 4:15–37.

Stephan, B. 1967. Die Schmuckfedern im Flugel von *Semioptera wallacii*. *J. Ornith.* 108:47–50.

Sterbetz, I. 1981. Comparative investigation in the reproduction behavior of monogamous, polygamous and unmated great bustard populations in south-east Hungary. *Aquila* 87:31–47.

Stiles, F. G. 1973. Food supply and annual cycle in the Anna's hummingbird. *Univ. Calif. Publ. Zool.* 97:1–109.

Stiles, F. G., and A. F. Skutch. 1989. A guide to the birds of Costa Rica. Cornell Univ. Press, Ithaca.

Stiles, F. G., and B. Whitney. 1983. Notes on the behaviour of the Costa Rican sharpbill (*Oxyruncus cristatus frater*). *Auk* 100:117–25.

Stiles, F. G., and L. L. Wolf. 1979. Ecology and evolution of lek mating behavior in the long-tailed hermit hummingbird. *Ornith. Monogr.* 27.

Stonor, C. R. 1936. The evolution and mutual relationships of some members of the Paradisaeidae. *Proc. Zool. Soc. London,* pp. 1177–85.

Sullivan, B. K. 1983. Sexual selection in the Great Plains toad (*Bufo cognatus*). *Behaviour* 84:258–64.

Taczanowski, L., and J. Stolzmann. 1881. Notice sur la *Loddigesia mirabilis* (Bounc.). *Proc. Zool. Soc. London,* pp. 827–34.

Tamm, S., D. P. Armstrong, and Z. J. Tooze. 1989. Display behavior of male calliope hummingbirds during the breeding season. *Condor* 91:272–79.

Tarboton, W. R. 1989. Breeding behaviour of Denham's bustard. *Bustard Studies* 4:160–69.

Thornhill, R., and J. Alcock. 1983. The evolution of insect mating systems. Harvard Univ. Press, Cambridge, Mass.

Titman, R. D., and N. R. Seymour. 1981. A comparison of pursuit flights by six North American ducks of the genus *Anas*. *Wildfowl* 32:11–18.

Trail, P. W. 1983. Cock-of-the-rock: jungle dandy. *Natl. Geog.* 164:831–39.

Trail, P. W. 1984. The lek mating system of the Guianan cock-of-the-rock: a field study of sexual selection. Ph.D. diss., Cornell Univ., Ithaca, N. Y.

Trail, P. W. 1985a. Courtship disruption modifies mate choice in a lek-breeding bird. *Science* 227(4688):778–80.

Trail, P. W. 1985b. A lek's icon. The courtship display of the Guianan cock-of-the-rock. *American Birds* 39:235–40.

Trail, P. W. 1985c. Territoriality and dominance in the lek-breeding Guianan cock-of-the-rock. *Natl. Geog. Res.* 1:112–23.

Trail, P. W. 1990. Why should lek breeders be monomorphic? *Evolution* 44:1837–52.

Trail, P. W., and E. S. Adams. 1989. Active mate choice at cock-of-the-rock leks; tactics of sampling and comparison. *Behav. Ecol. & Sociobiol.* 25:283–92.

Trail, P. W., and P. Donahue. 1991. Notes on the behavior and ecology of the red-cotingas (Cotingidae: *Phoenicircus*). *Wils. Bull.* 103:539–51.

Troy, S., and M. A. Elgar. 1991. Brush-turkey incubation mounds: mate attraction in a promiscuous mating system. *Trends in Ecol. & Evol.* 6:202–3.

Tsuji, L. J. S., D. R. Kozlovic, and M. B. Sokolowski. 1992. Territorial position in sharp-tailed grouse leks: the probability of fertilization. *Condor* 94:1030–31.

van Rhijn, J. G. 1973. Behavioural dimorphism in male ruffs *Philomachus pugnax* (L.). *Behaviour* 47:153–229.

van Rhijn, J. G. 1985. A scenario for the evolution of social organization in ruffs *Philomachus pugnax* and other charadriiform species. *Ardea* 73:25–37.

van Rhijn, J. G. 1991. The ruff: individuality in a gregarious wading bird. T. & A. D. Poyser, London.

van Someren, V. D. 1945. The dancing display and courtship of Jackson's whydah (*Coliuspasser jacksoni* Sharpe). *E. Afr. Nat. Hist. Soc. J.* 18:131–41.

van Someren, V. D. 1956. Days with birds. *Fieldiana: Zool.* 38:1–520.

Vellenga, R. E. 1970. Behavior of the male satin bowerbird at the bower. *Aust. Bird Bander* 8:3–11.

Vellenga, R. E. 1980. Distribution of bowers of the satin bowerbird at Leura, NSW, with notes on parental care, development and independence of the young. *Emu* 80:97–102.

Veselovsky, Z. 1978. Beobachtungen zur Biologie und Verhalten der grossen Kragenlaubvogels (*Chlamydera nuchalis*). *J. Ornith.* 119:74–90.

von Schantz, T., G. Göransson, G. Andersson, I. Fröberg, M. Grahn, A. Helgée, and H. Wittzell. 1989. Female choice selects for viability-based trait in pheasants. *Nature* 337:166–69.

Wagner, R. H. 1992. Extra-pair copulations in a lek: the secondary mating system of monogamous razorbills. *Behav. Ecol. & Sociobiol.* 31:63–71.

Warham, J. 1957. Notes on the display and behaviour of the great bowerbird. *Emu* 57:73–78.

Watts, C. R. 1968. Rio Grande turkeys in the mating season. *Trans. 33rd N. Am. Wild. & Nat. Res. Conf.,* pp. 205–10.

Watts, C. R., and A. W. Stokes. 1971. The social order of turkeys. *Sci. Amer.* 224:112–18.

Weatherhead, P. J. 1984. Mate choice in avian polygyny: why do females prefer older males? *Am. Nat.* 123:873–75.

Weathers, W. W., D. L. Weathers, and R. S. Seymour. 1990. Polygyny and reproductive effort in the malleefowl *Leipoa ocellata*. *Emu* 90:1–6.

Weeks, H. P. 1969. Courtship and territorial behavior of some woodcocks. *Proc. Indiana Acad. Sci.* 79:162–71.

Weidmann, U. 1990. Plumage quality and mate choice in mallards (*Anas platyrhynchos*). *Behaviour* 115:127–41.

Weller, M. W. 1967. Courtship of the redhead (*Aythya americana*). *Auk* 84:544–59.

Wiley, R. H. 1971. Song groups in a singing assembly of little hermits. *Condor* 72:28–35.

Wiley, R. H. 1973. Territoriality and non-random mating in the sage grouse, *Centrocercus urophasianus*. *Anim. Behav. Monogr.* 6:87–169.

Wiley, R. H. 1974. Evolution of social organization and life history patterns among grouse. *Quart. Rev. Biol.* 49:201–27.

Wiley, R. H. 1978. The lek mating system of the sage grouse. *Sci. Amer.* 238(5):114–25.

Wiley, R. H. 1991. Lekking in birds and mammals: behavioral and evolutionary issues. *Adv. Study Behav.* 20:201–91.

Williams, D. M. 1983. Mate choice in the mallard. Pp. 297–305, *in* Mate choice, P. Bateson (ed.). Cambridge Univ. Press, Cambridge, England.

Willis, E. O. 1966. Notes on a display and nest of the club-winged manakin. *Auk* 83:475–76.

Willis, E. O., D. Wechsler, and Y. Oniki. 1978. On behavior and nesting of McConnell's flycatcher (*Pipromorpha macconnelli*): does female rejection lead to male promiscuity? *Auk* 95:1–8.

Willson, M. 1984. Vertebrate natural history. Saunders, Philadelphia.

Wilson. E. O. 1975. Sociobiology: the new synthesis. Belknap Press, Cambridge, Mass.

Wittenberger, J. F. 1978. The evolution of mating systems in grouse. *Condor* 80:126–37.

Wittenberger, J. F. 1979. The evolution of mating systems in birds and mammals. Pp. 271–349, *in* Handbook of behavioural neurobiology, P. Marler and J. Vandburgh (eds.). Plenum Press, New York.

Wittenberger, J. F. 1981. Animal social behavior. Duxbury Press, Boston.

Wittzell, H. 1991. Directional selection on morphology in the pheasant, *Phasianus colchicus*. *Oikos* 61:394–400.

Wolf, L. L. 1970. The influence of seasonal flowering on the biology of some tropical hummingbirds. *Condor* 72:1–14.

Wrangham, R. W. 1980. Female choice of least costly males: a possible factor in the evolution of leks. *Z. Tierpsychol.* 54:357–67.

Wünschmann, A. 1966. Die Balz der Rotkropf-Kotinga *Pyroderus scutulatus*. *Gefied. Welt.* 90:46–48.

Zahavi, A. 1975. Mate selection—a selection for a handicap. *J. Theor. Biol.* 53:205–14.

Zuk, M. 1991. Sexual ornaments as animal signals. *Trends Ecol. & Evol.* 6:228–30.

Zuk, M., R. Thornhill, J. D. Ligon, and K. Johnson. 1990a. Parasites and mate choice in red jungle fowl. *Am. Zool.* 30:235–44.

Zuk, M., R. Thornhill, J. D. Ligon, K. Johnson, S. Austad, S. H. Ligon, N. W. Thornhill, and C. Costin. 1990b. The role of male ornaments and courtship behavior in female mate choice of red junglefowl. *Am. Nat.* 136:459–73.

Index

This index includes cited authors and vernacular names of birds mentioned in the text. Page references to illustrations are indicated by *italics*.